Mendeleev to Oganesson

Mendeleev to Oganesson

A Multidisciplinary Perspective on the Periodic Table

Edited by ERIC SCERRI AND
GUILLERMO RESTREPO

OXFORD
UNIVERSITY PRESS

OXFORD
UNIVERSITY PRESS

Oxford University Press is a department of the University of Oxford. It furthers the University's objective of excellence in research, scholarship, and education by publishing worldwide. Oxford is a registered trade mark of Oxford University Press in the UK and certain other countries.

Published in the United States of America by Oxford University Press
198 Madison Avenue, New York, NY 10016, United States of America.

Library of Congress Cataloging-in-Publication Data

Title: Mendeleev to Oganesson : a multidisciplinary perspective on the periodic table.
Description: New York, NY : Oxford University Press, [2018] | Includes index.
Identifiers: LCCN 2017034854 | ISBN 9780190668532
Subjects: LCSH: Periodic table of the elements. | Periodic law. | Chemistry—History—19th century. | Chemical elements.
Classification: LCC QD467 .M4294 2018 | DDC 546/.8—dc23 LC record available at https://lccn.loc.gov/2017034854

9 8 7 6 5 4 3 2 1

Printed by Sheridan Books, Inc., United States of America

CONTENTS

FOREWORD

THE ICON OF OUR profession has remained strong and deeply useful because of its nonreducible chemical nature, because of its flexibility in both definition and utilization. And it remains alive not only because of its essence, the way it organizes our elements, but also because of the natural challenges it provokes. Let me explain.

The periodic table has evolved as chemistry has evolved; its meanings are now multiple. That is its strength, not a putative weakness. Useful chemical concepts evolve by appropriation. So the line of Kekulé's time, a symbol for a then-unknown persistent connection between atoms, in time was assigned the meaning of a shared Lewis pair. This was further coopted by Pauling to stand for a covalent wavefunction. In a similar way, the periodic table of Mendeleev, which began as a summary of valence regularities of the elements in combination, was reinterpreted in the new atomic theory of Bohr (and by the quantum chemistry that followed) as the outcome of a sequential, quantum-mechanically determined filling of atomic orbitals. The table is alive because it can accommodate alternative interpretations that are of chemical utility, that suggest similarities and fruitful substitutions—in chemical reactions to be tried, in physical properties to be desired.

And what do I mean by the challenges it engenders? There were the initial ones, of atomic weights to be believed or not. Quandaries to be added on to, by asking why a table of chemical proclivities in compounds should have anything to do with filling the orbitals of isolated atoms. Should we, for instance, worry or not about the fact that the electronic configurations of the ground states of isolated Ni, Pd, Pt atoms are all different? No, for there is no ambiguity in the configuration taken on by the most common valence state, II, of those atoms in their compounds. And yes, we should think (if not worry) about what that ambiguity in Ni, Pd, Pt ground state atomic configurations tells us of how close the nd, (n-1)s levels are in the three transition series,

I was asked once by *Chemical & Engineering News* to pick an element to write about. I picked silicon. Not because I had done many calculations on silicon-containing compounds, but because... it was to me the most obvious example of that essential tension of identity, of similarity and difference, of being the same and not the same. Of course, just as the periodic table implies, silicon forms analogues of methane, ethane, and ethylene. And the Si-Si bond length in $H_2Si{=}SiH_2$ is shorter than in $H_3Si\text{-}SiH_3$, just as in the carbon analogues. But... there are no bottles of Si_2H_4; it's a fleeting molecule seen only in a matrix of solid Ne. And—hold on to your pants—quite unlike the carbon case, the enthalpy of the bond-breaking reaction of $H_2Si{=}SiH_2$ to two SiH_2 is actually less than that to $H_3Si\text{-}SiH_3$ to two SiH_3. And in a wide group of compounds, silicon goes five- and six-coordinate, whereas carbon resists that, with a vengeance.

Should we abandon the periodic table, look for a resemblance between C and Al or P? Nonsense. The periodic table gains in interest because (a) we see its limitations, and (b) we try to understand them. Not to substitute another numerology for that the table represents, but to reason out the "why?" that makes compounds of Si different from those of C.

We now have prime evidence for a multitude of exoplanets with conditions approaching those of Earth. Many low-probability accidents of evolution must intervene before conscious, intelligent life develops on these. But the chances of that happening have been much increased by our discovery of the exoplanets. A survivor and an optimist, I have no special desire to see the future; we will find a way, despite what we have done to the planet. Except for one thing—how I would like to learn, in that encounter with extraterrestrial intelligence that must come, how the ETs represent the periodic table! The regularities of atoms, ions, and molecules will be there in their world; but given the vagaries of evolution, the diverse ways we see things worked out in our world, our ET friends will not look the way we do. And the accidents of that piece of cultural evolution we call chemistry (that I am certain will transpire on their world) will have been constructed by them with a different mindset, with different symbols. I just long to see their "table."

Roald Hoffmann

Mendeleev to Oganesson

Introduction

THE PERIODIC TABLE IS ubiquitous in chemistry—as it is in science and in the popular imagination. These days "nerdism" has become a respectable way of being. While science may continue to get bad press, having an expertise in technical matters appears to be increasingly respected. We should all be grateful for the latter if not the former..

The periodic table is one of the areas in science that seems to come under the scrutiny of highly gifted children. Through the YouTube, and the Internet in general, we encounter younger and younger children who are drawn to learning the names of the elements in their correct atomic number sequence.[1] What is this scientific and cultural icon that seems to evoke such strong passions even at such early ages?

Much has been written on the periodic table in recent years, including technical articles and books as well as popular expositions aimed at a variety of audiences.[2] Why do we need another book on the subject? The answer is that as popular interest in the subject grows it becomes even more important for experts in different fields to provide an ever-deeper analysis of what it is we are all talking about.

This project began some years ago following a conference on the periodic table that was held in the idyllic Peruvian city of Cuzco.[3] The organizer, Julio Gutierrez Samanez, has had a lifelong interest in the periodic table, especially its mathematical aspects. It was at this meeting (which included the inevitable excursion to the nearby Inca temple at Machu Picchu) that Scerri and Restrepo conceived of the idea of a book of essays aimed at furthering the foundations of all aspects of the periodic table. We quickly decided that we should also invite authors who might not have attended the meeting. These included mathematicians, physicists, geologists, historians, chemists, semioticians, and philosophers from many different countries and with a variety of interests.

---

1 Some examples are available at https://www.youtube.com/watch?v=nM52VBSb9vk

2 A full list of all books published on the elements and the periodic table (and in several languages) can be found at www.ericscerri.com.

3 http://www.chemistryviews.org/details/event/2459831/Third_International_Conference_on_the_Periodic_Table.html. Also see the announcement in Foundations of Chemistry (2012) 14:105–106.

We naturally turned to Oxford University Press as a possible publisher given their recent record in, not to say virtual domination of, the field of books on the periodic table.[4] After the usual proposals and discussions the press accepted our project and we began to gather the essays you find collected here.

The 15 chapters of this book show aspects of the periodic system from different perspectives and disciplines, and include new translations of some documents by Mendeleev.

Paul J. Karol, a nuclear chemist, ponders the limits of the periodic table, and the stability of the heaviest elements, and provides a review of the subject in his chapter. Among other things, Karol discusses the nuclear liquid drop model and its different refinements, which he later combines with several electronic approaches. He also claims that nuclei with up to 172 protons are stable.

"Nuclear Lattice Model and the Electronic Configuration of the Chemical Elements" is by Jozsef Garai, a mechanical engineer by profession. The author argues that the periods of the periodic system are related to a lattice nuclear model in which protons and neutrons are regarded as spheres that adopt packing patterns of tetrahedra made of layers of nucleons. Garai argues that the number of protons in those layers is related to the periodicity patterns found in electronic configurations.

Philip J. Stewart, an ecologist by profession, reviews some graphical representations of the periodic system, including the short, medium-long, spiral, and left-step forms of the periodic table. He criticizes the reluctance of chemists to accept other non–medium-long periodic tables. He also points to what he regards as contradiction in regarding hydrogen, a non-metal, as being the head of the alkali metals while rejecting that helium, which is also a non-metal, should be at the head of alkaline-earth metals. Stewart stresses the role of amateurs in devising periodic tables and finally makes a plea for his own favorite spiral representation.

In "Similarity Structure of the Periodic System," Guillermo Restrepo discusses the differences between the concepts of periodic system, periodic law, and periodic table. The chapter is motivated by Mendeleev's distinction between basic and simple substance as well as Mendeleev's insistence on considering the properties of the compounds in the classification of elements. A total of seven similarity studies of the chemical elements based on experimental information are then reviewed, in an attempt to clarify the minimum number of properties needed to characterize chemical elements.

Ceulemans and Thyssen provide a discussion of the foundations of Madelung's rule and of its interpretations by exploring different approaches from

[4] J. Emsley, *Nature's Building Blocks: An A-Z Guide to the Elements*; E. Scerri, *The Periodic Table, Its Story and Its Significance*, OUP, 2007; *A Very Short Introduction to the Periodic Table*, OUP, 2011; *A Tale of Seven Elements*, 2013; M. Kaji, H. Kragh, G. Pallo, *The Early Reception of the Periodic Table*, OUP, 2015. M.V. Orna, M. Fontani, G. Fontani, *The Lost Elements, The Periodic Table's Shadow Side*, OUP, 2014.

mathematical physics. This is a topic that is of some importance to chemical education, which habitually attempts to indoctrinate students into being able to write the correct electronic configuration for any particular atom.

Jan C. A. Boeyens, a mathematical chemist, combines information on nucleons with number theory to devise sequences that are related to the periodic system. The author claims that all chemical phenomena displaying periodic trends should be amenable to mathematical analysis. Boeyens also concludes that the results of number theory are completely general and applicable to disciplines other than chemistry.

Eric Scerri and William Parsons are the authors of "What Elements Belong in Group 3 of the Periodic Table?" They claim that group 3 of the periodic table should include lutetium and lawrencium as opposed to lanthanum and actinium. The authors review the many arguments that have been proposed on both sides of the debate and conclude that although the evidence in favor of the proposed change is quite impressive there has yet to be a conclusive argument to satisfy all parties. They proceed to then provide what they consider such a conclusive argument by turning to a long-form presentation of the periodic table and the demand that the elements be strictly ordered according to increasing atomic number.

Paul Geerlings, a theoretical chemist, considers the relationship between density functional theory and chemical periodicity. Geerlings criticizes the quantum chemical approach to the periodic system based on isolated atoms and orbitals. Instead he presents an alternative approach based on electron density and on the amount of information among atoms. He adopts what may be termed a more "realistic" approach to the chemical elements by considering perturbed atoms rather than those in isolation.

Yu Liu, a semiotician, makes a semantic argument for having several periodic tables. With reference to two particular periodic tables published by John Newlands, the author examines the shifts of interests and meanings that Newlands wanted to address. Liu claims that different kinds of relations were depicted by Newlands including that of mereology, which in turn was related to the concept of a chemical period. Liu concludes that chemical periodicity is far from having a resolved meaning; rather, its meaning is constantly being negotiated.

Inorganic chemist and chemical educator Geoff Rayner-Canham discusses several classifications of transition metals and one that includes groups 4 to 11 of the periodic table. The author suggests a novel classification based on the aqueous chemistry of the compounds of elements. Rayner-Canham claims that the traditional approach through groups and periods is not appropriate to the similarities among transition elements. The author suggests a mixture of vertical and horizontal similarities and that one element may take part in more than one similarity class.

Bruce Railsback is the author of "The Earth Scientist's Periodic Table of the Elements and Their Ions: A New Periodic Table Founded on Non-Traditional Concepts," where chemical elements do not occupy a single position but rather several of them, depending on their oxidation states. The table especially includes information on chemical elements present in materials of geochemical interest. Railsback's table is divided into four blocks or neighborhoods according to the oxidized forms of the elements: hard, intermediate to soft cations, anions, and noble gases, which in turn are further divided into finer neighborhoods. Railsback suggests that the table can be used to highlight interactions between chemistry and various other disciplines.

The chemical historian Masanori Kaji analyzes the scientific and social conditions which gave rise to Mendeleev's system. He examines how Mendeleev's interests in analytical chemistry, isomorphism, specific volumes, indefinite compounds, and the classification of organic substances led him to develop his ideas of chemical element and to devise his periodic system. Kaji also shows how the creation of the Russian Chemical Society motivated Mendeleev to write his famous chemistry textbook, which in turn led him to discover the periodic system. This chapter also includes a complete English translation of Mendeleev's letter replying to accusations that he was not doing "real work."

The chemical historian and educator William B. Jensen discusses Abegg's rule of eight, which represented the first attempt to interpret the periodic system in electronic terms. Jensen reformulates and extends the rule to the concept of a valence manifold by using valence electrons and vacancies. Several controversial questions regarding the placement of elements in the periodic table are addressed. Jensen concludes hydrogen and helium should be separated from each other, that lanthanum and actinium should be grouped with the inner-transition, and that the zinc group elements are main-block elements.

The studies of the philosophical views of Mendeleev have been mainly based on translations of his scientific publications. However, as Michael Gordin states in his chapter, Mendeleev wrote two explicitly philosophical documents that Gordin translates for the first time into English. These documents are thoughts that Mendeleev appears to have hesitated from sharing openly. "The Unit," written under a pseudonym, reveals Mendeleev's views on spiritualism while "Worldview," a posthumously published document, contains his political and economic views in the midst of the first Russian revolution.

The final contribution in this multidisciplinary compilation is perhaps the most philosophical of all the studies presented. In it Mark Weinstein considers the methodology, and in particular the role of counter-evidence, in evaluating generalizations. He then examines how the table permits a reinterpretation of foundational epistemological notions of truth, and finally asks how the periodic table supports our commitment to the fundamental nature of reality.

We hope that these essays will further stimulate interdisciplinary as well as multidisciplinary studies on the central icon of chemistry—whose status is still widely debated. There is still widespread disagreement as to whether it should be regarded as a representation, a model, a law, and so on. Perhaps one should avoid even trying to categorize it according to the usual criteria that are imposed by philosophers of science; perhaps one should embrace the uniqueness of the periodic system or simply regard it as the paradigm or framework of modern chemistry.

FIGURE 0.1 Left to right: Rolando Alfaro, Elisa Seidner-Scerri, Eric Scerri, Alberto Vela, Pieter Thyssen, Alfio Zambon, Inelda Hefferlin, Ray Hefferlin, Ana María Gutierrez, Guillermo Restrepo, Margarita Viniegra, Heidi Rubio, Adolfo Gutierrez, Julio Gutierrez.

We end on a sad note in reporting that while this book was in press two of the authors, Masanori Kaji and Jan Boeyens passed away. They will be deeply missed by their families, friends and colleagues, including the editors and other authors represented in this volume. We also acknowledge the passing of Ray Hefferlin who spoke at the Cuzco meeting on the periodic table and who appears in the group photograph above.

CHAPTER 1 | Heavy, Superheavy . . . *Quo Vadis?*

PAUL J. KAROL
Department of Chemistry, Carnegie Mellon University, USA

URANIUM WAS DISCOVERED IN 1789 by the German chemist Martin Heinrich Klaproth in pitchblende ore from Joachimsthal, a town now in the Czech Republic. Nearly a century later, the Russian chemist Dmitri Mendeleev placed uranium at the end of his periodic table of the chemical elements. A century ago, Moseley used x-ray spectroscopy to set the atomic number of uranium at 92, making it the heaviest element known at the time. This chapter will deal with the quest to explore that limit and heavy and superheavy elements, and provide an update on where continuation of the periodic table is headed and some of the significant changes in its appearance and interpretation that may be necessary. Our use of the term "heavy elements" differs from that of astrophysicists who refer to elements above helium as heavy elements. The meaning of the term "superheavy" element is still not exactly agreed upon and has changed over the past several decades. "Ultraheavy" is occasionally used.

Interestingly, there is no formal definition of "periodic table" by the International Union of Pure and Applied Chemistry (IUPAC) in their glossary of definitions: the "Gold Book." But there are plenty of definitions in the general literature—including Wikipedia, the collaborative, free, internet encyclopedia which calls the "periodic table" a "tabular arrangement of the chemical elements, organized on the basis of their atomic numbers, electron configurations (electron shell model), and recurring chemical properties. Elements are presented in order of increasing atomic number (the number of protons in the nucleus)." IUPAC's first definition of a "chemical element" is: "A species of atoms; all atoms with the same number of protons in the atomic nucleus." Their definition of atom: "the smallest particle still characterizing a chemical element. It consists of a nucleus of positive charge (Z is the proton number and *e* the elementary charge) carrying almost all its mass (more than 99.9%) and Z electrons determining its size." Our approach here will be to start with some necessary historical context, followed by a look at the nucleus, without which there would be no atom, and then an examination of the electrons, without which there would be no atom.

Early thoughts on heavy elements trace back at least to John Newlands, an English analytical chemist and one of the earliest contributors in developing a periodic table. Following Mendeleev's proposal, Newlands noted that "All our views on the subject of the ordinal numbers to be attached to various elements are liable to be corrected by the light of future experience. There may be elements having atomic weights as much above uranium as uranium is above hydrogen" (Newlands 1878). Since neither electronic structure of atoms nor nuclear theory had been launched yet, speculations on new elements could proceed without restraint. And did.

In the nearly one-and-a-half centuries since Mendeleev, conjecture about the number of possible elements has been erratic, as illustrated in the historical summary in Table 1.1, which is based partially on the work of Kragh and Carazza (1994) and Kragh (2013 *q.v.*). Obviously, very early speculations about heavy elements are beyond incorrect physics, they were wild guesses. Neither atomic nor nuclear structure was launched. As to more recent speculations, all the proposed physics may be amiss since the leaping extrapolation from $Z = 118$ to $Z \sim 175$ involves enormous relativistic effects that are of unknown/unproven applicability. To this day, relativity and quantum mechanics are not unified. Calculations depend on approximations which are likely unjustifiable as Z increases since the corrections depend very strongly on increasing Z to a power argued to be at least the square, if not the fourth power.

To an extent, much of the speculation was fueled by the discovery of radioactivity revealing the phenomena of transformations and isotopy. Ernest Rutherford and Marie Curie helped clarify the process of radioactive decay, although the physics behind the phenomenon was still a total mystery. Their efforts eventually revealed the uranium decay series.

The series begins with ^{238}U, shown on the right in Figure 1.1 below and, through a now-known chain of alpha-decays (downward arrows) and beta-decays (upward arrows), terminates in stable ^{206}Pb on the left in the figure.

Two other series, the thorium series and the actinium series, were also established. One series began with thorium (Fig. 1.2) and the other (Fig. 1.3) with a lighter isotope of uranium—at the time called "actino-uranium."

The history of these efforts is well-covered in Marjorie Malley's "Radioactivity" (2011).

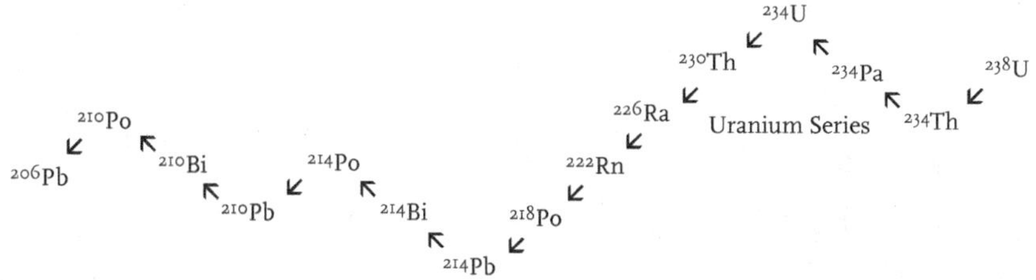

FIGURE 1.1 Uranium decay series commencing with ^{238}U and terminating at ^{206}Pb. Downward arrows are alpha decays and upward arrows represent beta decays.

TABLE 1.1 History of predictions about the upper limit of the periodic table

PROPOSED BY	YEAR	Z LIMIT
J. Newlands	1878	A → 480
E. Mills	1884	92
V. Meyer	1889	100
C. Baskerville	1901	93
S. Losanitsch	1906	>92
W. Tilden	1910	92
N. Bohr	1922	hundreds or thousands
S. Rosseland	1923	92
S. Rosseland and N. Bohr	1923	137
A. Sommerfeld	1924	137
R. Swinne	1926	110
J. Jeans	1926	thousands
M. Snyder	1926	143
E. Meyer	1927	92
W. Gordon	1928	119
W. Nernst	1928	>92
W. Kossel	1928	<137
H. Flint and O. Richardson	1928	97
W. Glaser and K. Sitte	1928	90.5
J. Jeans	1928	95
S. Stone	1930	A<340
G. Lemaître	1931	millions
A. Eddington	1932	137
V. Narliker	1932	92
O. Koblic	1934	93
G. Gamow	1942	several times > 92
F. Werner and J. Wheeler	1958	147
B. Fricke and W. Greiner	1971	172
A. Migdal	1974	> $(137)^{3/2}$ ~ 1600
J. Berger et al.	2001	≈300
V. Nefedov	2006	164
A. Khazan	2007	155
J. Emsley	2011	128
W. Brodziński and J. Skalski	2013	126

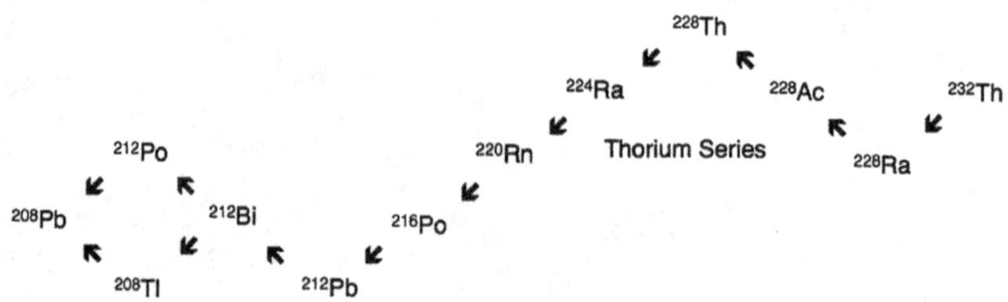

FIGURE 1.2 Thorium decay series commencing with ^{232}Th and terminating at ^{208}Pb. Downward arrows are alpha decays and upward arrows represent beta decays.

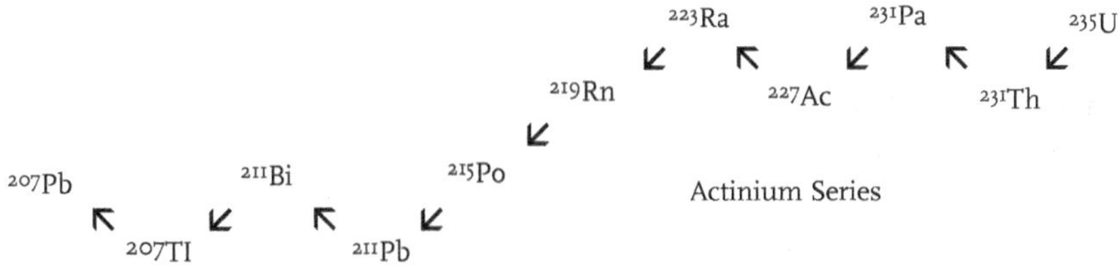

FIGURE 1.3 Actinium decay series commencing with ^{235}U and terminating at ^{207}Pb. Downward arrows are alpha decays and upward arrows represent beta decays.

James Jeans, a British physicist and astronomer, contributor to the field of modern cosmology, and author of "The Kinetic Molecular Theory of Gases," hypothesized that elements with atomic numbers even a thousand times greater than that of uranium exist in stars (1928). The need for such superheavy elements was, at the time, a missing ingredient to explain the huge energy production exhibited by the sun.

First, the Atomic Nucleus

In 1929, a bit before the discovery of the neutron, the Russian physicist George Gamow, working in Niels Bohr's institute, published the first theory of nuclear structure (1930). In it he referred to a *liquid drop model* based on the theory of capillarity. The drop model plays an essential role in discussing the behavior, observed and predicted, for heavy nuclei. Gamow developed an expression for the energy of such a drop, consisting (for simplicity) of alpha particles in which the necessary attractive forces were short-ranged (as with capillary action in a macroscopic liquid) but of unknown strength. By matching predictions of the nuclear drop model energies with Aston's 1927 published mass measurements, Gamow was able to demonstrate a primitive agreement. There was a major problem with the prediction though: At nuclear masses of 120 or higher, instability toward radioactive decay was predicted despite the perceived absence of instability through masses of about twice that value. The fact that nuclear masses were approximately twice the atomic numbers was accommodated, as usual for these times, by assuming that many more protons and alpha particles in the nucleus were accompanied by a sufficient number of electrons within the nucleus itself to account for the eventual balance and mass number, A (equaling N + Z, number of neutrons plus number of protons).

Only a few years after the discovery of the neutron by British physicist James Chadwick in 1932, Gamow's nuclear drop model was adapted by German theoretician Carl von Weizsäcker into an extended form of the liquid drop model leading to the semi-empirical mass formula (1935), a very successful stratagem for understanding major bulk characteristics of atomic nuclei.

On July 5, 1934, Czechoslovak newspapers reported that an element with atomic number 93 and atomic weight 240 had been discovered in Joachimsthal pitchblende by Odelon Koblic, Director of the National Uranium and Radium

Plant in Joachimstal (1934). The logic behind Koblic's search was very reasonable. An element heavier than uranium had to exist in order to explain the chain of radioactivities that included actinium (10-day ^{225}Ac) in a fashion identical to the three uranium series already quite well-characterized. But within the year, a retraction of the evidence was published as no characteristic x-radiation could be found to confirm the new substance. Neptunium was discovered in 1940 and the now-known series appears below in Figure 1.4. Its parent isotope (half-life of 2.1 million years) is too short-lived to be naturally occurring: any original neptunium is extinct.

As we begin to consider understanding the extension of the periodic table beyond uranium, we must deal with the stability of the nucleus itself without which, of course, there would be no elements. With a stable nucleus, the second consideration is the stability of the electronic structure, a requirement to have atoms and then, *ipso facto*, an element, according to definition. Beginning just with stability of the nucleus is our chosen approach.

The nuclear liquid drop model is a very simple and effective recipe for many major features of nuclear stability. From early measurements of nuclear size using alpha particle scattering measurements, the nuclear radius, r, was seen to be proportional to the cube root of the mass number, $A^{1/3}$. From this observation one may infer that the nuclear force between the constituents of the atomic nucleus (initially thought to be only protons and electrons, eventually recognized to be protons and neutrons) had to be short-ranged leading to a saturation effect qualitatively similar to the forces between molecules in ordinary liquid matter. Increasing the numbers of nuclear particles, collectively termed *nucleons*, would increase the volume of the nucleus, meaning the total number of "bonding" interactions would likewise increase. The total "bonding" energy would then be proportional to the nuclear volume ($\frac{4}{3}\pi r^3 \propto A$) and the proportionality coefficient would depend on the strength of the nuclear force, unknown, and therefore just represented as a constant-to-be-determined, a_{vol}. Recalling the behavior of ordinary liquids, as long as the volume is small, there is a considerable surface tension owing to the absence of "bonding" external to the surface of the drop. The magnitude of that surface effect reduces the total bonding energy by an amount proportional to the number of (bonding

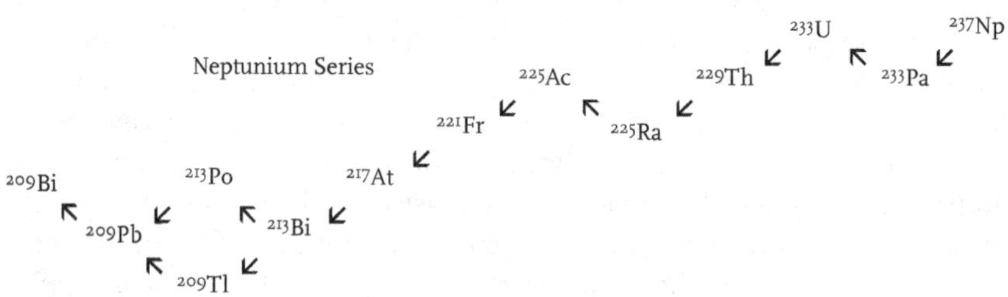

FIGURE 1.4 Neptunium decay series commencing with ^{237}Np and terminating at ^{209}Bi. Downward arrows are alpha decays and upward arrows represent beta decays.

unsaturated) particles on the nuclear surface. That, in turn, varies with the surface area. The strongest total bonding will occur when the reduction due to the surface effect is minimal, namely with a spherical shape. The surface area of a sphere ($4\pi r^2$), and therefore the dependence of the surface binding energy reduction, consequently depends on $A^{2/3}$. To summarize to this point, the total binding energy, E_B, of a spherical drop of liquid can be written as

$$E_B = a_{vol}A - a_{surf}A^{2/3}$$

However, within the nucleus are Z positively charged protons that are mutually repelling each other, reducing the net attractive bonding strength accordingly. Electrostatic (coulomb) repulsion energy among particles is proportional to the product of the charges and inversely proportional to the distance between them. A collection of Z protons is each repelled by the other Z-1 protons and the average distance of separation is proportional to the size of the nucleus, r, which depends on $A^{1/3}$. Proceeding to the binding energy of a *charged* liquid drop, we have

$$E_B = a_{vol}A - a_{surf}A^{2/3} - a_{coul}\frac{Z(Z-1)}{A^{1/3}}$$

In this case, the coefficient a_{coul} is actually known since it incorporates classical electrostatic energy constants.

If we ceased our development of the nuclear liquid drop model at this point, the composition (Z,A) of any nucleus corresponding to the maximum binding energy and therefore the highest stability for that A would be such that Z = 0 or 1. Not very encouraging. But missing from the development at this stage is an important quantum effect. Modern quantum mechanics had already been launched by the time of the nuclear liquid drop model's construction. The constituents of the atomic nucleus were known to be protons and neutrons, both of which obey the same Pauli Exclusion Principle so essential to the electronic structure of atoms: two protons per quantized energy state and two neutrons per quantized energy state within the nucleus: spin up, spin down. As an example, considering four nucleons; having two protons and two neutrons in the lowest state for each species is one possibility. Having two protons in the lowest state and a third in the next higher state (as required by the Exclusion Principle) with one neutron in the lowest state would be higher in energy than the previous configuration. The mirror image alternative having two neutrons in the lowest state and a third in the next higher state with one proton in the lowest state is accommodated by exactly the same energy increase.

Generalizing this picture, the most favorable configuration, considering just the effect of the Exclusion Principle and quantized levels, is that any departure from a composition that has equal numbers of protons and neutrons is higher in energy than that *symmetric* composition. A detailed consideration of the quantum effect of symmetric composition of nucleons, two different spin ½ particles, gives us the next term in the total binding energy, a term that reduces the energy for compositions where Z and N are not equal.

$$E_B = a_{vol}A - a_{surf}A^{\frac{2}{3}} - a_{coul}\frac{Z(Z-1)}{A^{\frac{1}{3}}} - a_{sym}\frac{(N-Z)^2}{A}$$

The symmetry effect is most important for light nuclei since it is inversely proportional to A. It explains why the most stable light nuclei have $N = Z$, or at least nearly so. Illustrating this with a few examples we have the most stable nuclei with mass numbers A = 4, 16 and 40 being ^{4}He, ^{16}O, and ^{40}Ca.

Although the details of the nuclear force were unknown, Weizsäcker was able to use measured masses to get values for the four coefficients in the expression for E_B. Since the energy equivalent of a mass equals mc^2 where c is the speed of light:

$$M(Z,N)c^2 + E_B = ZM_Hc^2 + ZM_nc^2$$

In this expression M(Z,N) represents the *atomic* mass of the isotope of element Z with N neutrons and M_H represents the *atomic* mass of hydrogen. M_n is the neutron mass. *Atomic* masses are used because they are what is determined in mass spectrometer measurements and the number of electrons, equal to Z, is conserved on both sides of the above equation. Consequently, mass measurements allow evaluation of E_B for each isotope. Best values for a_{vol}, a_{surf}, and a_{sym} can then be extracted from fitting the data.

An additional property that emerges from the nuclear liquid drop model is the ability to get a firm grasp of the energetic considerations of radioactive transformations. For example, the decay of ^{238}U into ^{234}Th plus an alpha particle ^{4}He involves a decay energy expressible as

$$M(^{234}\mathrm{Th})c^2 + M(^{4}\mathrm{He})c^2 - M(^{238}\mathrm{U})c^2$$

Substitution converts this into the decay energy in terms of binding energies:

$$E_B(^{234}\mathrm{Th}) + E_B(^{4}\mathrm{He}) - E_B(^{238}\mathrm{U})$$

for which either the experimentally determined values can be used or the model predictions from the nuclear liquid drop picture. The model's predicted value is 4.1 MeV and the experimental value is 4.3 MeV. Interestingly, what emerges is that any element with a mass above about 140 is unstable with respect to alpha decay. Energy available or unavailable (negative values) for alpha decay along the smoothed trend of stable compositions is shown in Figure 1.5.

For lighter systems the negative energies shown below suggest that fusion of an alpha particle with a nucleus is energetically feasible; it is effectively the reverse of emission. Why alpha-decay isn't prevalent just above its threshold in practice is resolved by the well-known dichotomy in ordinary chemistry: thermodynamic (energetic) considerations versus kinetic (rate) considerations.

As far back as 1911, the strong connection between the energy release in alpha-decay and the rate at which the transformation occurred was recognized. That relationship was first embodied in the Geiger-Nuttall Law which gave the

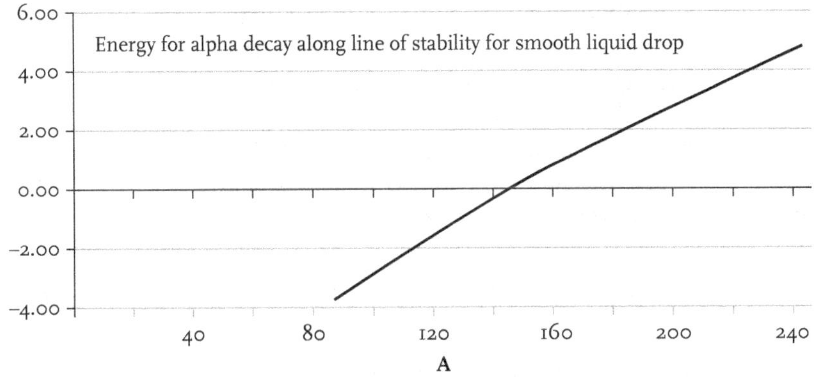

FIGURE 1.5 Energy available for alpha decay as a function of nuclide mass number A for the most stable nuclide at that mass number. Negative energies means the decay is thermodynamically forbidden.

variation of the alpha-decay half-life as log $t_{1/2}$ being proportional to $Z/\sqrt{E_{decay}}$. Thus a small increase in decay energy could be reflected as a very large increase in the rate of decay and vice versa. In 1928, George Gamow explained the Geiger-Nuttall empirical relationship by devising a theory of alpha-decay involving the quantum mechanical tunneling of the alpha particle through the nuclear electrostatic barrier (Gamow 1928). Following this model, it is easy to understand why alpha-decay around A = 140 with very little energy available for decay would proceed exceptionally slowly, as observed in the rare cases where such nuclear behavior has been measured. (^{144}Nd has an observed half-life of 2.4 quadrillion years.) Alpha-decay is spontaneous above A~140 but not necessarily instantaneous.

The rate of decay plays an essential role in looking at very massive systems but we will postpone consideration of this aspect of nuclear behavior until later.

The liquid drop model can also be employed to recognize that there are limits to how many or how few neutrons a nucleus can tolerate before the binding of an additional neutron or proton goes to zero. Predicted limits on composition result. ^{400}U, for instance, will not persist but will hemorrhage neutrons spontaneously and nearly instantaneously until stability is achieved again at around ^{300}U. (The most massive uranium isotope known is ^{242}U.)

A much more illustrative approach to nuclear stability is to look at the average binding energy *per nucleon*, E_B/A. For a bulk liquid, this average would be constant, independent of the size/volume of the bulk liquid under consideration. But as the volume of the liquid approaches zero (nuclear size), the surface effect becomes important and the average binding energy decreases significantly. That is exactly what is observed. On the other hand, as the volume of our nuclear system increases, as the number of particles increases, the binding energy per particle also begins to drop. This is understandable from the contribution of the coulomb repulsion effect in which, as we go to higher and higher elements, more and more protons counteract the net effect of the short-ranged attractive nuclear force. Our quest here is to explore what happens as heavier and heavier systems are considered.

Quantitatively, the simplified version of the liquid drop model here when fitted to mass-derived data gives what is frequently called the *semi-empirical binding energy formula* with units of MeV:

$$E_B = 14.1A - 13A^{\frac{2}{3}} - 0.6\frac{Z(Z-1)}{A^{\frac{1}{3}}} - 20\frac{(A-2Z)^2}{A}$$

A graph of the average binding energy per particle is displayed quantitatively for the most stable nucleus at each mass number A in Figure 1.6. As implied in the above expression, for fixed A, the maximum binding energy can be found as a function of Z and would correspond to the most stable element, Z_A, at that mass number. The approximate expression that emerges is

$$Z_A \approx \frac{79.2A}{160 + 1.2A^{2/3}}$$

Although we have omitted several important aspects of nuclear structure from our treatment so far, the above illustration serves to demonstrate some interesting features of the most stable nuclei. There is a peak in stability around A = 60 corresponding to nickel and iron isotopes. As we get to much heavier nuclei, fission into two more tightly bound fragments becomes energetically favorable. Fission was discovered by German nuclear chemists Otto

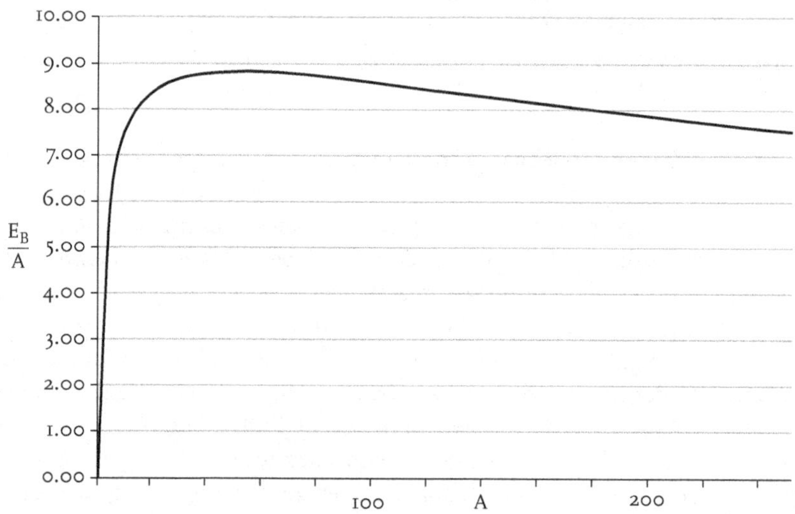

FIGURE 1.6 The liquid drop model average binding energy E_B per nucleon along the valley of stability as a function of A, the number of nucleons in the nucleus.

Hahn and Fritz Strassmann in 1938. Within the year, Austrian physicists Lise Meitner and (her nephew) Otto Frisch came up with an explanation of fission based on the liquid drop model of the nucleus. If ^{238}U hypothetically splits exactly in half, the product ^{119}Pd's are not themselves stable and so Figure 1.6 is inaccurate for that nucleus. Our simple formula gives Z_A for A = 119 as close to ^{119}Sn, not ^{119}Pd. Nevertheless, the energy released in splitting, calculated from the binding energies, still exceeds 150 MeV. That energy is typical for what is available to fission for heavy nuclei. As with alpha-decay, fission is disfavored by a Coulomb barrier, the energy needed to overcome the electrostatic repulsion associated with the two large ≈ Z/2 fragments. The liquid drop model provides a very straightforward scheme for estimating the composition at which the barrier to fission vanishes implying instantaneous fission, that is, instability.

A nucleus undergoing fission, splitting into two fragments, has to distort its spherical shape—stretching, forming a neck, and breaking in two. The process affects two features of the drop model through the shape change: the surface area will increase (since the single original sphere has the smallest area) and the electrostatic repulsion will decrease (since the *average* distance of separation between protons will increase). For *small* distortions from the original spherical shape, the surface energy and the electrostatic coulomb energy will change from their original spherical values, $E^{\circ}_{surf} = a_{surf}A^{2/3}$ and $E^{\circ}_{coul} = a_{coul}Z(Z-1)A^{-1/3}$, according to

$$E_{surf} = E^{\circ}_{surf}\left(1+\frac{2}{5}\beta^2\right)$$

$$E_{coul} = E^{\circ}_{coul}\left(1-\frac{1}{5}\beta^2\right)$$

in which β is a measure of the distortion. Total volume is conserved as is the total contribution of neutron-proton asymmetry energy of the two fragments which maintain the same relative composition as the fissioning parent. Distortion of the spherical shape on the way to fission leads to a total change of energy:

$$\Delta E_{dist} = \frac{2}{5}E^{\circ}_{surf}\beta^2 - \frac{1}{5}E^{\circ}_{coul}\beta^2$$

The path to accomplish fission requires no additional energy when $\Delta E_{dist} = 0$ or

$$\frac{E^{\circ}_{coul}}{2E^{\circ}_{surf}} = 1$$

which, upon reference to the semi-empirical equation for the liquid drop model becomes

$$\frac{E^{\circ}_{coul}}{2E^{\circ}_{surf}} = \frac{0.6Z(Z-1)/A^{1/3}}{2(13A^{2/3})} \approx \frac{Z^2}{43A}$$

re-expressible as

$$\left(\frac{Z^2}{A}\right)_{crit} = 43$$

Translation: for heavy nuclei, when the composition is such that the *fissility factor* $(Z^2/A)/(Z^2/A)_{crit}$ reaches or exceeds 1, fission will occur spontaneously and instantaneously. For lower values of Z^2/A, there is a rate-determining barrier to fission that must be tunneled through by the forming fission fragments. As with alpha-decay, the less tunneling (lower barrier), the more likely quantum mechanical leakage can occur. ^{238}U has a fissility factor of about 0.84. Classically, the electrostatic coulomb barrier would prevent fission from occurring. But the isotope does undergo spontaneous fission albeit very "slowly" (with a very long half-life value). Here in Figure 1.7 is the behavior of the drop model barrier toward fission along the region of most stable nuclei. Also shown are a few measured barriers for some heavy nuclei.

There are now three features introduced that work against very heavy stable nuclear systems:

alpha-decay
fission
binding energy going to zero.

(But in our simple picture in the liquid drop model the latter does not occur until A is roughly 3800.) Chemistry Nobel laureate Glenn Seaborg and W. Loveland hypothesized that the liquid drop model, when its opposing surface and electrostatic effects balance, "provides a limit to Z of about 100-125 elements,

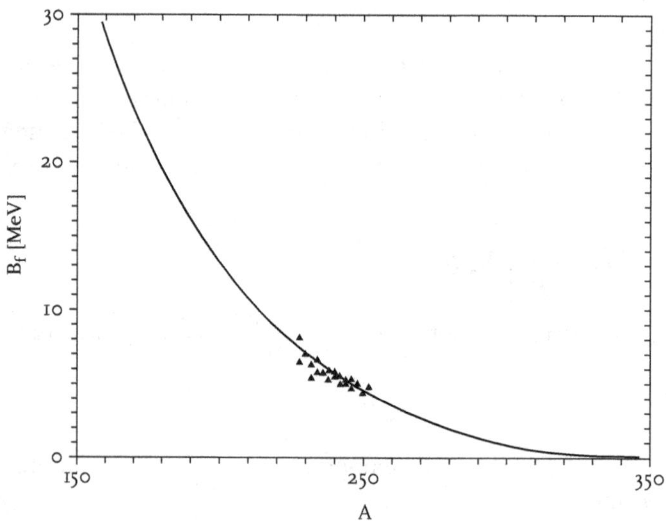

FIGURE 1.7 The nuclear liquid drop model barrier (smooth curve) toward fission as a function of mass number along the line of maximum stability. Data points are for some measured values.

at which point fission is not only spontaneous, but likely instantaneous" (1990).

There has been a claim that an isotope of $_{122}X$, eka-thorium with a mass number around 292 and an exceptional half-life of the order of a hundred million years has been found in natural samples of thorium at ultra-trace levels (Marinov et al. 2010). Identification was done with mass spectrometry but the technique has been uniformly criticized by experts in the field. Furthermore, the chemical isolation of this species prior to the mass measurement assumed that eka-Th would follow a chemistry similar to that of thorium despite the high likelihood that strong relativistic effects would drastically negate those expectations as we shall see later when electronic structure of superheavy elements is addressed.

What is missing from our (overly) simplified description of nuclear behavior are some additional features associated with the detailed assignment of protons and neutrons to quantized states in the nucleus and also a quintessential look at the *rate* of the two "destructive" phenomena, alpha-decay and fission. After all, if they are very slow, we can disregard their influence to a good approximation. No such luck though.

An addition to the semi-empirical formula based on the liquid drop model is accommodation of a *pairing effect* between like particles favoring, thereby, even numbers. That is, an aspect of the nuclear force gives an extra binding strength for proton-proton pairs and for neutron-neutron pairs within the nuclear drop. There is much evidence for this, but most striking is the comparison of abundances of 335 stable isotopes as evidenced in Table 1.2.

Although there are many interesting and important consequences of the pairing (even-odd) effect, we need not consider them here.

A remaining feature is the very momentous influence wielded by the quantum states in which the protons and neutrons find themselves. Not surprisingly, as with electronic states in atoms, there is a shell structure associated with the arrangements of the spin-1/2 nucleons. The evidence for shells is overwhelming. But perhaps the best illustration is to compare the most complete liquid drop model predictions with the precise data for all known isotopes, stable and radioactive. This appears for neutrons in Figure 1.8. A similar result pertains to protons.

For the neutrons, there is indubitably an increase in stability associated with neutron numbers N = 28, 50, 82, and 126. These are associated with

TABLE 1.2 Abundances of stable isotopes by composition

Z	N	NUMBER OF STABLE ISOTOPES
even	even	201
even	odd	69
odd	even	61
odd	odd	4

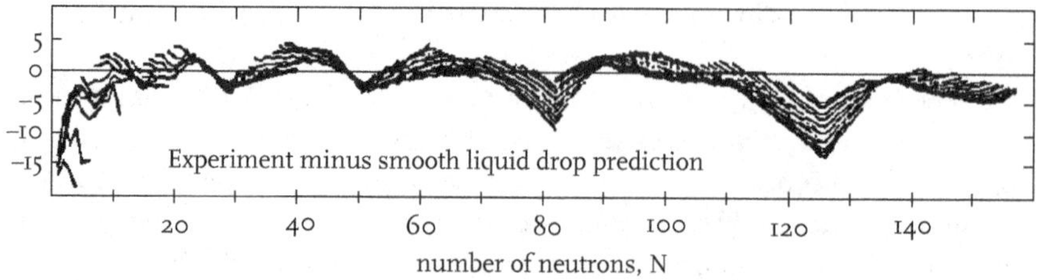

FIGURE 1.8 The difference between the smooth nuclear liquid drop model prediction of total energy compared to the measured values as a function of the number of neutrons in the nucleus. Extra stability compared to the liquid drop prediction is indicated by the dips in the data at certain values of N, magic neutron numbers.

TABLE 1.3 Closed shell populations (*magic numbers*) for protons and neutrons

Z	2	8	20	28	50	82	
N	2	8	20	28	50	82	126

closed, stable shells. Collating all such data for existing systems provides us with the closed shell populations for both protons and neutrons in Table 1.3.

The blank space for the next Z shell closure is at Z = 126 is a compelling observation. But stay tuned.

The analogous shell structure of electrons in atoms emerged from the early days of quantum mechanics in the 1920s following Neils Bohr and Erwin Schrödinger. Similarly, existence of nuclear shell structure was unmistakable early in the development of nuclear physics. It was initially attempted by the Russian physicist Dmitri Ivanenko, a student of George Gamow, and modified subsequently by several others. But the models did not successfully reproduce the observations until 1949. Closed-shell behavior was matched by the scheme devised by German physicists Maria Goeppert-Mayer and J. Hans Jensen, who were awarded a share of the 1963 Nobel Prize in Physics for their achievement. The key to their accomplishment was recognizing the role of *spin-orbit coupling* in ordering of the energy states. Spin-orbit coupling considerations will also further impact the ordering of electronic states and chemical periodicity as we extend the march toward heavier atoms.

With shells affecting nuclear stability by as much as 12 MeV, they can have very profound influence on transformation energetics and consequently on half-lives, extending them beyond the simple liquid drop predictions. For electronic systems, the shell structure emerges from bound states in the coulomb attractive potential including an accounting of the Pauli Exclusion Principle and electron-electron repulsions. In Figure 1.9, the emergence of nuclear shells is reproduced under three distinct situations. On the left are the levels for spin-1/2 particles such as neutrons or protons in a three-dimensional harmonic oscillator potential with central symmetry. Gaps appear at the total particle numbers indicated: 2, 8, 20, 40,...In the center is the result for a

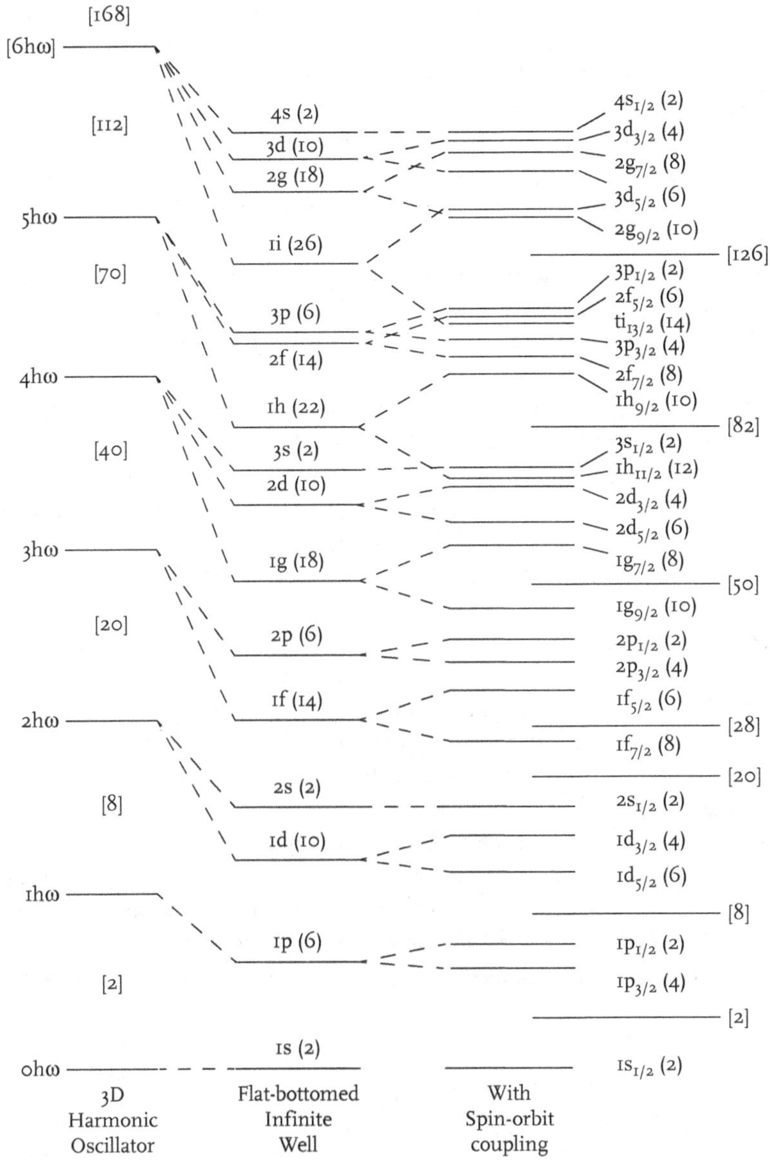

FIGURE 1.9 Energy levels in three dimensional potential wells with different shapes. On the left is that for a parabolic well corresponding to the harmonic oscillator. The center arrangement pertains to a sharp-edged infinite potential well including angular momentum. The right set of levels corresponds to what happens when spin-orbit coupling between orbital angular.

spherically symmetric flat-bottomed attractive well with infinitely repulsive walls and including quantized angular momentum. The various angular momentum states *s, p, d,*... give rise to shell gaps at 2, 8, 20, 34,... momentum and intrinsic angular momentum of the nucleons are introduced. Gaps in the energy spacing, interpreted as indications of closed shells, are indicated by total occupation numbers for the class of nucleon under consideration.

The scheme on the right is what led Mayer and Jensen to come up with a successful explanation of shell structure in atomic nuclei. Introduction of spin-orbit coupling splits angular momentum levels of $\ell > 0$ into $\ell \pm 1/2$ sublevels where, for example, the $1p$ partitions into $1p_{3/2}$ and $1p_{1/2}$. Shell gaps appear at the occupation numbers identical with the previous table in agreement with observation.

Bulk behavior of nuclei embodied in the liquid drop model is often referred to as the *macroscopic effect* whereas the detailed characteristics of nuclear states' shell considerations within the liquid drop environment are considered *microscopic effects.*

Recent studies (Steppenbeck et al. 2013) suggest that stability may arise for more extreme compositions, away from stability. The "exotic" nucleus ^{54}Ca shows evidence of a stable shell (closed) at neutron number 34. Explanations involve the inclusion of three-body forces previously ignored. To what extent these new considerations affect predictions concerning conventional superheavy elements and the possibility of "exotic" superheavy elements has not yet been explored.

Departures from a symmetrical, spherical shape (attractive nuclear potential) further affect the energies of particles bound to the system. Around 1950, nuclei with non-zero quadrupole moments were measured for some rare-earth elements, clearly indicating that non-spherical shapes were at play. An important extension of the liquid drop model then was to allow non-spherical shapes of the nucleus shown in Figure 1.10.

A heuristic example of how such options play into our picture is the three-dimensional harmonic oscillator again, but allowing for oblate and prolate spheroidal distortions of the type shown schematically above. Results of calculations (Fig. 1.11) signal that dramatic changes in energies and occupation numbers corresponding to shell gaps can occur with shape distortions when

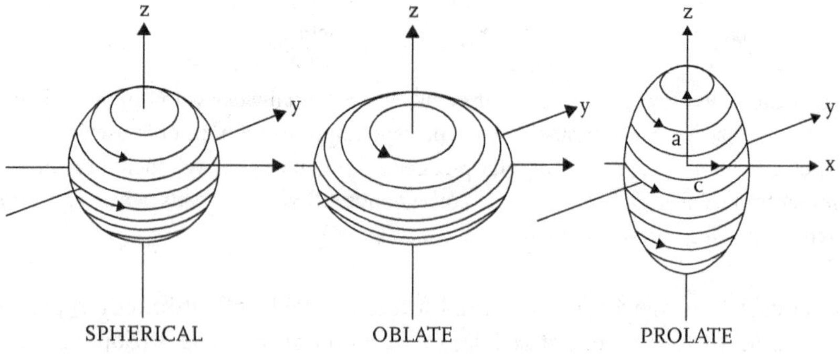

FIGURE 1.10 Distortion of nuclear shapes from the symmetric spherical shape on the left to either an oblate geometry in the center or a prolate geometry on the right.

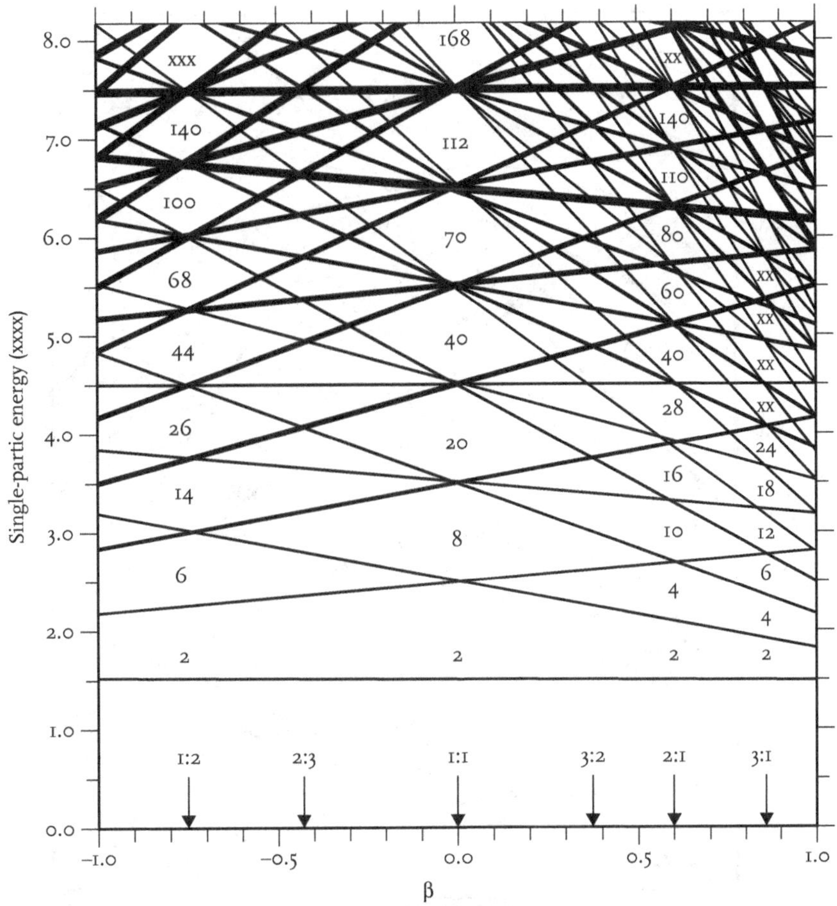

FIGURE 1.11 The consequence of taking a symmetrical, three dimensional parabolic well (harmonic oscillator) and imposing distortions of oblate (distortion parameter $\beta < 0$) or prolate ($\beta > 0$) on removing the degeneracy of the nuclear states.

accompanied by a lowering in the total energy of the system. In the chemical world, this is similar to the Jahn-Teller effect.

The same approach employed for the 3D harmonic oscillator was used for the more realistic nuclear potential as a function of distortion in extensive calculations by Swedish physicist Sven Gösta Nilsson. Included were non-spheroidal shapes and also spin-orbit coupling and centrifugal force effects in the new shapes (Nilsson 1955). The diagrams become very complex and just a portion of the result is shown below in Figure 1.12 for lower energy states up to the gap that is associated with 50 particles in the spherical (distortion parameter $\beta = 0$) potential well.

How influential shell structure is on stability of heavy nuclei can be seen by comparing the macroscopic prediction for rates of penetrating through a liquid drop barrier to measured half-lives. Figure 1.13 shows the experimental

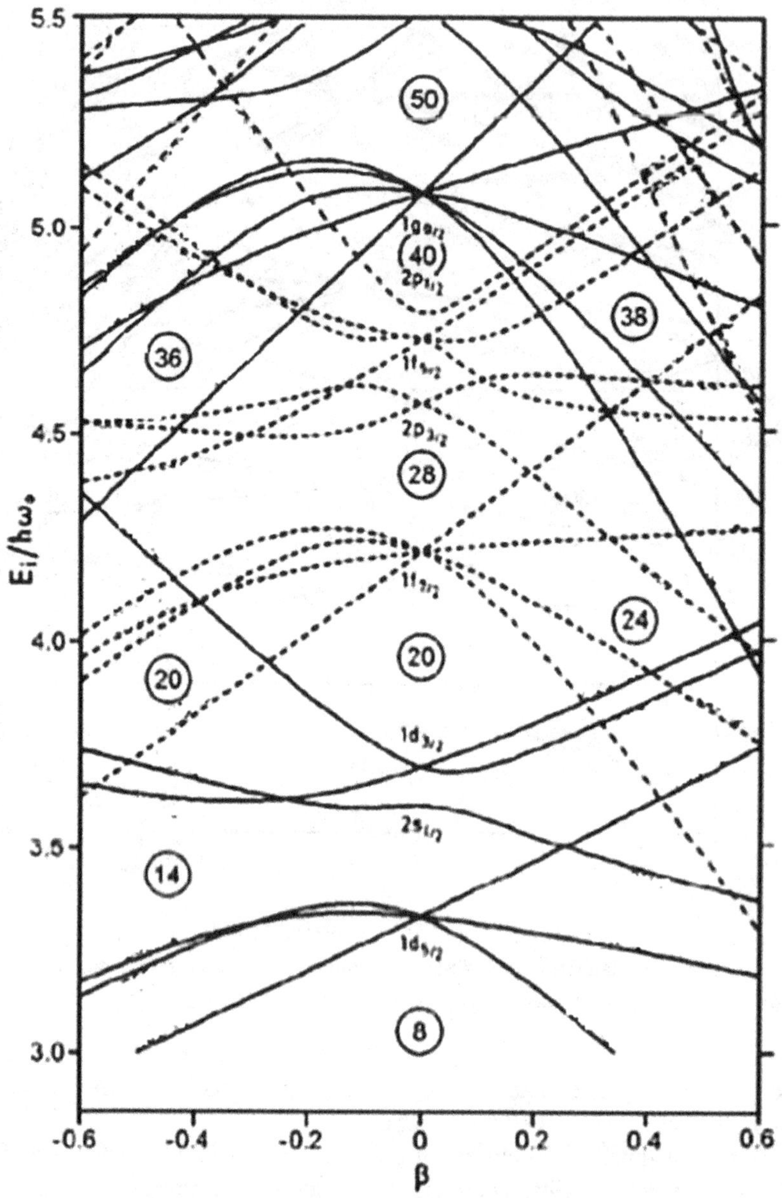

FIGURE 1.12 As in Fig. 11 but for nucleons in an infinite well with spin-orbit coupling (Fig. 9, right) as a function of distorting the shape of that well in either an oblate (distortion parameter $\beta < 0$) or prolate ($\beta > 0$) sense.

fission half-lives of a number of trans-uranium isotopes as a function of the relative fissility parameter (x in the graph). The macroscopic liquid drop model predictions for half life "through" the fission barrier (dashed curve) and the approximate time to form the electronic cloud about a nucleus (horizontal dashed line) are also indicated.

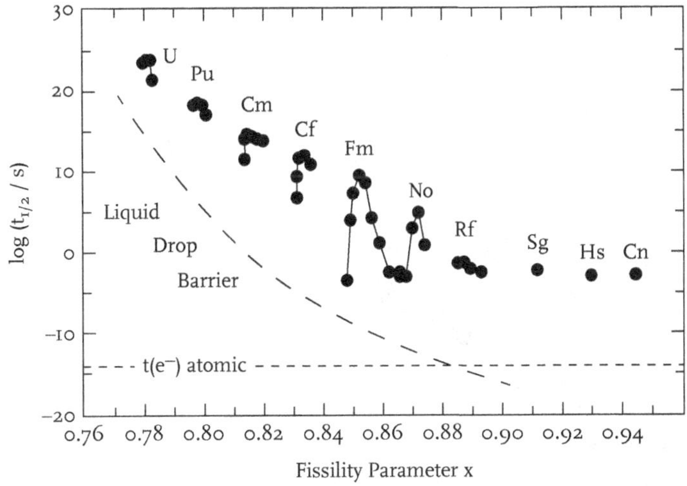

FIGURE 1.13 The probability of fission for a heavy element, as entailed in its fission half-life, as a function of the fissility parameter, $x = (Z^2/A)/(Z^2/A)_{crit}$. Data points are measured half-lives; downward sloping smooth dashed curve is the macroscopic prediction of the nuclear liquid drop model; horizontal dashed line is the minimum lifetime for a chemically viable species.

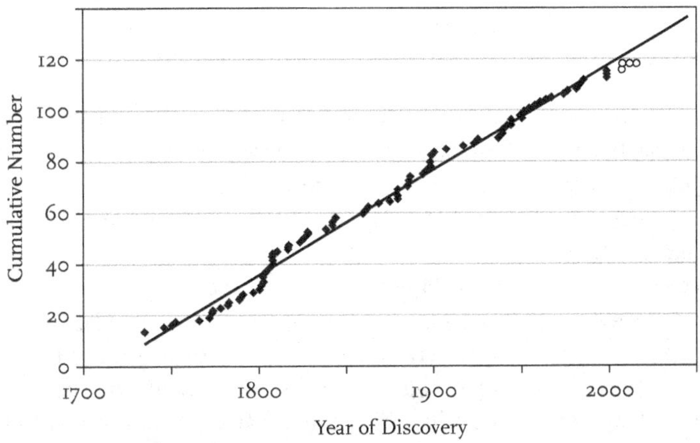

FIGURE 1.14 The cumulative number of known nuclei as a function of year of discovery over the past two and a half centuries.

The liquid drop model, without further modification due to microscopic effects, predicts that by Z = 104 the lifetime of the nucleus would be insufficient for establishment of bound electrons needed to discuss atomic structure and placement on the periodic table. But the actual data reveal that even as the relative fissility, x, approaches 1 at the right edge of the figure, half-lives toward fission are ten orders of magnitude longer than the minimum for atom formation. The effect is entirely due to microscopic considerations; that is, to the influence of shell structure on energetics and fission barrier characteristics.

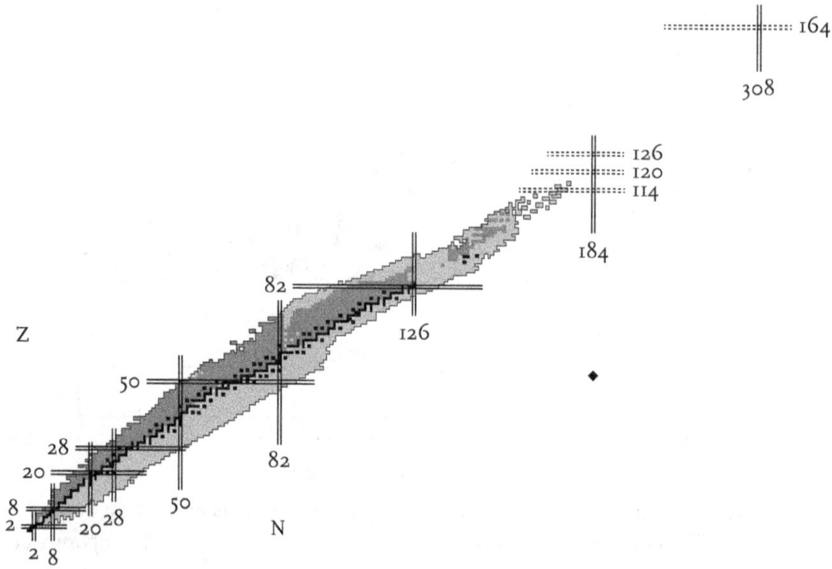

FIGURE 1.15 On a grid where neutron number is the horizontal axis and proton number is the vertical axis, a display of known nuclides; stable as black squares and radioactive species in grey. Closed shells for both neutrons and protons are indicated with their occupation numbers including extrapolated predictions for the superheavy element regions at the upper right. The isolated black diamond (at N = 184) pinpoints the hypothetical ^{236}Te.

The cumulative rate of discovery of new elements, plotted in Figure 1.14, has been relatively constant (straight line) at about one discovery per two-and-a-half years since the mid 18th century. The trend modified from (Karol 2002) by adding recent newer elements, is pictured. The discoveries are not necessarily sequential.

Since closed shells are so very influential in stability properties, where do nuclear theories place the next closed proton and neutron shells? Stability predictions for heavy and superheavy nuclei currently focus on the influence of the macroscopic-microscopic non-spherical shape considerations. Mosel and Greiner (1969) among others have suggested that the next island of stability is way off at the distant (and most likely unreachable) Z = 164. But such possibilities are not the only ones under exploration. There have also been assessments of yet more exotic departures from spherical geometries: triaxial (pear) nuclear shapes (Brodzinski and Skalski 2013) and bubble nuclei (Berger et al. 2001). Among radioactive decay possibilities that might prove detrimental are cluster decays (emitting fragments larger than alpha particles) and the re-emerging importance of electron capture. The German theoretical physicist Walter Greiner comments that shape-dependent models have revealed that the magic shells in the superheavy region are very isotope dependent (Greiner 2012). There is a strong dependence on the parameters chosen and the particular model employed. Forces predict the magic numbers Z = 114, 120 and 126 (see

Fig. 1.15). The corresponding magic neutron numbers are predicted to be N = 172 and, to a lesser extent, N = 184. The charge distribution of the Z = 120, N = 184 nucleus indicates a partially hollow inside core. Just how influential esoteric closed shells can be is demonstrated by considering that the most neutron-rich isotope of tellurium is ^{142}Te. Instability by neutron *drip* is predicted to occur once ^{178}Te is reached. Yet theoretical calculations imply that ^{236}Te with N = 184 will be stable toward spontaneous neutron loss (Tarasov et al. 2013).

An extrapolation of the various searches for the superheavy closed shell system(s) to the next closed shell brings the distant, hypothetical island of stability at around Z ≈ 164 (Grumman et al. 1969) into focus. This is admittedly a huge leap of faith. Is that island a mirage? Will we ever get there? Is it legitimate to talk about the chemistry of Z = 164, less just of an atom?

Electronic States of Atoms

Energies

Erwin Schrödinger gave us the recognizable wave mechanical basis for electronic structure in one-electron systems in 1926. Most are familiar with his equation for the allowed energy states of one electron attracted to a nucleus of Z units of charge *e* as $E_n = -RZ^2/n^2$ where R is the Rydberg constant (in energy units), $m_e e^4/8h^2\varepsilon_0^2 = 13.6$ eV. Planck's constant is *h*, the rest mass of the electron is m_e (actually, the very slightly different reduced mass should be used), and the permittivity of the vacuum toward the coulomb force is ε_0. The principal quantum number, *n*, can take any integer value (1, 2, etc.). In one-electron systems, the electron's energy is independent of its orbital angular momentum $[\ell(\ell+1)]^{1/2}h/2\pi$ which itself is also quantized as $\ell = 0, 1, \ldots (n\text{-}1)$. Energy states are labeled as 1*s*, 2*s*, 2*p*, and so on, as is well known. The electron also has a quantum mechanical "intrinsic" angular momentum, traditionally called its "spin" and represented by the quantum number ½. Orbital angular momentum and intrinsic angular momentum can couple to give a total angular momentum, *j*, which has quantum numbers $\ell \pm ½$. Total angular momentum could be visualized as an electron spin pointing in the same or the opposite direction of the angular momentum as the only two possibilities. The "orientation" of the total angular momentum vector is also quantized.

Having reviewed those fundamentals (quite superficially), we can move on to a modification of our understanding. First, we will introduce an abbreviation for the Rydberg constant by using

$$\alpha = \frac{e^2}{2\varepsilon_0 h} = 0.007297\ldots \approx \frac{1}{137}$$

Dimensionless α is called the *fine structure constant* and transforms the one-electron energy equation into $E_n = -m_e c^2\alpha^2 Z^2/2n^2$ unchanged, except in appearance, from the previous form that also gave $-13.6\ Z^2/n^2$ eV. The fine structure constant embodies many physical features. It is the ratio of the speed of a Bohr model hydrogenic 1s electron to the speed of light; it is a measure of the strength of the electromagnetic interaction of a charged particle; it is a measure of the fraction of an electron's rest mass energy that corresponds to the total energy of the hydrogen atom in its ground state. Richard Feynman referred to its value as "one of the greatest damn mysteries of physics." Now why did we introduce α and what has it to do with pursuing the *extremum* of the periodic table?

Even the Bohr planetary model of the atom foresees a difficulty with large Z. The velocity of a 1s electron in Bohr's model can be written as $v = Z\alpha c$. You can see that for $Z > 137$ a speed greater than that of light is predicted and is physically impossible.

Reminding ourselves that the total energy, including now rest mass energy m_ec^2, for a single electron about a point nucleus of Z unit charges may be written as

$$E_n = m_ec^2 - \frac{1}{2}m_ec^2\frac{Z^2\alpha^2}{n^2} = m_ec^2\left(1 - \frac{Z^2\alpha^2}{2n^2}\right) \quad \textit{Non-relativistic, one electron}$$

the expression under relativistic treatment becomes

$$E_{n,j} = m_ec^2\left[1 + \left(\frac{Z\alpha}{n - \left(j + \frac{1}{2}\right) + \sqrt{\left(j + \frac{1}{2}\right)^2 - Z^2\alpha^2}}\right)^2\right]^{-\frac{1}{2}} \quad \textit{one-electron, relativistic}$$

Accordingly, even in the hydrogen atom, the degeneracy of the $n = 2$ level's states is removed. Two different energy levels occur, one with a degenerate pair $2s_{1/2} = 2p_{1/2}$ and the remainder as $2p_{3/2}$, although the differences are quite small: the former energy (without the rest energy m_ec^2) is -3.40132 and the latter is -3.40127, eV. It is instructive to take a more detailed look at the $1s_{1/2}$ result under the condition that Z becomes large. In this case, for one electron and $Z >> 1$ but less than $1/\alpha$

$$E_{1,\frac{1}{2}} = m_ec^2\sqrt{1 - Z^2\alpha^2}$$

At $Z = 100$, $E_{1,1/2} = 0.68\ m_ec^2$ or 347 keV. (Exclusive of the electron rest mass energy of 511 keV this is -164 keV.) For $Z > 137$, $1—Z^2\alpha^2$ becomes negative and the 1s electron energy becomes imaginary: physically meaningless. The wave function for the *1s* becomes that of an unbound electron.

Often, the relativistic effect is interpreted as due to the increase in mass of the electrons as their speeds approach that of light. When that increase is not too large, perturbation methods can be used to accommodate relativistic effects. For the 1s electron, the mass is nearly twice the rest mass as the eighth period commences indicating the perturbation approach is no longer justified.

Originally, Dirac thought that relativistic corrections would not have much influence on atomic properties since the valence electron velocities remained small compared to *c*. But that turned out to be a premature judgment.

There is much more than just a change in mass when relativistic effects are put into play. Spin-orbit coupling has its origin here for example. An electron moves in the coulomb field of the nucleus and relativity shows that a magnetic field appears which can interact with the electron's intrinsic magnetic moment. Quantum mechanically, this translates into spin-orbit coupling. Going further, replacing the aforementioned local interaction with the more accurate non-local interaction introduces another consideration, the *Darwin term*. Additionally, there is an interaction with the weaker magnetic moment of the proton. These corrections are dealt with in more detail by Cohen-Tannoudji, Diu, and Laloë (Cohen-Tannoudji et al. 1977).

Shapes (Powell 1968)

Electronic structure of atoms involves not only the energy of the electrons but also the geometry of their orbitals, in other words, where they are likely to be found. Relativistic effects change orbital shapes as follows:

(a) Orbitals with the same value of j and m_j have the same angular distribution. The $p_{1/2}$ distribution is spherically symmetrical, like the $s_{1/2}$. The set of $d_{3/2}$ orbitals has the same angular distribution as the set of $p_{3/2}$; the set of $f_{5/2}$ orbitals has the same shape as those of $d_{5/2}$.

(b) The $s_{1/2}$ distribution is spherically symmetrical. For other states with $m_j = 1/2$, the angular distribution is stretched out along the z axis but has no nodes. The $p_{3/2}$, $m_j = 1/2$ state, for example, has an angular dependence of $\cos^2\theta + 1/3$ so it no longer resembles two touching ellipsoids. The probability density angular distribution for the $p_{3/2}$, $m_j = 1/2$ state is shown in Figure 1.16 on the right compared to the p_o for the no spin-orbit splitting shape on the left.

(c) The state with the highest m_j value for its j—the $p_{3/2}$, $m_j = 3/2$, the $d_{5/2}$, $m_j = 5/2$ the $f_{7/2}$, $m_j = 7/2$ states—always has a doughnut-shaped (toroid) distribution; no nodes.

(d) Intermediate m_j states have distributions resembling two-lobed, three-lobed, etc., toroids; no nodes.

Many Electron Atoms in Superheavy Elements

For the very heavy elements, as the periodic table is extended, theory currently plays a quintessential role. With rare exception nowadays, theory is the only source of chemical insight. For example, electronic configurations, chemical composition, stability, geometries, ionization potential (IP), and electron affinities

FIGURE 1.16 Comparison of the shape of p-orbitals in the non-relativistic case (left) and with relativistic considerations (right).

(EA), can be obtained only theoretically. Consequently, relativistic quantum mechanical calculations are crucial for predicting and/or interpreting extraordinarily difficult experiments with single atoms. Yet only in a comparison of the observed behavior with that predicted theoretically can science assess the correctness, importance, and magnitude of relativistic effects.

Following a suggestion of J. R. Oppenheimer, theoreticians L. Schiff, H. Snyder, and J. Weinberg solved the problem of electrons in a square well used to simulate the coulomb field (Schiff et al. 1940). They used relativistic quantum mechanics and showed that the solutions would turn complex when Z exceeded ½·137. I. Pomeranchuk and J. Smorodinski (1945) extended that one-electron effort to a coulomb potential with a non-point nucleus and, using reasonable values for the nuclear dimensions, arrived at an upper Z limit of 175. In 1969, W. Pieper and W. Greiner continued these calculations and looked at the energies of the single particle electronic states as a function of Z for uniform density nuclei of finite size. Their results are shown in Figure 1.17 for $100 \leq Z \leq 250$. Note the spin-orbit splitting (between $2p_{1/2}$ and $2p_{3/2}$, for example) increases very rapidly with Z and furthermore, that the $3p_{1/2}$ actually becomes more stable than the $2p_{3/2}$.

Fricke et al. (1971) used still more realistic nuclear potential wells and self-consistent field approaches to extend the orbital calculations to many-electron systems and noted again that a physical limit is given by the total binding energy (including rest energy) of $-mc^2$ because at this value the level drops into the continuum of filled states of negative energy. This occurs at approximately $Z = Z_{critical} \approx 175$. It is not yet clear whether this really happens or—what is deemed more likely—that quantum electrodynamical effects, such as vacuum

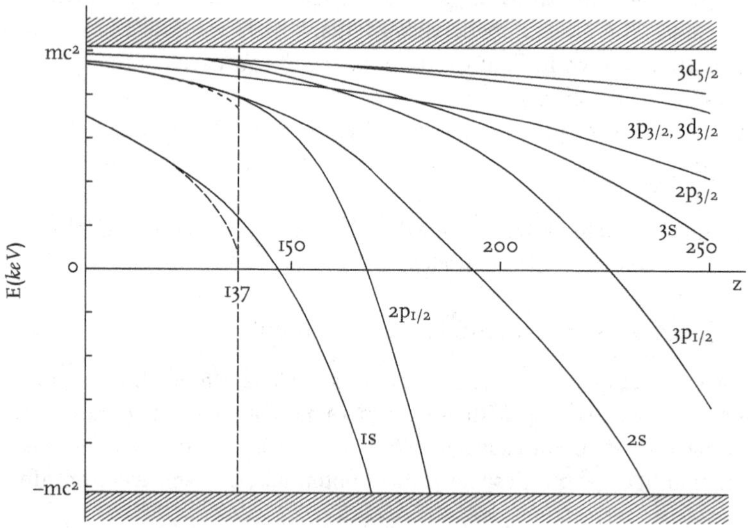

FIGURE 1.17 Relativistic orbital energies for core levels in Z = 100-250 atoms. The conventional zero of energy (unbound electron), when the rest mass is added, appears as the upper limit in the diagram.

polarization and fluctuation, become so large that the single particle Dirac equation breaks down completely. One possibility is that the levels do not drop into the negative contribution but they will asymptotically reach the lower continuum for very large Z. That would imply that in nature a maximal field strength would exist. Another possibility is that spontaneous electron-positron production occurs at the critical Z-value.

There are (at least) three main effects influencing orbital electronic structure of atoms when relativistic effects are taken into consideration. These and their domains are

Spin-orbit splitting $(p, d, f$ shells)

Relativistic contraction and stabilization (s, p)

Relativistic expansion and destabilization (d, f)

Furthermore, as P. Pyykkö (2011) points out, "while nuclear charges are well defined, the concept of an electron configuration... is only approximate... the true many-electron wave function can be a superposition of a large number of effective configurations."

Several groups have now calculated orbital energies of atoms and ions in which relativistic effects have been accommodated. A key feature among these effects is the spin-orbit splitting of states with ℓ - ½ and ℓ + ½, the former being stabilized—lowered in energy—with respect to the latter. The phenomenon is most significant for $\ell = 1$ orbitals increasingly much less so for $\ell > 1$. An illustration, meant to be semi-quantitative, is derived from the tables of orbital energies published by Desclaux (1973). Table 1.4 shows the $p_{3/2}$-$p_{1/2}$, $d_{5/2}$-$d_{3/2}$, and $f_{7/2}$-$f_{5/2}$ splittings calculated for element 119, eka-francium. (Recall from earlier that for the hydrogen atom, the 2p splitting was approximately 0.00005 eV.)

The spin-orbit splitting has a most profound effect on the periodic behavior of the elements since it shuffles, to an extent, not only the energies of the orbitals, but also their geometric placement about the nucleus. That is, the $p_{1/2}$ is not only lowered in energy (stabilized), but also correspondingly contracted in size (as is the unsplit $s_{1/2}$).

To illustrate, Pershina and Fricke (1993) published the radial density distribution in $_{108}$Ha for the valence 7s state under both the non-relativistic and relativistic restraints. The contraction of the s-orbital due to relativistic considerations is apparent in the Figure 1.18 display adopted from their calculations.

TABLE 1.4 Spin-orbit energy differences for some $n = 5$ states

$_{119}$EKA-Fr ORBITAL	SPLITTING (eV)
5p	333.2
5d	45.7
5f	10.6

To see general trends, the root-mean-square radius of the 2*s* orbital, for example, decreases with increasing nuclear charge Z, of course, but much more rapidly than would occur if treated non-relativistically. Those radii in pm are shown in Figure 1.19 below for the group I alkali metals (Desclaux 1973).

Relativistic effects on atomic properties are already influential at only moderately high Z (Pyykkö 2011). For $_{87}$Fr, Table 1.5 shows the calculated ratio of relativisitic to non-relativistic orbital radii. Pershina (2011) has calculated that the 20% correction for francium's 7s at the beginning of the seventh period has increased to a 25% correction by copernicium (Z = 112) in the same period of the periodic table.

The periodic trend in atomic radii for a given column is not monotonic but shows the effect of size contraction in Figure 1.20, again for the group I elements

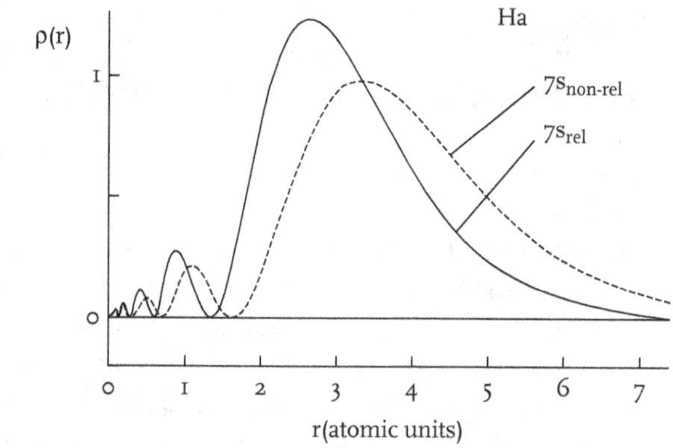

FIGURE 1.18 An illustration of the relativistic contraction of s-orbitals in hahnium. The dashed curve is the radial density probability distribution for the 7s orbital calculated non-relativistically in arbitrary units. The solid curve is the same 7s density distribution calculated relativistically showing an approximate 25% shrinkage in the electron's distance from the nucleus.

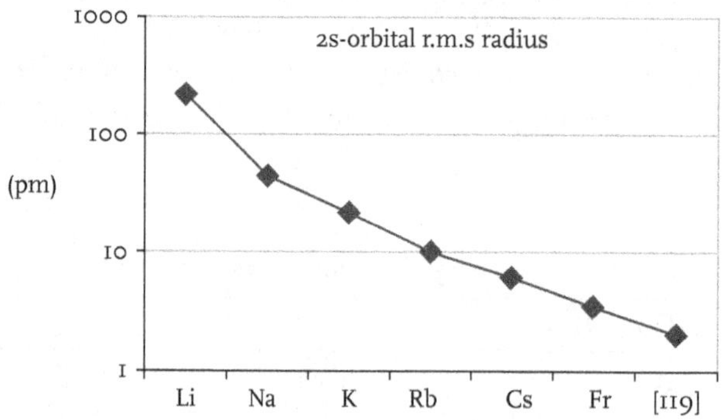

FIGURE 1.19 Relativisitically calculated r.m.s. radii for the 2s atomic orbital as n increases from 2 to 7 for the group I (alkali metal) elements.

TABLE 1.5 The ratio of relativistic to non-relativistic calculated orbital r.m.s. radii in Z = 87, francium

n	$ns_{1/2}$	$np_{1/2}$	$np_{3/2}$	nd (AVERAGE)	nf (AVERAGE)
1	0.757				
2	0.729	0.691	0.946		
3	0.775	0.761	0.943	0.969	
4	0.799	0.794	0.954	0.984	1.043
5	0.811	0.807	0.972	1.026	
6	0.804	0.794	1.028		
7	0.803				

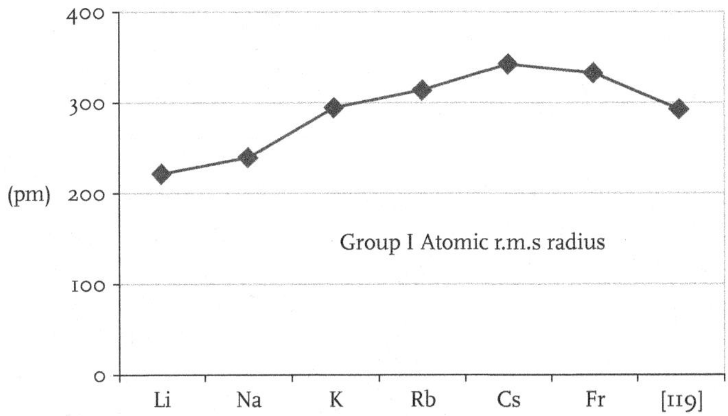

FIGURE 1.20 Atomic (root mean square) radii in picometers for the group I (alkali metals) taken from Desclaux (1973).

(Desclaux 1973). A consequence of atomic radius shrinkage is that the density of the bulk elements is predicted to increase. The densest elements actually measured are the Pt-group in the sixth period of the periodic table with densities slightly greater than 20 g/cm³. For the 8th period, the density is calculated to peak at more than twice those values (Fricke et al. 1971).

Reciprocally, the other orbitals ($p_{3/2}$, d, f, and g) are screened more effectively by the inner s and $p_{1/2}$ orbitals and are simultaneously destabilized—raised in energy—and expanded in extent. Splitting increases for each orbital across a period of the periodic table as shown below in Figure 1.21.

Splitting also increases in magnitude down a column as evidenced in the calculated values for group I elements compared in Table 1.6 showing a very steep dependence on Z and yet, the calculations have not approached the periodic table limit, Z ≈ 175, anticipated).

The final set of trends is associated with the decreasing significance of the spin-orbit effect as the principal quantum number increases, moving to orbitals

TABLE 1.6 4p Splitting for alkali elements

ATOM	$4p_{3/2,1/2}$ SPLITTING (eV)
$_{37}$Rb	1.0
$_{55}$Cs	14.3
$_{87}$Fr	171
$_{119}$eka-Fr	1121

TABLE 1.7 Spin-orbit splitting for the *p*-orbitals in Z = 119

$_{119}$EKA-Fr ORBITAL	SPLITTING (eV)
2*p*	15556
3*p*	3825
4*p*	1121
5*p*	333
6*p*	84.6
7*p*	13.6

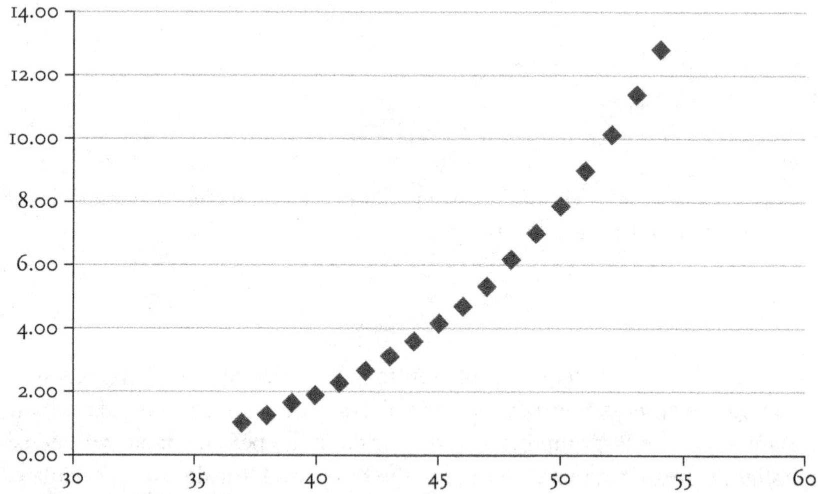

FIGURE 1.21 The splitting (in eV) between the $4p_{3/2}$ and $4p_{1/2}$ across the 5th period of the Periodic Table from Rb through Xe.

further removed from the nucleus, as appears for the n*p* orbital energies for eka-francium in Table 1.7.

A look at the spin-orbit influence on electron configuration is had by viewing what should be a familiar sequence of orbital energies for group I metal Rb where the conventional (Magdelung Rule) energy sequence is

1*s*<2*s*<2*p*<3*s*<3*p*<3*d*<4*s*<4*p*<5*s*

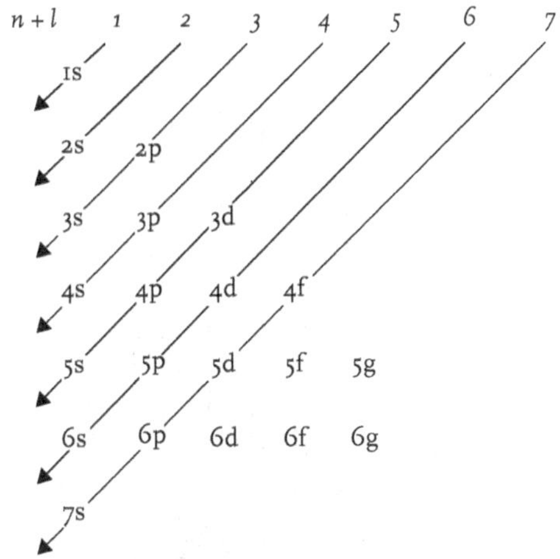

FIGURE 1.22 Aufbau sequence of filling orbitals according to the conventional n + l rule up to the beginning of the seventh period of the periodic table.

That periodic table's aufbau sequence, in which each period corresponds to one value of $n + \ell$, was first suggested in 1927 by Charles Janet, a French engineer, and put on a sounder footing by the German physicist Erwin Madelung in 1936 from analysis of atomic spectra. In 1962, Vsevolod Klechkowski, the Russian chemist, advocated a theoretical explanation for the $n + \ell$ rule based on the Thomas-Fermi model of the atom. The sequence is diagramed schematically in Figure 1.22 but its strict *aufbau* property has been arguably problematic.

If we ignore the strong influence of relativistic spin-orbit coupling, then the Mendeleev-Seaborg extended periodic table would follow the traditional *spdf* (*s*hell-*p*artitioned *d*isplay *f*ormat) scheme (Karol 2002) shown in Figure 1.23. The 8th period, at the bottom of the periodic table would run from Z = 119 (eka-Fr) through Z = 168 (eka-Og). This period would be the initial appearance of the g-block elements, termed *octadecanoids* by Goldansii (1970). Seaborg (1969), on the other hand, referred to a *superactinide series*, occupying the 6*f* and 5g shells.

But the progression of the periodic table is not that simple, by far.

Aufbau trends are relatively "normal" up to about Z ≈ 120. As Fricke and Soff (1977) and Pyykkö (2011) note, the sequence becomes quite seriously chaotic as the various (*f*, *g*) transition series come into play as shown in Figure 1.24. Above Z ≈ 160, where the transition series seem to be completed, the more normal sequence reappears.

The significance of these results for the chemistry of elements above the 8th period's alkaline earth eka-radium (Z = 120) is that exact assignment of electron configuration becomes problematic as evidenced by three choices of

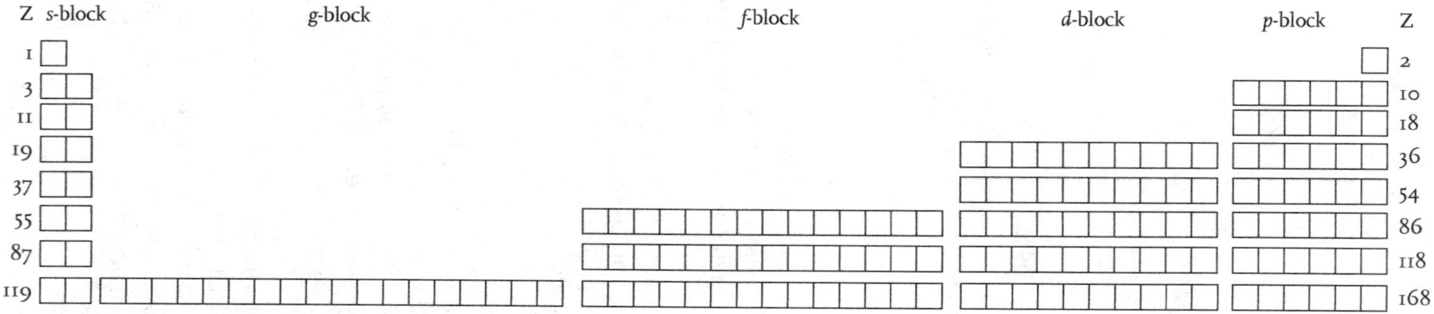

FIGURE 1.23 The spdf long display of the Periodic Table following the n + l rule.

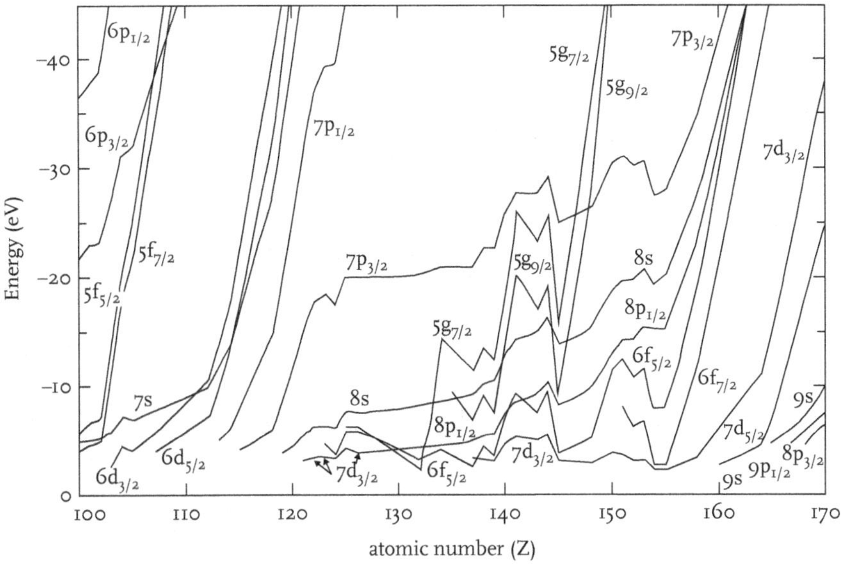

FIGURE 1.24 Calculated energies for various atomic states in the superheavy region. Note that some neighboring energy states can be separated only slightly in energy, implying probable mixing if not imprecise assignment.

periodic table displayed here in Figures 1.23, 1.25, and 1.26. Even the most disjointed result has been somewhat smoothed over by relying on configurations of ions rather than atoms to decide on placement. For example, element 121's ground state configuration could be written as $[118]8s^28p_{1/2}^{1}$ or $[118]8s^25g^1$. The former is the theoretically more stable arrangement. However, Pyykkö notes that $[125]^{6+}$ has a predicted configuration of $[118]5g^1$ and that most subsequent ions also have 5g-electron ground states from which the conclusion was that elements 121–138 may be viewed as consisting of a g-block despite the fact that some early members of the block would have other valence orbitals preferred.

Pyykkö cautions that even though the atomic number Z can be known for superheavy elements, the concept of an electron configuration is only approximate. "True" electron configurations can be a superposition of many effective electron configurations. For elements $121 \leq Z \leq 131$, *8p*, *7d*, *6f*, and 5g orbitals may all be involved (Pyykkö 2011). Nevertheless, the trend in the later parts of the periodic table aufbau is represented approximately by the sequence

$$8s < 5g \leq 8p_{1/2} < 6f < 7d < 9s < 9p_{1/2} < 8p_{3/2}.$$

Predictions of ionization energies (Fig. 1.27) and oxidation states have also been made with the intriguing suggestion that oxidation numbers as high as 12+ might be attainable.

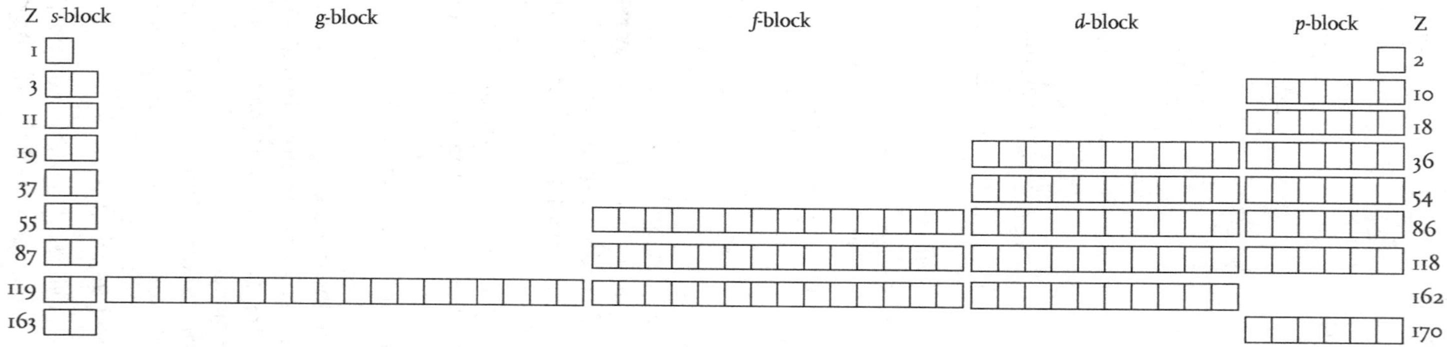

FIGURE 1.25 Extended periodic table calculated according to relativistic treatment of spin-orbit splitting which stabilizes the 9s orbital sufficiently to break the conventional sequence (Fricke).

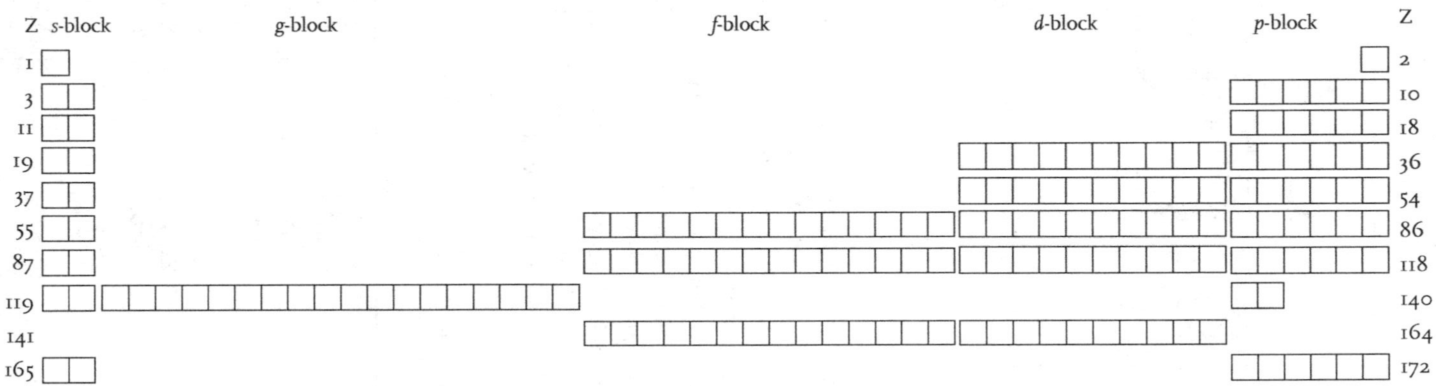

FIGURE 1.26 The extended Periodic Table including the effect of spin-orbit relativistic effects stabilizing the $8p_{1/2}$ and 9s levels (Pyykkö).

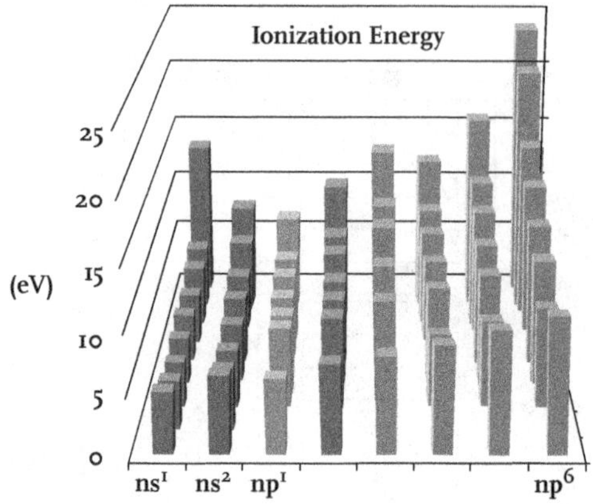

FIGURE 1.27 Ground state first ionization energies for the s- and p-block elements in the first through ninth rows of the Periodic Table (data from Fricke 1971) from the back to the front, respectively.

Quo Vadis?

While predictions of nuclear stability currently seem to project viable compositions up to elements $Z \approx 172$, production is already problematic at $Z \approx 120$, the region of the "island of stability" where half-lives might be long enough to enable chemical studies. It remains to be seen whether or not new synthesis technologies will allow continuation of the journey. This is certainly a conceivable possibility. The vast unexplored zone beyond $Z \approx 120$ is likely strewn with physicochemical surprises that will allow, if not provoke, modifications to relativistic treatment of electronic structure at a fundamental level and where perturbation approaches will fail. These need not wait for a contiguous extension of atomic number species, for even a few islets in the ocean of instability

TABLE 1.8 Synthetic routes reported by Emsley (2011) to produce elements with $Z \geq 119$

Z	TARGET	PROJECTILE	FUSED PRODUCT
119	${}_{99}$Es-254	${}_{20}$Ca-48	[${}_{119}$X-302]
	${}_{97}$Bk-249	${}_{22}$Ti-50	[${}_{119}$X-299]
120	${}_{94}$Pu-244	${}_{26}$Fe-58	[${}_{120}$X-302]
	${}_{92}$U-238	${}_{28}$Ni-64	[${}_{120}$X-302]
	${}_{96}$Cm-248	${}_{24}$Cr-54	[${}_{120}$X-302]
	${}_{98}$Cf-249	${}_{22}$Ti-50	[${}_{119}$X-299]
122	${}_{92}$U-238	${}_{30}$Zn-66	[${}_{122}$X-304]
	${}_{92}$U-238	${}_{30}$Zn-70	[${}_{122}$X-308]
	${}_{68}$Er(nat'l)	${}_{54}$Xe-136	[${}_{122}$X-many]
124	${}_{92}$U-238	${}_{32}$Ge(nat'l)	[${}_{124}$X-many]
126	${}_{90}$Th-232	${}_{36}$Kr-84	[${}_{126}$X-316]
127	${}_{73}$Ta(nat'l)	${}_{54}$Xe-136	[${}_{127}$X-many]

will suit the purpose. Even now, theoretical predictions for the changed appearance of the Periodic Table are considerable and await experimental validation.

Epilogue

We have not reviewed how much the methods for synthesis and characterization have improved over the years. These aspects are covered amply in two recent reviews (Hamilton et al., 2013 and Türler and Pershina 2013). Unsuccessful attempts at synthesizing new elements beyond Z = 118, commencing the eighth period of the Periodic Table have been reviewed by Emsley (2011) and are tabulated, with some additional references, in Table 1.8.

References

Berger, J., Bitaud, L., Dechargé, J., Girod, M., and Dietrich, K. 2001. Superheavy, hyperheavy and bubble nuclei. Nucl. Phys. A685, 1c.

Brodzínski, W. and Skalski, J. 2013. Predictions for superheavy elements beyond Z = 126. Phys. Rev. C 88, 044307.

Cohen-Tannoudji, C., Diu, B., and Laloë, F, 1977. Quantum Mechanics, vol 2, chapt XII, John Wiley & Sons, New York.

Desclaux, J.P. 1973. Relativistic Dirac-Fock expectation values for atoms with Z=1 to Z=120. Atom Data Nucl. Data Tables 12, 311.

Emsley, J. 2011. *Nature's Building Blocks: An A-Z Guide to the Elements*. Oxford, New York.

Fricke, B., Greiner, W., and Waber, J. 1971. The continuation of the periodic table: the chemistry of superheavy elements, Theor. Chim. Acta 21, 235.

Fricke, B. and Soff, G. 1977. Dirac-Fock-Slater calculations for the elements Z = 100, fermium, to Z = 173. Atom. Data Nucl. Data Tables 19, 83–95.

Gamow, G. 1928. Zur Quantentheorie des Atomkernes. Z. Phys. 51, 204.

Gamow, G. 1930. Mass defect curve and nuclear constitution. Proc. Roy. Soc. A 126, 632.

Gol'danskii, V.I.: 1970. The periodic system of D. I. Mendeleev and problems of nuclear chemistry. J. Chem. Ed. 47, 409.

Greiner W. 2012. Superheavy nuclei and beyond: hypermatter and antimatter, J. Phys., Conference Series 337 (2012), 012002.

Grumman, J., Mosel, U., Fink, B., and Greiner, W. 1969. Investigation of the stability of superheavy nuclei around Z = 114 and Z = 164. Z. Phys. 228, 371–386.

Hamilton, J.H., Hofmann, S., and Oganessian, Y.T. 2013. Search for superheavy nuclei. Ann. Rev. Nucl. Part. Sci. 63, 383–405.

Jeans, J.H. 1928. *Astronomy and Cosmology*. Cambridge University Press, London. pp. 98–101.

Karol, P. 2002. The Mendeleev-Seaborg periodic table: through Z = 1138 and beyond. J. Chem. Ed. 79, 60.

Khazan, A. 2007. Upper limit of the periodic table and synthesis of superheavy elements. Prog. Phys. 2, 104.

Koblic, O. 1934. A new radioactive element beyond uranium. Nature 134, 55.

Kragh, H. and Carazza, B. 1994. From time atoms to space-time quantization: the idea of discrete time, ca 1925–1936. Stud. Hist. Phil. Sci. 25, 447.

Kragh, H. 2013. Superheavy elements and the upper limit of the periodic table: early speculations. Eur. Phys. J. H38, 411.

Losanitsch, S. 1906. *Die Grenzen des Periodischen Systems der chemischen Elemente.* Dositije Obradowits, Belgrade.

Malley, M. 2011. *Radioactivity*. Oxford University Press, New York.

Marinov, A., Rodushkin, I,. Kolb, D., Pape, A., Kashiv, Y., Brandt, R., Gentry, R.V., Miller, H.W. 2010. Evidence for the possible existence of a long-lived superheavy nucleus with atomic mass number A=292 and Atomic Number Z~122 in natural Th. Int. J. Mod. Phys. E 19, 131.

Mosel, U. and Greiner, W. 1969. On the stability of superheavy nuclei against fission. Z. Phys. 222, 261–282.

Narliker, V.V. 1932. The highest atomic number. Nature 129, 402.

Newlands, J.A.R. 1878. On relations among the atomic weights of the elements. Chem. News. 37, 255.

Nilsson, S.G. 1955. Binding states of individual nucleons in strongly deformed nuclei, Dan. Mat. Fys. Medd. 29, No. 16, 1.

Pershina, V. and Fricke, B. 1993. Relativistic effects in physics and chemistry of element 105. IV. Their influence on electronic structure and related properties. J. Chem. Phys. 99, 9720–9729.

Pershina, V. 2011. Relativistic electronic studies on the heaviest elements. Radiochim. Acta 99, 459.

Pieper, W. and Greiner, W. 1969. Interior electron shells in superheavy nuclei. Z. Phys. 218, 327.

Pomeranchuk, I. and Smorodinsky, J. 1945. On the energy levels of systems with Z > 137. J. Phys. (USSR) 9, 97.

Powell, R.E. 1968. Relativistic quantum chemistry. J. Chem. Ed. 45, 558–563.

Pyykkö, P. 2011. A suggested periodic table up to Z ≤ 172, based on Dirac-Fock calculations on atoms and ions. Phys. Chem. Chem. Phys. 13, 161.

Schiff, L., Snyder, H., and Weinberg, J. 1940. On the existence of stationary states of the mesotron field. Phys. Rev. 57, 315.

Seaborg, G. T. 1969. Prospects for Further Considerable Extension of the Periodic Table. J. Chem. Educ.46, 626.

Seaborg, G.T. and Loveland, W.D. 1990. *The Elements Beyond Uranium.* Wiley, New York.

Steppenbeck, D., Takeuchi, S., Aoi, N., Doornenbal, et al., 2013. Evidence for a new nuclear "magic number" from the level structure of ^{54}Ca. Nature 502, 207.

Tarasov, V., Gridnev, K., Gridnev, D., Tarasov, D., Schramm, S. Viñas, X., and Greiner, W. 2013. Stability peninsulas on the neutron drip line. Intl. J. Mod. Phys. E 22, 1350009.

Türler, A. and Pershina, V. 2013. Advances in the production and chemistry of the heaviest elements. Chem. Rev. 113, 1237.

Weizsäcker, C. von. 1935. Zur Theorie von Kernmassen. Z. Phys. 96, 431.

CHAPTER 2 | Nuclear Lattice Model and the Electronic Configuration of the Chemical Elements

JOZSEF GARAI
Department of Civil Engineering, University of Debrecen, Hungary

1. *Introduction*

1.1 Preface

In the earliest days of science researchers were arguing philosophically what might be the reasonable explanation for an observed phenomenon. The majority of the contemporary scientific community claims that these arguments are useless because they do not add anything to our understanding of nature. The current consensus on the aim of science is that science collects facts (data) and discerns the order that exists between and among the various facts (e.g., Feynman 1985). According to this approach the mission of science is over when the phenomenon under investigation has been described. It is left to the philosophers to answer the question what is the governing physical process behind the observed physical phenomenon. Quantum mechanics is a good example of this approach, "It works, so we just have to accept it." The consequence is that nearly 90 years after the development of quantum theory, there is still no consensus in the scientific community regarding the interpretation of the theory's foundational building blocks (Schlosshauer et al. 2013). I believe that identifying the physical process governing a natural phenomenon is the responsibility of science. Dutailly (2013) expressed this quite well: A "black box" in the "cloud" which answers our questions correctly is not a scientific theory, if we have no knowledge of the basis upon which it has been designed. A scientific theory should provide a set of concepts and a formalism which can be easily and indisputably understood and used by the workers in the field.

In this study the main unifying principle in chemistry, the periodic system of the chemical elements (PSCE) is investigated. The aim of the study is not only the description of the periodicity but also the understanding of the underlying physics resulting in the PSCE.

1.2 The periodic pattern of the chemical elements emerged

By 1860 about 60 elements had been identified, and this initiated a quest to find their systematic arrangement. Based on similarities, Döbereiner (1829) in Germany suggested grouping the elements into triads. Newlands (1864; 1865) in England arranged the elements in order of increasing atomic weights and based on the repetition of chemical properties, proposed the "Law of Octaves." Listing the elements also by mass, Mendeleyev (1869; 1872) in Russia and Meyer (1870) in Germany simultaneously proposed a 17-column arrangement with two partial periods of seven elements each (Li-F and Na-Cl) followed by nearly complete periods (K-Br and Rb-I). Mendeleev gets greater credit for this discovery because he published the results first; he also rearranged a few elements out of strict mass sequence in order to fit better to the properties of their neighbors and corrected mistakes in the values of several atomic masses. Additionally, he predicted the existence and the properties of a few new elements by leaving empty cells in his table. Mendeleev's periodic table did not include the noble gases, which were discovered later. Argon was identified by Rayleigh in 1895(a,b). The remaining noble gases were discovered by Ramsay (1897) who positioned them in the periodic table in a new column. Anton van den Broek (1911; 1913) suggested that the fundamental organizing principle of the table is not the weight but rather the nuclear charge, which is equivalent to the atomic number. The extended 18-column table was slightly modified based on Moseley's experiments (1913), and he rearranged the table according to atomic number. The discovery of the transuranium elements from 94 to 102 by Seaborg (1951) further expanded the table. He also reconfigured the table by placing the lanthanide/actinide series at the bottom.

There is no "standard" or approved periodic table. The only specific recommendation of the International Union of Pure and Applied Chemistry (IUPAC), which is the governing body in chemistry, is that the PSCE should follow the 1 to 18 group numbering (Leigh 2009) (Fig. 2.1).

1.3 Sequences of the Table

The organizing pattern of the PSCE can be described by three sequences. These digital descriptions of the table are the fundamental [$S_{fundamental}$], the periodic [$S_{\Delta z}$], and the atomic number [S_Z] sequences (Fig. 2.1) (Garai 2008). The fundamental sequence of the table is:

$$S_{fundamental} = \{1, 2, 2, 3, 3, 4, 4, \ldots\} \quad [1]$$

The number of elements [$\Delta Z(n)$] within the period or the length of the periods has the sequence of

$$S_{\Delta Z} = \{2, 8, 8, 18, 18, 32, 32 \ldots\}. \quad [2]$$

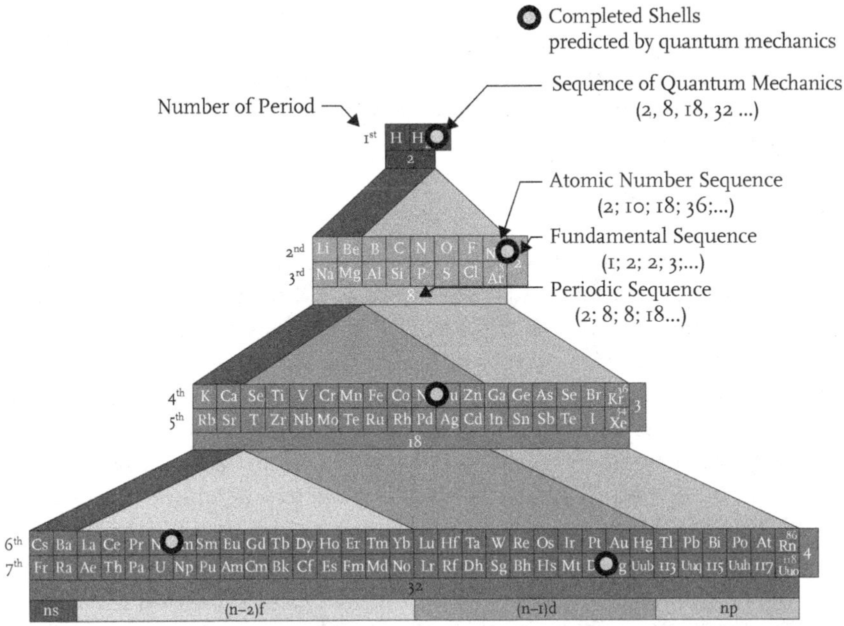

FIGURE 2.1 The sequences of the periodic table. The sequence derived from quantum theory is also shown. It can be seen that the sequence derived from quantum theory does not match with the sequences of the PSCE, indicating that the theory should not be considered as a viable physical explanation for the periodicity.

The atomic number or the nuclear charge of the elements [Z(n)] in a completely developed period follows the sequence

$$S_Z = \{2, 10, 18, 36, 54, 86, 118\ldots\}. \quad [3]$$

1.4 Pauli Exclusion Principle

In order to explain the observed light emission patterns observed in atoms, Pauli postulated the exclusion principle: "In an atom there cannot be two or more equivalent electrons for which the values of all four quantum numbers coincide. If an electron exists in an atom for which all of these numbers have defined values, then this state is occupied" (Pauli 1925; 1964). The principle was originally postulated for electrons but it has been generalized and now includes other particles. The Pauli Exclusion Principle is one of the cornerstones of quantum physics and thus it is at the basis of modern physics. It is connected to spin statistics dividing the world into fermions and bosons. Particles with half-integer spin (fermions) are described by antisymmetric wavefunctions and particles with integer spin (bosons) are described by symmetric wavefunction (Pauli 1946). The rise of the periodic table is one of the important outcomes of the Pauli Exclusion Principle.

In his Nobel lecture in 1946 (Pauli 1946) Pauli stated that "I was unable to give a logical reason for the exclusion principle or to deduce it from more general assumptions. I had always the feeling and I still have it today, that this is a deficiency." After many decades, there is still no physical explanation for Pauli's postulate (Kaplan 2013). The Pauli Exclusion Principle is the theoretical basis for the periodic table. Thus the physical explanation for the periodicity of the chemical elements is also lacking.

1.5 Physical Explanations

The Aufbau principle of how the electronic configuration of the atom is built up was originally postulated by Niels Bohr and Wolfgang Pauli. They stated that "The orbitals of lower energy are filled in first with the electrons and only then the orbitals of high energy are filled." The energy levels calculated from the hydrogenic model, or the main quantum number, follow the sequence of $2n^2$. Only the first two entries of this quantum number sequence are consistent with the periodic sequence. In order to explain the periodic sequence by quantum theory the (n, l) rule or the Madelung energy ordering (Karapetoff 1930; Madelung 1936) that applies to neutral atoms in their ground state has been suggested. The majority of general and physical chemistry books present the symmetry of the PSCE as satisfactorily explained by either the electronic structure of the elements (Bohr 1922; Hund 1925; Slater 1930; Condon & Shortley 1935; Landau & Lifschitz 1977; Schwabl 2001; Atkins & Atkins 2001) or by quantum mechanics (Hartree 1928a,b,c 1957; Fischer 1977; Johnson 2005). Some authors present a quantum justification of the rule (Demkov and Ostrovsky 1971; Ostrovsky 1981, 2001). Not all of the elements comply with the (n, l) rule; therefore, general acceptance is lacking (Scerri 2004; Boeyens 2008).

Based on the conflicting interpretations it can be concluded that the most important periodic sequence of the chemical elements has not been satisfactorily explained (Scerri 1998; Schwarz and Wang 2010; Boeyens 2013). One example of this conclusion might be the positions of hydrogen and helium. Hydrogen has one 1s electron but one electron is also needed to attain inert configuration; this can be placed either in the 1st group or in the 17th group. Based on chemical behavior, hydrogen is neither a halogen nor an alkali metal but rather both;thus the position of hydrogen in the PSCE is uncertain (Scerri 2007). With its $1s^2$ electron configuration, helium is the other element with uncertain position. Helium should be in the 2nd group. However, its chemical properties are equivalent to an inert gas, and it is placed into group 18. Besides helium, the outermost electron configuration of group 18th is $2p^6$. No theoretical explanation for the shift in the electron configuration from $2p^6$ to $1s^2$ or vice versa is offered.

Based on these deficiencies, it has been suggested that quantum mechanics is unable to explain the most important aspects of the periodic table (Scerri 1998; Boeyens 2008; Schwarz and Wang, 2010; Boeyens, 2013). The substantial number of articles in the current literature (e.g., Sneath 2000; Giunta 2001; Kragh 2001; Ostrovsky 2001; Scerri 2001; Dudek et al. 2002; Baum 2003; Ostrovsky 2003; Moore 2003; Scerri 2003; Friedrich 2004; Kibler 2004; Schunck & Dudek 2004; Scerri 2004; Schwarz 2004; Rouvray & King 2005; Bent 2006;

Rouvray & King 2006; Wang et al. 2006; Restrepo & Pachon 2007; Schwarz 2007; Weinhold & Bent 2007; Boeyens & Levendis 2008; Wang & Schwarz 2009) discussing the fundamental problems of the PSCE, nearly one-and-a-half centuries after its invention, also indicates that there is no complete physical understanding of the symmetry expressed by the PSCE.

Despite the remaining outstanding questions, there is a general agreement that the elements should be arranged in an increasing atomic number order in the PSCE. Invariant symmetry relating to the nuclear charge can only be possible if the positions of the nucleons are invariant too.

2. *Nuclear Models*

The most widely accepted models for the nucleus are the shell (Mayer 1949; Haxel et al. 1949; Rainwater 1950; Mayer & Jensen, 1955), the liquid drop (Bohr 1936; Feenberg 1947) and the cluster (Hafstad & Teller 1938) models. The shell model assumes a gas phase (Fermi) for the nucleus and is able to explain the independent quantum characteristics of the nucleons. The liquid drop model is able to explain the observed saturation properties of the nuclear forces, the low compressibility of the nucleus, the well-defined nuclear surface, the binding energies, and—most importantly—the fission phenomena. The clustering of the alpha particle model was deduced from the fact that certain large nuclei emit alpha particles and the stability and the abundance of the 4n-nuclei are significantly higher. These models are able to describe successfully certain selected properties of the nucleus; however, none of them are able to give a comprehensive description. The basic assumptions of the models are contradictory; therefore, it is impossible to combine them and develop a hybrid model.

The assumed phase of the nucleus in these models is gas, liquid, or semi-solid. None of these phases has an invariant nucleon position; therefore, these models cannot maintain the symmetry present in the PSCE. The preservation of symmetry requires a "solid phase" nucleus, in which the positions of the nucleons are preserved. "Solid phase" or lattice models have not been considered for many decades as a viable option because of the uncertainty principle and the lack of diffraction. In the 1960s, the discovery of quarks and neutron star research satisfactorily answered these objections and opened the door for lattice models. The first nuclear lattice models were presented by Linus Pauling in 1965(a,b,c).

The lattice models can easily reproduce the various shell, liquid-drop, and cluster properties (Cook & Dallacasa 1987; Cook & Hayashi 1997). Asymmetric fission and heavy-ion multifragmentation are some of the phenomena that the traditional models of nuclear structure theory cannot explain, yet can be reproduced by lattice models (Gupta et al. 1996, 1997).

Significant effort has been made to find correlation between lattice positions and quantum numbers, with partial success for FCC structure (Pauling 1965d; Cook & Dallacasa 1987). The symmetries of Schrödinger's equation also correspond to FCC geometry (Wigner 1937). The common features of the developed lattice models are that the protons and the neutrons are the same size and that

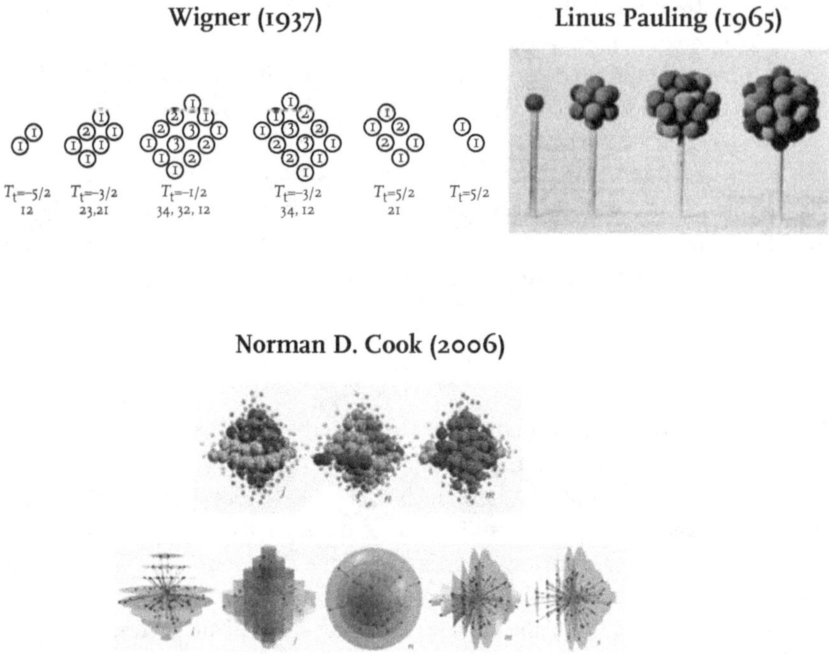

FIGURE 2.2 Previous lattice models of the nucleus. These investigations expanded the FCC nuclear lattice structure spherically.

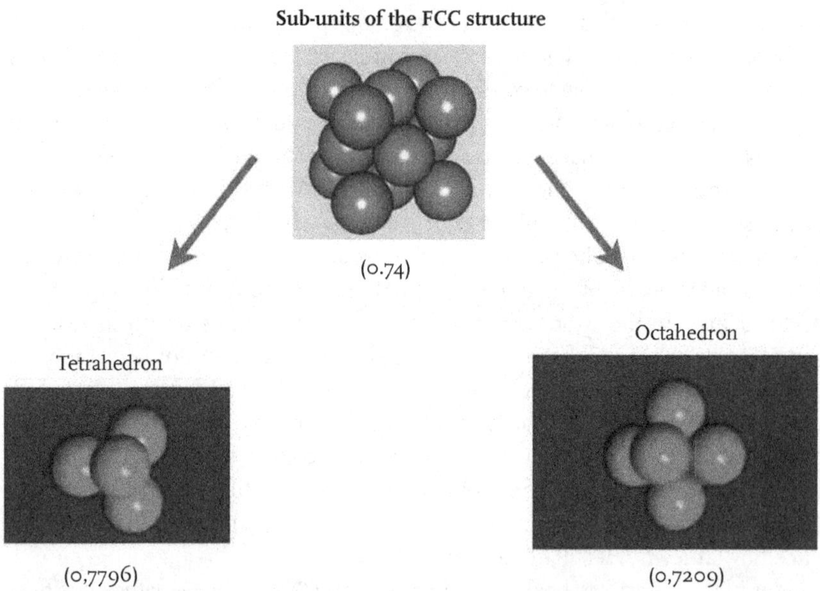

FIGURE 2.3 The subunits of the face centered cubic structure and their densities. The density of the FCC structure is 0.74. The density of the subunits, tetrahedron and octahedron are 0.7796 and 0.7209, respectively. The ratio of the tetrahedron and octahedron subunits in the FCC structure is 2:1.

they are alternately arranged in the closest packing array (Anagnostatos, 1973; Canuto & Chitre 1974; Lezuo 1974; Cook 1976; Matsui et al. 1980; Dallacasa 1981). These assumptions are reasonable.

The radii of protons and neutrons differ only slightly (Schery et al. 1980). The same proton and neutron magic numbers indicate the same structural development for both protons and neutrons, which is consistent with an alternate arrangement. The equal spheres will most likely utilize the available space in the most efficient way, which is a closest packing arrangement. Previous investigations (Pauling 1965d; Cook & Dallacasa 1987), which expanded the FCC structure spherically, had partial success in finding correspondence between lattice positions and quantum states (Fig. 2.2). The structure of the closest packing arrangements consists exclusively of tetrahedron and octahedron subunits. In order to find structural symmetry patterns in the FCC lattice, the expansion of these sub-units should be investigated (Garai 1999; 2003) (Fig. 2.3).

3. Tetrahedron FCC Lattice Model

Assuming that the bonding energy correlates with the nucleon density, then the densest structure should be formed preferentially. The densities of the tetrahedron and octahedron units in FCC are 0.7796 and 0.7209, respectively (Garai 2010) indicating preferential tetrahedron formation.

The first sequence of the nuclear structural development is completed by the nucleus of helium, which contains four nucleons. The closest packing arrangement of the four nucleons is a tetrahedron. Forming a tetrahedron subunit nucleus in the first completed period is an additional support for tetrahedron nucleus formation. Calculations of potential models, constrained by the hadron spectrum for the confinement of the relativistic quark (Goldman et al. 1988; Maltman et al. 1994) and the colored quark exchange model (Robson 1978), are also consistent with a tetrahedron forming the He nucleus. The expansion of this tetrahedron seed of four nucleons is investigated here.

The tetrahedron shape of equal spheres arranged in FCC packing can be formed from layers of equilateral triangles packed in two-dimensional closest-packing arrangement as shown in Figure 2.4a. Starting with one sphere and increasing the length of the side of the triangles by one additional sphere, the number of nucleons in each triangle plane will be 1, 3, 6, 10, 15, 21, 28, 36… (Fig. 2.4a). Stacking these layers, the numbers of spheres in two consecutive layers are 4, 16, 36, 54…. Assuming that protons and neutrons are alternately arranged in the lattice the number of spheres should be divided by two. This gives the proton numbers to 2, 8, 18, 32…respectively (Fig. 2.4b). These numbers are identical with the number of possible states of the principle quantum numbers. If the tetrahedron is extended on two faces, then a shell-like structure can be formed (Fig. 2.4c), which is consistent with the physical interpretation of the principle quantum number.

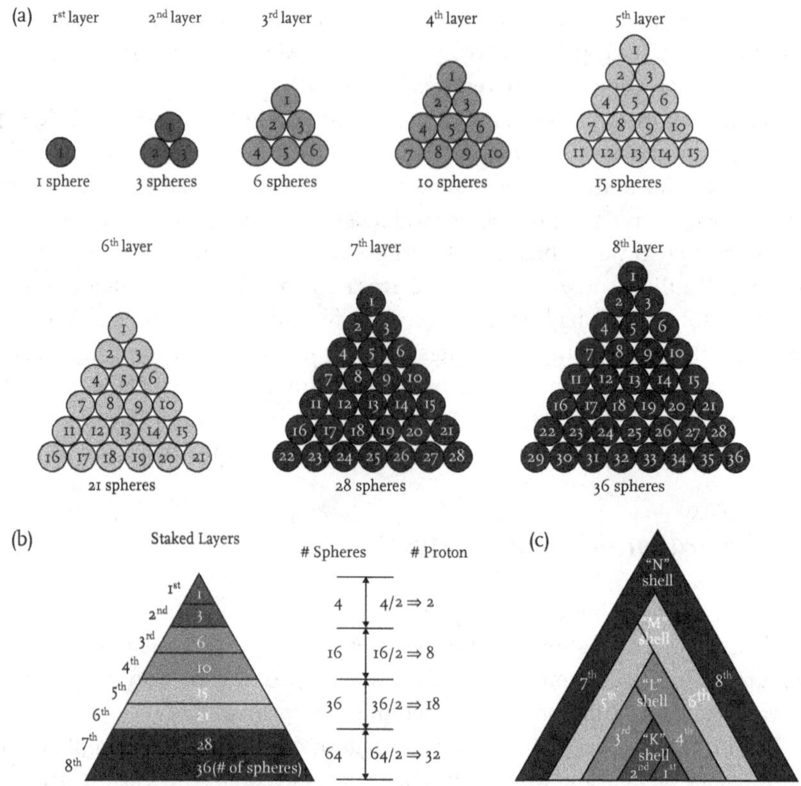

FIGURE 2.4 Representing both protons and neutrons with equal spheres and arranging them in FCC structure, the number of protons in the outer layers of a tetrahedron formation is the same as the number of possible states of the principle quantum numbers. (a) The number of spheres in a two dimensional closed packing arrangement in equilateral triangles. (b) The number of spheres in two consecutive layers of the tetrahedron formation. Assuming a proton:neutron ratio of one, the outer layers of the tetrahedron contain the same number of protons as predicted by quantum theory. (c) The same tetrahedron formation can be developed by adding the new layers to alternate sides.

Investigating how many spheres can be accommodated in one row in the outer shell gives the total number of spheres in one row of 4, 12, 20, 28..., which corresponds to proton numbers 2, 6, 10, 14... (Fig. 2.5a). These proton numbers are identical with the number of states determined by the angular momentum quantum numbers corresponding to s, p, d, and f orbitals. The rows in the outer layers of the tetrahedron are one unit distance away from each other; thus the identical agreement of the number of nucleons with the angular momentum quantum numbers is not only a numerical agreement but also an agreement with the physical meaning defined by quantum theory.

The number of different positions of protons in one row of the outer shell is the same as the number of magnetic quantum numbers (Fig. 2.5b). The lattice positions also reproduce the multiplicities. Thus the number of positions

Cross-Sections of Vertex of Tetrahedron

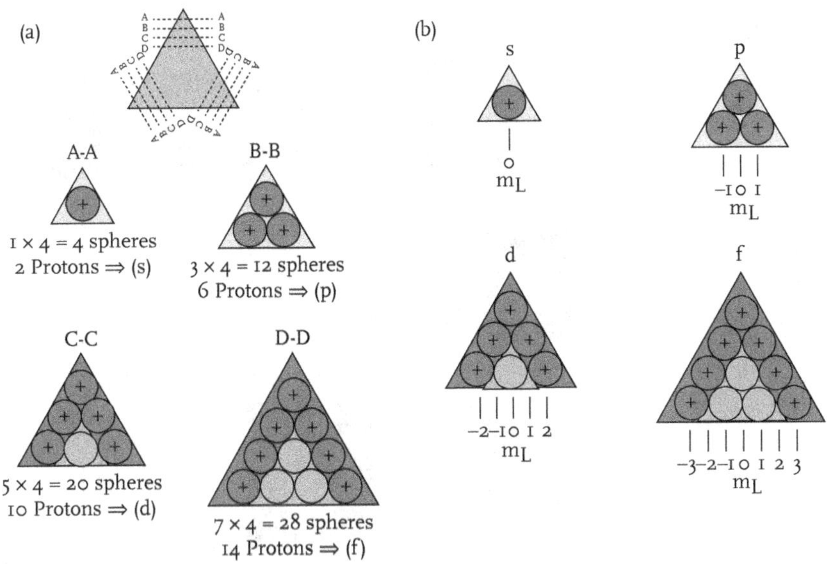

FIGURE 2.5 If a tetrahedron has been developed from a core tetrahedron, which contains four spheres, then the number of protons in one layer of the outer shell of the tetrahedron is equivalent with the number of states of the angular momentum quantum number or the corresponding sub-shell. The number of different positions of the protons in one layer of the shell is the same as the number of magnetic quantum numbers. The red circles represent the outer shell nucleons. The tetrahedron has four vertexes; therefore, the number of spheres is multiplied by four. (a) Number of spheres in one layer of a vertex of the tetrahedron. (b) Number of the different proton positions in one layer of a vertex of a tetrahedron.

in an FCC lattice is identical to the quantum numbers if a tetrahedron seed is expanded. The lattice positions not only reproduce all of the quantum numbers but also bear the same physical meaning.

It has been assumed that a higher nuclear density is preferable to a lower one. The density of structures built in FCC arrangement can be described by the ratio of the tetrahedron and octahedron sub-units. The higher ratio corresponds to higher nuclear density. It can be shown that if a tetrahedron is expanded by rotating 90 degrees at each expansion, then the density of this joint (or double) tetrahedron is higher in comparison to a single tetrahedron. The higher density suggests that the initial single tetrahedron should be developed by alternately rotating 90 degrees at each expansion of the tetrahedron. Three-dimensional images of the completed tetrahedrons corresponding to the elements He, Ne, Ar, Kr, Xe, and Ra are shown in Figure 2.6(a–f). The nuclear FCC tetrahedron lattice structure expanded by rotating 90 degrees reproduces not only the quantum numbers but also the periodicity of the chemical elements.

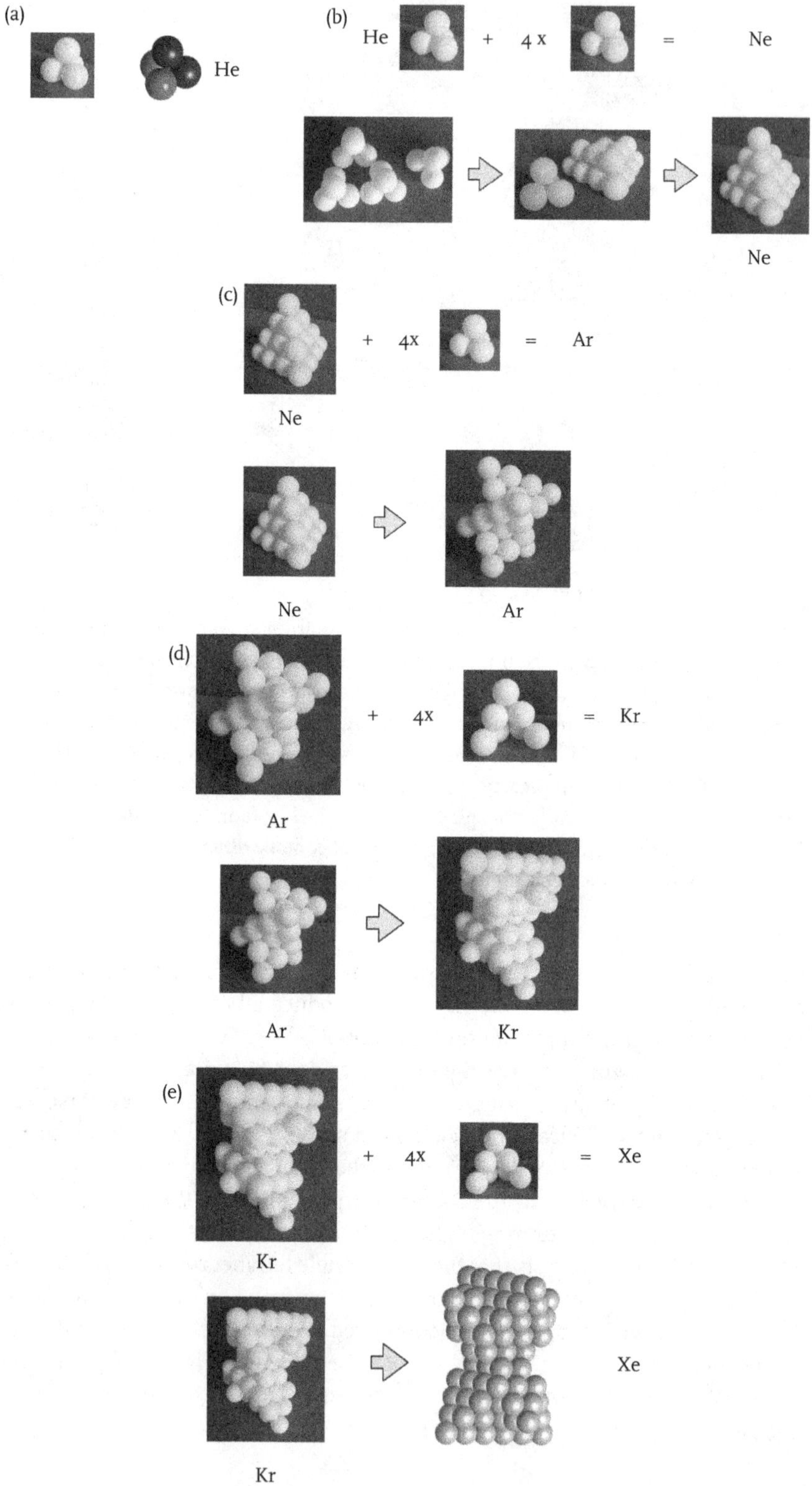

FIGURE 2.6 3D images show the completed nuclear structures of the noble gases. (a) Helium (b) Neon (c) Argon (d) Krypton (e) Xenon (f) Radon

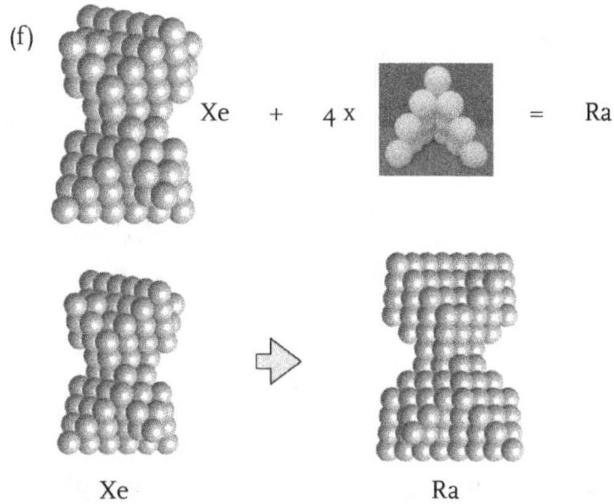

FIGURE 2.6 Continued

This study focuses on the relationship between the nucleus and the PSCE; therefore, the additional characteristics of the nucleus supporting the tetrahedron lattice model are only listed here.

- The expansion of a tetrahedron into four dimensions with angles of 109.5 degrees reproduces the original tetrahedron symmetry for every fourth nucleus. This is consistent with the observed zero magnetic momentum for each even-even nucleus.
- The disintegration of a structure should produce fragments of its basic lattice. The fragmentation of the tetrahedron FCC nuclear lattice is consistent with nuclear fission. The preferred alpha decay of the nuclear structure is consistent with the disintegration of an FCC lattice, which is built up from tetrahedron sub-units.
- The measured bulk density of the nucleus is consistent with the density of the FCC lattice arrangement (Cook and Dallacasa 1987; Cook 2010).
- The charge density distribution of the individual elements (Hofstadter 1961) is also consistent with the shape of the "double" tetrahedron FCC nuclear model.

4. *Mathematical Description of the Periodic System*

The proposed tetrahedron nuclear lattice model can be used to derive the sequences of the periodic table. The three noticeable sequences in the periodic table are the fundamental {1, 2, 2, 3, 3, 4, 4...}, the periodic {2, 8, 8, 18, 18, 32, 32...}, and the atomic number {2, 10, 18, 36, 54, 86, 118...} sequences (Fig. 2.1).

4.1 Fundamental Sequence

The relationship between the periods (n) and the sequence numbers (m) can be described as

$$m = \frac{2n + (-1)^n + 3}{4} \qquad [4]$$

4.2 Periodic Sequence

The number of nucleons in the kth layer of a tetrahedron can be calculated by the triangular number [Tr(k)] (Abramowitz & Stegun 1964; Beiler 1964) (Fig. 2.4a):

$$Tr(k) = \frac{k}{2}(k + 1). \qquad [5]$$

In each structural step of its development, the tetrahedron is expanded by one layer in two directions (Fig. 2.4c) giving the relationship between the tetrahedron layers and the sequence numbers as

$$k = 2m. \qquad [6]$$

The number of nucleons in the outer shell of the tetrahedron [Tr(m)] is the sum of the two consecutive triangular numbers:

$$Tr(m) = Tr(k) + Tr(k - 1) = Tr(2m) + Tr(2m - 1) = 4m^2 \qquad [7]$$

The number of charges in the completely developed shell is

$$\Delta Z(n) = \frac{Tr(m)}{2} = 2m^2 \qquad [8]$$

Substituting the sequence number from Eq. 4 gives:

$$\Delta Z(n) = \frac{1}{8}[2n + (-1)^n + 3]^2 \qquad [9]$$

which is the same formula given by Tomkeieff (1951, 1954).

4.3 Atomic Number Sequence

A formula giving the total number of charges in the nucleus with completely developed shells can be derived in a similar manner. The total number of nucleons in a tetrahedron with k layers can be determined by its tetrahedral number [Th(k)] (Abramowitz & Stegun 1964; Conway & Guy1996):

$$Th(k) = \frac{k}{6}(k + 1)(k + 2) \qquad [10]$$

Substituting the sequence number from Eq. 6 gives the number of nucleons in a tetrahedron for sequence (m) as

$$Th(m) = \frac{m}{3}(2m + 1)(2m + 2) = \frac{4m^3}{3} + 2m^2 + \frac{2m}{3} \qquad [11]$$

The tetrahedron nucleus is developed by alternately expanding the tetrahedrons rotated by 90 degrees (Fig. 2.4d). The number of nucleons in the tetrahedron is

$$Th^{double}(n) = 2Th(m) - Tr^{even\text{-}period}(m) - 4 \qquad [12]$$

The formula

$$\frac{(-1)^n + 1}{2} \qquad [13]$$

can be used to generate 0 for odd periods and 1 for even periods, and

$$Tr^{even\text{-}period}(m) = \frac{(-1)^n + 1}{2} Tr(m)$$

Equation 12 can be rewritten then as

$$Th^{double}(n) = 2Th(m) - \frac{(-1)^n + 1}{2} Tr(m) - 4 \qquad [15]$$

The number of charges in the nucleus in a completely developed sequence is

$$Z(n) = \frac{Th^{double}(n)}{2}. \qquad [16]$$

Combining Eqs. 4, 7, 11, 12, 15, and 16 gives the number of nuclear charges for any period. The atomic number sequence of the periodic table can then be described as:

$$Z(n) = \frac{1}{3}\{4m^3 - [(-1)^n - 1]3m^2 + 2m - 6. \qquad [17]$$

Substituting m from Eq. 4 gives the atomic number sequence (Garai 2008)

$$Z(n) = \frac{1}{48}[2n + (-1)^n + 3]^3 - \frac{1}{16}[(-1)^n - 1][2n + (-1)^n + 3]^2 + \frac{1}{6}[2n + (-1)^n + 3] - 2 \qquad [18]$$

Any physical models proposed for the explanation of PSCE has to explain and reproduce the sequences of the table.

For comparison the sequence of the Principle Quantum Numbers [PQN(n)] is given as:

$$S_{PQN} = \{2, 8, 18, 32\ldots\}. \qquad [19]$$

The first two entries are the same in both the sequence of the principle quantum number and the sequence of the length of the period. The rest of the entries in these sequences show similarity but they are not identical. The sequence of the principle quantum number can be given as

$$PQN(n) = 2n^2 \qquad [20]$$

This formula (Eq. 20) emerging from quantum theory does not agree with any of the sequences of the periodic table (Eqs.4, 9, 18) indicating that quantum mechanics cannot give a viable explanation for the sequences of the periodic table.

5. *Pauli Exclusion Principle*

The chemical properties of the elements are defined by the electronic structure of the outermost, or valence, shell. The identity of an atom, including its electronic structure is determined by the nucleus—or more specifically by the number of protons (Broek 1911, 1913). The interaction between the atomic particles can be described as:

$$\text{Neutrons} \overset{\text{Strong Force Effects}}{\Leftrightarrow} \text{Protons} \overset{\text{Electrostatic Effects}}{\Rightarrow} \text{Electrons.}$$

The negatively charged electrons in an atom are captured by the electrostatic attraction of the positively charged protons. The electrostatic attraction between the two differently charged particles is described by an inverse square law. The attraction between a proton and a captured electron is the function of the distance between the two charges. The energy of a captured electron in its ground state depends on the distance between the two particles. Electrons with different energies in their ground states should be separated by different distances from the protons capturing them. This distance must be invariable otherwise the electronic structure of the atom would not be stable. In order to maintain the configuration of the protons and the distances between the protons and the captured electrons the position of the protons must be invariant. This condition is fulfilled by a nuclear lattice. The lattice structure of the nucleus ensures that if a lattice point is occupied then another proton cannot have the same position. Thus the nuclear lattice model offers a feasible physical explanation for the Pauli Exclusion Principle, which requires that the protons in the nucleus cannot coexist in the same location. We conclude that in order to comply with the Pauli Exclusion Principle the protons in the nucleus should be arranged in a lattice.

6. *Possible Physical Explanation for the Periodicity of the Electronic Structure*

The most important feature in the development of the electronic structure of the chemical elements is the {2, 8, 8, 18, 18, 32, 32...} pattern which represents both the number of elements in each period of the periodic table and the number of electrons in each completed shell. The periodicity pattern is different for the ground state neutral atoms and for the positive atomic ions (Brakel 2000; Goudsmith & Richards 1964). The simplest manifestation of periodicity, the ground state neutral atoms, are investigated here. The periodic sequence is described by Equation 9, and it is attributed to the number of protons in the outer shell of the tetrahedron nuclear structure. It will be investigated later how protons in the nuclear lattice can affect the electronic structure of the elements.

The interaction between the proton and electron is the result of the exchange of virtual photons. Based on the energies of the electrons in the atoms, virtual photon or quantum electrodynamic treatment is not necessary when the classical approach is sufficient to describe the interaction. The classical approach is supported by the success of the Bohr model, which correctly describes the orbit of electron in the Hydrogen atom (Bohr 1913a,b,c).

One of the outcomes of the nuclear lattice model is that the nucleus cannot be considered a point charge. The lattice position of a given proton must be considered when the attraction on an electron is calculated. The relative attractive force [$F_{e\text{-}P}(Z)$] of a proton [P(Z)] on an electron [e(Z)] is characterized by the distance between the charge center of the nucleus and the proton [$d_{NCC\text{-}P(Z)}$]

$$F_{e-p}(Z) = f(d_{NCC-P(Z)}). \qquad [21]$$

where Z is the corresponding atomic number defined as follows: The first proton in the hydrogen nucleus is P(1), the next proton added into the structure to form the nucleus of Helium is P(2) and the last proton added to the nuclear structure of element Z-1 to form the nucleus of element Z is P(Z). The electrons are labeled accordingly. Thus electron(1) is captured by P(1) and electron(Z) is captured by P(Z). Depending on the lattice positions of proton(Z) the attraction can be either stronger or weaker than the attraction of proton(Z-1). The relative attraction of proton Z-1 and Z can be defined by comparing the distances between the charge center of the nucleus and the position of the proton:

$$d_{p-NCC}(Z-1) < d_{p-NCC}(Z) \text{ then } F_{P-e}(Z-1) < F_{P-e}(Z) \qquad [22]$$

or vice versa. In order to overcome the repulsion force of the existing valence electrons, the relative attractive force on the captured electron(Z) must be stronger than the relative attraction force on the electrons already occupying the valence shell. Thus the following inequality must be satisfied:

$$F_{P-e}(Z) > F_{P-e}(Z-1) \Leftrightarrow d_{p-NCC}(Z) > d_{p-NCC}(Z-1). \qquad [23]$$

This inequality is consistent with the Aufbau principle, which states that orbitals with the lowest energy are filled first. When this condition (Eq. 23) is not satisfied then it is predicted that the electron(Z) is unable to join to the valence shell and starts to form a new valence shell outside of the existing one. Thus a new electron shell starts to form when Equation. 24 is satisfied.

$$F_{P-e}(Z) < F_{P-e}(Z-1) \Leftrightarrow dp-NCC(Z) < dp-NCC(Z-1) \qquad [24]$$

Based on stability considerations the first proton in the new nuclear shell should be positioned at the middle of the face of the tetrahedron and this new outer layer of the structure should develop toward the edges. It can be seen by visual inspection that in a given nuclear shell Equation 23 or the Aufbau principle is satisfied (Fig. 2.7a).

The distance of the last proton completing a shell [n] and the distance of the first proton in the new shell [n+1] are both calculated from the nuclear charge center. The position of the last proton in layer n is assumed to be at the vertex of the tetrahedron and that the he first proton in layer n+1 is positioned at the

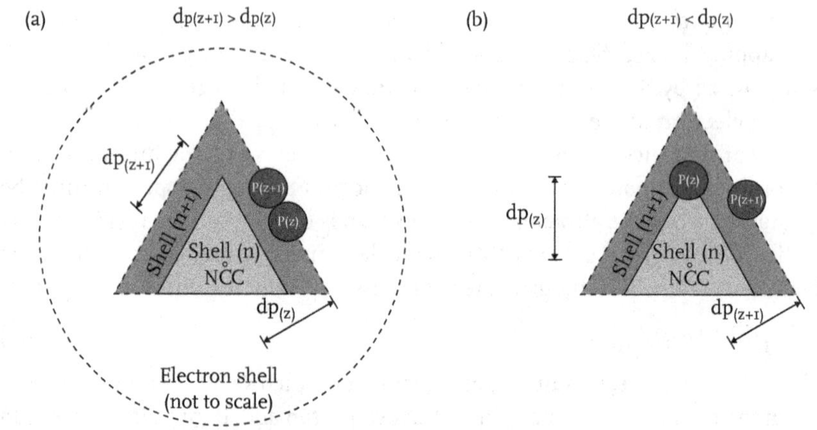

FIGURE 2.7 Schematic figure in 2D showing the geometry of the nucleus. (a) The distance between the nuclear charge center (NCC) and the proton increases as the protons occupy the lattice positions in the same layer. (b) When a new layer starts to build up in the nucleus, the proton is closer to the nuclear charge center than the proton completing the previous shell. Thus proton [P(Z+1)] starting the period of n+1 has weaker attraction on the captured electron than the proton [P(Z)] completing the period n has. The captured electron (Z+1) is unable to overcome the repulsion of the existing shell because of its weaker attraction and starts to built up a new valence shell outside of the existing one.

surface, closest to the charge center (Fig. 2.7b). It is assumed that the charge center of the nucleus coincides with the mass center. For the completed structures of the tetrahedrons He and Ne, and "double" tetrahedrons Ar, Xe, and Ra nucleus, the mass center coincides with the charge centers. When a new nuclear shell is started by adding a proton, then the charge center shifts toward the proton, resulting in a smaller distance between the charge center and the proton than the distance between the proton and the mass center. Thus, using the mass center instead of the charge center is a conservative estimate of the inequality of $d_{P\text{-}NCC}(Z) < d_{P\text{-}NCC}(Z\text{-}1)$.

The distance between the first proton in the shell and the mass center is calculated first. For simplification the length of the edge of the basic tetrahedron is one unit. Thus the diameter of the nucleons is also one unit. The height of a tetrahedron $[h_\Delta]$ with unit length is

$$h_\Delta = \sqrt{\frac{2}{3}}. \qquad [25]$$

Assuming a unit mass for the nucleons the distance between the vertex of the tetrahedron and the mass center is:

$$d_{\Delta vertex} - MC = \frac{3}{4} h_\Delta = \sqrt{\frac{3}{8}} \qquad [26]$$

The distance between the face of the tetrahedron and the mass center is then

$$d_{\Delta surface} - MC = \frac{1}{4} h_{\Delta} = \sqrt{\frac{1}{24}} \quad [27]$$

The tetrahedron is expanded by one-one layers on both sides; thus the length of the side of the tetrahedron is increased by two units in each of the periods (Fig. 2.7). The distance between the center of a sphere (nucleon lattice position), placed on the surface of the tetrahedron, and the mass center in period n+1 is then

$$d_{\Delta surface - MC}(n+1) = (2n+1)\frac{1}{4} h_{\Delta} = (2n+1)\sqrt{\frac{1}{24}} \quad [28]$$

where n is the number of the period. The distance between a sphere placed at the vertex of the tetrahedron in period n can be calculated then as

$$d_{\Delta Vertex - MC}(n) = (2n-1)\frac{3}{4} h_{\Delta} = (2n-1)\sqrt{\frac{3}{8}} \quad [29]$$

It can be seen that

$$d_{\Delta vertex - MC(n)} > d_{\Delta surface - MC(n+1)} \quad [30]$$

where n = 2, 3, 4, 5... Arranging the nucleons in an FCC lattice and building up a tetrahedron inequality in Equation 24 is fulfilled when a new layer in the nuclear structure starts to form (Fig. 2.7b). Thus

$$\begin{aligned} &d_{p-NCC}(Z=2) > d_{p-NCC}(Z=3);\ d_{p-NCC}(Z=10) > d_{p-NCC}(Z=11); \\ &d_{p-NCC}(Z=18) > d_{p-NCC}(Z=19);\ d_{p-NCC}(Z=36) > d_{p-NCC}(Z=37); \\ &d_{p-NCC}(Z=54) > d_{p-NCC}(Z=55);\ d_{p-NCC}(Z=86) > d_{p-NCC}(Z=87). \end{aligned} \quad [31]$$

The remaining rest of the pairs of the elements ([Z]⇔[Z+1]) satisfy the relationship of Equation 23 and $d_{P-NCC}(Z) > d_{P-NCC}(Z-1)$.

As a result of the inequalities of Eq. 31, the relative attraction force on the newly captured electron is weaker than the force on the electrons in the existing shell, leading to the formation of a new electron shell for the following elements:

$$\begin{aligned} &F_{P-e}(He) > F_{P-e}(Li);\ F_{P-e}(Ne) > F_{P-e}(Na);\ F_{P-e}(Ar) > F_{P-e}(K) \\ &F_{P-e}(Kr) > F_{P-e}(Rb);\ F_{P-e}(Xe) > F_{P-e}(Cs);\ F_{P-e}(Rn) > F_{P-e}(Fr) \end{aligned} \quad [32]$$

Based on the geometrical consideration presented, it is suggested that the cyclical structural development of the nuclear structure results in an interruption of the development of valence shells because the relative attraction of the proton on the captured electron weakens when a new layer starts to built in the nucleus (Garai 2011). The cycles of the weaker relative attractions predicted from the geometry of the nuclear structure are identical to the length of the periods in the periodic table. We conclude that the electronic configuration of the chemical elements is the consequence of the structural development of the nuclear lattice.

7. Conclusions

The symmetry pattern of the periodic system of the chemical elements emerges from the nuclear charge. This invariant pattern can be maintained only if the positions of the nucleons are also invariant. Thus the nucleons should form a lattice structure. The lattice arrangement of the nucleons is consistent with the Pauli Exclusion principle and offers a feasible physical explanation for the exclusion principle. Representing protons and neutrons with equal spheres, arranging them alternately in an FCC lattice, and developing a tetrahedron, rotating by 90 degrees at each expansion, results in an identical symmetry pattern expressed by the PSCE. The lattice positions of the nucleons not only reproduce the quantum numbers but they also bear the same physical meaning—indicating that the model presented should be considered a credible physical explanation for quantum theory.

Mathematical solutions for the sequences of the periodic table are derived based on the presented nuclear lattice mode. Investigating the structural development of the nucleus it was shown that the periodicity of the electronic structure is the natural outcome of the nuclear lattice geometry. The tetrahedron lattice model presented here is the first nucleus model which is able to reproduce all of the sequences of the periodic table. It offers a credible physical explanation for the identical symmetry of the nucleus and the electronic structure, for the Pauli Exclusion Principle, and for the Aufbau principle.

References

Abramowitz, M., Stegun, I.A. (1964). Handbook of Mathematical Functions; National Bureau of Standards Applied Math. Series 55, 828.

Anagnostatos, G.S. (1973). The geometry of the isomorphic model of the closed-shell nuclei. Can. J. Phys. 51, 998–1011.

Atkins, F.P., Atkins J.P. (2001). Physical Chemistry, 7th ed.; Oxford University Press, New York

Baum, R.M. (2003). Celebrating the periodic table. Chem Eng News 81, 28–29. Beiler, A.H. (1964). Re-creations in the Theory of Numbers. Dover,: New York.

Bent, H. (2006). New Ideas in Chemistry from Fresh Energy for the Periodic Law. AuthorHouse, Bloomington, IN.

Boeyens, J.C.A. (2008). Chemistry from First Principles. Springer, Dordrecht.

Boeyens, J.C.A., Levendis, D.C. (2008). Number Theory and the Periodicity of Matter. Springer, Dordrecht.

Boeyens, J.C.A. (2013). The Chemistry of Matter Waves. Springer,. Bohr, N. (1913a). On the constitution of atoms and molecules, Part I. Phil. Mag. 26, 1–25.

Bohr, N. (1913b). On the constitution of atoms and molecules, Part II: Systems containing only a single nucleus, Phil. Mag. 26, 476–502.

Bohr, N. (1913c). On the constitution of atoms and molecules, Part III: Systems containing several nuclei, Phil. Mag. 26, 857–875.

Bohr, N. (1922). The Theory of Atomic Spectra and Atomic Constitution. Cambridge University Press, Cambridge.

Bohr, N. (1936). Neutron capture and nuclear constitution. Nature 137, 344–348.

Brakel van, J. (2000). Philosophy of Chemistry. Leuven University Press, Leuven, Belgium.

Broek, A. (1911). The number of possible elements and Mendeléeff's "cubic" periodic system. Nature 87, 78.

Broek, A. (1913). Intra-atomic charge. Nature 92, 372–373.

Canuto, V., Chitre, S.M. (1974). Crystallization of dense neutron matter. Phys. Rev. D, 9, 1587–1613.

Condon E.U., Shortley, G.H. (1935). The Theory of Atomic Spectra. Cambridge University Press, Cambridge. Conway, J.H., Guy, R.K. (1996). The Book of Numbers. Copernicus Press, New York.

Cook, N.D. (1976). An FCC Lattice model for nuclei. Atomkernenergie 28, 195–199.

Cook, N.D., Dallacasa, V. (1987). Face-centered-cubic solid-phase theory of the nucleus. Phys. Rev. C 35, 1883–1890.

Cook, N.D., Hayashi, T. (1997). Lattice models for quark, nuclear structure and nuclear reaction studies. J. Phys. G: Nucl. Phys. 1997, 23, 1109–1126.

Cook, N.D. (2010). Models of the Atomic Nucleus, 2nd ed. Springer, Berlin. ed.

Dallacasa, V. (1981). FCC (face centered cubic) lattice model for nuclei. Atomkernenergie 37, 143144.

Demkov, Y.N., Ostrovsky, V.N. (1971). n + l filling rule in the periodic system and focusing potentials. Zh. Eksp. Teor. Fiz. 62, 125–132; Errata 63, 2376 (1972); Sov. Phys.-JETP 35, 66–69. (1972)

Döbereiner, J.W. (1829). Versuch einer Gruppirung der elementaren Stoffe nach ihrer Analogie, Pogg. Ann. Phys. Chem. 15, 301–307.

Dudek, J., Gozdz, A., Schunck, N., Miskiewicz, M. (2002). Nuclear tetrahedral symmetry: possibly present throughout the periodic table. Phys. Rev. Lett. 88, 252502.

Dutailly, J.C. (2013). Quantum Mechanics revisited, arXiv:1301.0885 [math-ph].

Feenberg, E. (1947). Semi-empirical theory of the nuclear energy surface. Rev. Mod. Phys. 19, 239–258.

Feynman R.P. (1985). QED, The Strange Theory of Light. Princeton University Press, Princeton, NJ.

Fischer, C.F. (1977). The Hartree-Fock Method for Atoms. A Numerical Approach. Wiley-Interscience, New York.

Friedrich, B. (2004). Hasn't it? A commentary on Eric Scerri's paper "Has quantum mechanics explained the periodic table?" Fund. Chem. 6, 117–132.

Garai, J. (1999). The Structure of the Nucleus, Yearly Progress Report. Dept. of Geosciences, SUNY at Stony Brook, New York.

Garai, J. (2003). The double tetrahedron structure of the nucleus, Physics Archive, arXiv:nucl-th/0309035.

Garai, J. (2008). Mathematical formulas describing the sequences of the periodic table. Int. J. Quant. Chem. 108, 667–670.

Garai, J. (2010). Upper bound on the disordered density of sphere packing and the Kepler Conjecture arXiv:1001.1714v1 [cond-mat.mtrl.sci, cond-mat.dis-nn].

Garai, J. (2011). Physical model explaining the periodic pattern of the chemical elements. arXiv:1101.4478v1 [physics.gen-ph].
Giunta, C.J. (2001). Argon and the periodic system: the piece that would not fit. Fund. Chem. 3, 105–128.
Goldman, T., Maltman, K.R., Stephenson Jr., G.J., Schmidt, K.E. (1988). The importance of nucleon substructure in nuclear ground states. Nucl. Phys. 481, 621–667.
Goudsmith, S.A., Richards, P.I. (1964). The order of electron shells in ionized atoms, Proc. Natl. Acad. Sci. USA 51, 664–671.
Gupta, S.D., Pan, J., Tsang, M.B. (1996) Temperature determination from the lattice gas model. Phys. Rev. C 54, R2820.
Gupta, S.D., Pan, J., Kvasnikova, I., Gale, C. (1997). Similarities between the lattice gas model and some other models of nuclear multifragmentation. Nucl. Phys. A 621, 897–910. Hafstad, L.R., Teller, E. (1938). The alpha-particle model of the nucleus. Phys. Rev., 54, 9, 681–692.
Hartree, D.R. (1928a). The wavemechanics of an atom with a non-Coulomb central field, Part I Theory and Methods, Proc. Cambridge Phil. Soc. 24, 89–110 (1928).
Hartree, D.R. (1928b). The wavemechanics of an atom with a non-Coulomb central field, Part II Some results and discussion, Proc. Cambridge Phil. Soc. 24, 110–132.
Hartree, D.R. (1928c). The wavemechanics of an atom with a non-Coulomb central field, Part III Term values and intensities in series optical spectra, Proc. Cambridge Phil. Soc. 24, 426–437.
Hartree, D.R. (1957). The Calculation of Atomic Structures. Wiley, New York.
Haxel, O., Hans, J., Jensen, D., Suess, H.E. (1949). On the "magic numbers" in nuclear structure. Phys. Rev. 75, 1766.
Hofstadter, R. (1961). The electron-scattering method and its application to the structure of nuclei and nucleons. Nobel Lecture, December 11.
Hund, F. (1925). Concerning the interpretation of complex spectra, especially the elements scandium to nickel, Zeitschrift für Physik 33, 345–371.
Johnson, R.D. (2005). NIST Computational chemistry comparison and benchmark database, NIST Stand. Ref. Database, 101.
Kaplan, I.G. (2013). Exclusion principle. Can it be proved? Found. Phys. 43, 1233–1251.
Karapetoff, V. (1930). A chart of consecutive sets of electronic orbits within atoms of chemical elements. J. Franklin Inst. 210, 609–614.
Kibler, M.R. (2004). Classifying chemical elements and particles. In: Rouvray D.H. and R.B. King (eds.), The Periodic Table: Into the 21st Century, pp. 297–329. Nova Science, New York.
Kragh, H. (2001). The first subatomic explanation of the periodic system. Fund. Chem. 3, 129–143.
Landau, L.D., Lifschitz, L.M. (1977). Quantum Mechanics. Pergamon Press, New York.
Leigh, J. (2009). Periodic tables and IUPAC. Chem. Int. IUPAC, 31, 1.
Lezuo, K.J. (1974). A nuclear model based upon the close-packing of spheres. Atomkernenergie 23, 285.

Madelung, E. (1950). Die Mathematischen Hilfsmittel des Physikers, 6th ed. Springer, Berlin, 611.

Maltman, M., Stephenson Jr., G.J., Goldman, T. (1994). A relativistic quark model of the nuclear substructure in the A = 3 system. Phys. Lett. B 324, 1–4.

Matsui, T., Otofuji, T., Sakai, K., Yasuno, M. (1980). Possible structures of pion condensate in symmetric nuclear matter. Prog. Theor. Phys. 63, 1665–1687.

Mayer, M.G. (1949). On closed shells in nucleu. II. Phys. Rev. 75, 1969–1970.

Mayer, M.G., Jensen, J.H.D. (1955). Elementary Theory of Nuclear Shell Structure. Wiley, New York.

Mendelejev, D. (1869). Ueber die Beziehungen der Eigenschaften zu den Atomgewichten der Élemente, Z. Chem. 12, 405–406 (1869); abstr trans into German from Zhur. Russ. Fiz. Khim. Obshch. 1, 60–77.

Mendelejev, D. (1872). Die periodische Gesetzmässigkeit der chemischen Élemente, Ann. Chem. Pharm., 1872, Suppl. 8, 133–229.

Meyer, L. (1870). Die natur der chemischen Elemente als Function ihrer Atomgewichte, Ann. Supplementband, VII, 351–364 (Heft 3).

Moore, J.W. (2003). Modern Quantum Look at the Periodic Table of Elements. In: Brändas, E. and Kryachko, E.S. (eds.), Fundamental World of Quantum Chemistry Kluwer Academic, Dordrecht, Vol. 2, pp. 631–674.

Moseley, H.G.J. (1913). The high-frequency spectra of the elements. Phil. Mag. 26, 1024–1034.

Newlands, J.A.R. (1864). On relations among the equivalents. Chem. News. 10, 94–95.

Newlands, J.A.R. (1865). On the law of octaves. Chem. News. 12, 83.

Ostrovsky, V.N. (1981). Dynamic symmetry of atomic potential. J. Phys. B 14, 4425–4439.

Ostrovsky, V.N. (2001). What and how physics contributes to understanding the periodic law. Found. Chem. 3, 145–182.

Ostrovsky, V.N. (2003). Physical explanation of the periodic table. Ann. N.Y. Acad. Sci. 988, 182–192.

Pauli, W. (1925). Über den Zusammenhang des Abschlusses der Elektronengruppen im Atom mit der Komplexstruktur der Spektren. Z. Physik 31, 765–783.

Pauli, W. (1964). Collected Papers, 2 vols. Interscience, New York.

Pauli, W. (1946). Exclusion principle and quantum mechanics. Nobel Lecture, December 13.

Pauling, L. (1965a). The close-packed-spheron theory and nuclear fission. Science 150, 297–305.

Pauling, L. (1965b). Structural basis of neutron and proton magic numbers in atomic nuclei. Nature 208, 174.

Pauling, L. (1965c). The close-packed-spheron model of atomic nuclei and its relationship to the shell model. Proc. Natl. Acad. Sci. 54, 989–994.

Pauling, L. (1965d). Structural significance of the principal quantum number of nucleonic orbital wave functions. Phys. Rev. Lett. 15, 499.

Rainwater, J. (1950). Nuclear energy level argument for a spheroidal nuclear model. Phys. Rev. 79, 432.

Ramsay, W. (1897). An Undiscovered Gas. Report of the 67th Meeting of the British Association for the Advancement of Science, 593–601.

Rayleigh, L. (1895a). Argon. Royal Inst. Proc. 14, 524–538.

Rayleigh, L., Ramsay, W. (1895b). Argon, a new constituent of the atmosphere. Phil. Trans. 186A, 187–241.

Restrepo, G., Pachon, L. (2007). Mathematical aspects of the periodic law. Fund. Chem. 9, 189–214.

Robson, D. (1978). Many-body interactions from quark exchanges and the tetrahedral crystal structure of nuclei. Nucl. Phys. A, 308, 381–428.

Rouvray, D.H., King, R.B., eds. (2005). The Mathematics of the Periodic Table. Nova Science, New York.

Scerri, E.R. (1998). How good is the quantum mechanical explanation of the periodic table? J. Chem. Ed. 75, 1384–1385.

Scerri, E.R. (2001). The periodic table: the ultimate paper tool in chemistry. In: Klein, E. (ed.), Tools and Modes of Representation in the Laboratory Sciences, pp. 163–177. Kluwer Academic, Dordrecht.

Scerri, E.R. (2003). Löwdin's remarks on the Aufbau principle and a philosopher's view of ab initio quantum chemistry. In: Brändas, E. and Kryachko, E.S. (eds.), Fundamental World of Quantum Chemistry, Vol. 2, pp. 675–694. Dordrecht: Kluwer.

Scerri, E.R. (2004). The best representation for the periodic system: the role of the n + l rule and of the concept of an element as a basic substance. In: Rauvray, D. H., King R B (ed.), The Periodic Table: Into the 21st Century, ch. 5. Research Studies Press: Philadelphia, PA, Baldock/Hfsh & Institute of Physics Publishing.

Scerri, E.R. (2007). The Periodic Table. Its Story and Its Significance. Oxford University Press, New York.

Schery, S.D., Lind, S.D., Zafiratos, C.D. (1980). Evidence for near equality of neutron and proton radii in the N = 82 isotones. Phys. Lett. B 97, 25–28.

Schlosshauer, M., Kofler, J., Zeilinger, A. (2013). A snapshot of foundational attitudes toward quantum mechanics. arXiv:1301.1069 [quant-ph] (2013); Stud. Hist. Phil. Mod. Phys. 44, 222–230.

Schunck, N., Dudek, J. (2004). Nuclear tetrahedral symmetry. Int. J. Modern. Phys. E-Nucl. Phys. 13, 213–216.

Schwabl, F. (2001). Advanced Quantum Mechanics. Springer, Berlin.

Schwarz, W.H.E. (2004). Towards a physical explanation of the periodic table (pt) of chemical elements. In: E.J. Brändas and E.S. Kryachko (eds.), Fundamental World of Quantum Chemistry, Vol. 3, pp. 645–669. Dordrecht: Kluwer.

Schwarz, W. (2007). Recommended questions on the road towards a scientific explanation of the periodic system of chemical elements with the help of the concepts of quantum physics. Found. Chem. 2, 139–188.

Schwarz, W.H.E., Wang, S.-G. (2010). Some solved problems of the periodic system of chemical elements. Int. J. Quant. Chem. 110, 1455–1465.

Seaborg, G.T. (1951). The Transuranium Elements: Present Status. Nobel Lecture, December 12.

Slater, J.C. (1930). Atomic shielding constants. Phys. Rev. 36, 57–64.

Sneath, P.H.A. (2000). Numerical classification of the chemical elements and its relation to the periodic system. Fund. Chem. 2, 237–263.

Tomkeieff, M.V. (1951). Length of the period of the periodic system. Nature 167, 954.

Tomkeieff, M.V. (1954). The periodic table and its applications. Nature 173, 393.

Wang, S.G., Qiu, Y.X., Fang, H., Schwarz, W.H.E. (2006). The challenge of the so-called electron configurations of the transition metals. Chem. Eur. J. 12, 4101–4114.

Wang, S.G., Schwarz, W.H.E. (2009). Icon of chemistry: the periodic system of chemical elements in the new century. Angew Chem. Int. Ed. 48, 3404–34–15.

Weinhold, F.J., Bent, H.A. (2007). News from the periodic table: an introduction to "periodicity symbols, tables, and models for higher-order valency and donor–acceptor Kinships." Chem. Educ. 84, 1145.

Wigner, E. (1937). On the consequences of the symmetry of the nuclear Hamiltonian on the spectroscopy of nuclei. Phys. Rev.w 51, 106–119.

CHAPTER 3

Amateurs and Professionals in Chemistry

The Case of the Periodic System

PHILIP J. STEWART
Department of Plant Sciences, University of Oxford, UK

AMATEURS HAVE MADE VALUABLE contributions to various sciences, including astronomy, geology, biology, and engineering. In chemistry they have been drawn to the periodic system of the elements, with its deceptive simplicity, its density of information, its aesthetic potential, and its implication of deep order in the universe. They have suggested novel ways of representing it visually, in particular spirals and lemniscates in two or three dimensions. However, in the course of a century and a half, professional chemists have generally ignored the amateur versions of the table, and contented themselves with a couple of utilitarian tabulations.

Edward Mazurs, one of the two great historians of the periodic system, surveyed approximately 700 graphic representations produced between 1862 and 1972. He was obsessed with classification, and he counted 146 different types.[1] It seems astonishing that out of all these only two have ever attained any lasting and widespread currency among professionals.

Dmitri Mendeleev's short form was rapidly taken up by chemists and remained the standard for half a century. It was compact and easy to read, and by the clever device of combining three or four groups of elements as column VIII it concealed the difference in length between what we now call the p block and the d block. Indeed it confused the two; Mendeleev's predicted properties of scandium (in the d block) were based on those of boron (in the p block). His inability to deal with the f block did not attract attention because as yet so few

1 Edward Mazurs, Types of Graphic Representation of the Periodic System of Chemical Elements. (La Grange, IL: Published by author, 1957).

Idem. Graphic Representations of the Periodic System during One Hundred Years. (Tuscaloosa: University of Alabama Press, 1974).

lanthanoids were known and that the early actinoids behave rather like the first members of the d-block.

In the interwar years, the standard medium-long form gradually displaced Mendeleev's short form, and since the 1940s it has become ubiquitous. Mazurs classified it as his type IIC2-4, and he referenced it 67 times—more than any other type (pp. 175–180)—but he thought so poorly of it that he accorded less than a page to discussing it. He pointed out that the first to envisage the medium-long form was Mendeleev himself, who described it but drew only a fragmentary version of it, because he did not like the gaps within the rows.[2]

Interruptions are inevitable in any table; the rows must end somewhere. In Mendeleev's table, with 12 rows, there were 11 interruptions, but no gaps wider than one column except after hydrogen and for the "rare earth" elements. By the device of placing what he called "the transition elements" (the groups headed by Fe, Co, Ni, and—hesitatingly—Cu) in a single column, he was able to make columns I to VII represent highest oxidation states (while VIII transitions back from Os^{VIII} to Ag^{I} and Au^{II}).

In the standard medium-long table, there are 7 rows, but because the lanthanoids and actinoids are footnoted there are 9 interruptions; there are also 7 gaps, including the one between Ba/Ra and Hf/Rf, making for a very lopsided and discontinuous table. In the column numbering, the link with highest oxidation state is lost. For professional chemists, to whom the table is just a tool and who do not care about aesthetics, this does not matter, but one must suppose that Mendeleev would not have liked it. In fact his short form, with its display of valency, would be superior for beginners, who do not need to go to the lowest periods.

Alternative Forms

To show the sequence of elements without interruptions, it is necessary to represent it as a spiral in two or three dimensions. This was in fact the solution adopted by three of the pioneers of the system. The first was the geologist and mineralogist Alexandre Béguyer de Chancourtois (1820–1886). In 1862 he submitted an article to the French Académie des Sciences, proposing to depict the elements as a sequence in order of atomic weight, and the following year he published the diagram.[3] The elements were to be disposed along a helix wound round a vertical cylinder ruled with 16 lines parallel to the axis. The elements were arranged so that similar ones occupied the same column. The system worked well enough as far as calcium, after which what are now called

2 Dmitri Mendeleev, "On the relation of the properties to the atomic weights of the elements," in William Jensen, ed., Mendeleev on the Periodic Law. (Mineola, NY: Dover Publications, 2002), 34n 2.

3 Alexandre Béguyer de Chancourtois. Vis tellurique: classement naturel des corps simples ou radicaux obtenu au moyen d'un système de classification hélicoidal ou numérique. (Paris: Published by Author, 1863).

the transition elements complicated things. Béguyer de Chancourtois did not limit himself to elements but included a number of radicals.

It is hard to say how far Gustavus Hinrichs (1836–1923) identified himself as a chemist. Hinrichs was a Dane who emigrated to the United States in 1861; within a year he was appointed professor of natural philosophy, chemistry, and modern languages at the University of Iowa. He was something of a numerologist and produced a classification of the planets based on the ratios between their distances from the sun, comparing them to the ratios between musical tones. Turning his attention to the chemical elements Hinrichs looked for similar ratios between their atomic numbers. This resulted in a diagram published in 1867 that depicted the elements in order of atomic weight, grouped according to properties.[4] This was not strictly a flat spiral, because it was divided into two unequal halves, "trigonoid" and "tetragonoid" (roughly nonmetals and metals, respectively), but the principle was clear.

The first almost satisfactory representation of the system as a flat spiral was published in 1870 by a young German mineralogist, Heinrich Adolf Baumhauer (1848–1926), but like other pioneers he had problems with the elements beyond barium.[5] His was by no means the last spiral version. Others have continued to appear ever since. Most notable was the one with a circular outline published by Ingo Hackh; it was the first spiral to make use of Moseley's atomic numbers, and it treats the longer periods as separate pairs of coils—but it is difficult to read the groups.[6]

Mendeleev himself steadfastly refused to depict the system as a spiral, although he wrote "In reality the series of elements is uninterrupted, and corresponds, to a certain degree, to a spiral function."[7] His objection was that he could not express this function mathematically. This refusal may have helped to deprive him of the Nobel Prize, since the gap in which he might have suspected the missing group of noble gases coincided with the interruption between rows of his table.[8] It was an amateur, Lieutenant-Colonel W. Sedgewick of the Royal Engineers in British India, who first suggested, on the basis of an almost Lucretian model of the atom, the existence of elements of zero valency with theoretical atomic weights of about 20, 40, 80, and 120.[9]

The concept of blocks of elements, characterized by the subshell of their outer electrons, became possible only after Bohr had introduced the theory of electronic structure, though he himself did not refer to them as "blocks." In his

---

[4] Gustavus Hinrichs. *Programm der Atomechanik* [sic], *oder die Chemie einer Mechanik der Panatome.* (Iowa City, 1867).

[5] H. Baumhauer. *Die Beziehungen zwischen dem Atomgewichte und der Natur der chemische Elemente.* (Braunschweig: Vieweg und Sohn, 1870).

[6] Ingo Hackh, *Das synthetische System der Atome.* (Hamburg: Hephaestos, 1914), 5.

[7] W. B. Jensen (ed.), *Mendeleev on the Periodic Law: Selected Writings, 1869–1905.* (Cincinnati, OH: University of Cincinnati, 2002), 56.

[8] P.J. Stewart, "A Ccentury on from Dmitrii Mendeleev: Tables and spirals, noble gases and Nobel Prizes." *Found. Chem.*, 9 (October 2007), 235.

[9] W. Sedgewick, *Force as an Entity with Stream, Pool And Wave Forms: An Engineer's Practical Way of Explaining the Facts Ascertained by Science.* (London: Sampson Low, 1890), 64.

adaptation of the Bayley-Thomsen table he did not separate them, nor did he delimit them consistently. His periods were respectively of 2, 8, 8, 18, 18, 32, and [6] elements. H and He were separated from the rest of the s block, which was combined with the p block. Bohr outlined only eight 3d elements (Sc to Ni), eight 4d (Y to Pd), but nine 5d elements (La and Lu to Pt) and anticipated the same number for 6d (Ac to U plus five spaces) He outlined 13 4f elements (Ce to Yb) and allowed for the same number in 5f.[10]

It was left to the French polymath, Charles Janet (1849–1933), to find a more satisfactory way to delimit what he called *les nappes*—layers—the word used for geological strata in French. This engineer, entrepreneur, geologist, inventor, and entomologist turned his attention to the periodic system at the age of 78. After a couple of false starts in 1927, based on an eccentric theory of the atomic nucleus, he discussed three possible versions in April 1928: the first omitted hydrogen, the second placed hydrogen and helium over fluorine and neon.[11] By November he had decided for the third, which gave him a perfectly regular system by separating four *nappes* and moving hydrogen and helium to the head of the first one, giving him four pairs of periods of 2, 8, 18, and 32.[12] Janet started from a helix traced on nested cylinders with diameters 2, 6, 10, and 14—one for each layer or block. He then took this through a succession of beautiful spiral or lemniscate transformations, to reach what has come to be called his "Left-Step Periodic Table."

There were three striking innovations in Janet's system. First, in terms of the medium-long table, he moved the column that we now call the s block from the left to the right-hand side. Second, in his November 1928 paper, despite the noble-gas behavior of helium, he moved hydrogen and helium to the top of the s block (having in one of his April versions placed them over fluorine and neon), to make four pairs of periods of increasing length. Third, he anticipated Seaborg by twenty years in recognizing the actinoids in their proper place, pushing the last period to a hypothetical element 120. If this last element is ever created it must surely be called janetium. An additional detail: unlike most authors, Janet made the compartments for the individual elements higher than they were wide, giving the table as a whole a width to height ratio of 3.25—relatively low for a long form, though still too wide for convenience. His spiral, on the other hand, is circular in outline and fits comfortably on a page, as is true of spirals in general (see Fig. 3.1),

It is remarkable that Janet arrived at his table without reference to quantum theory, basing it purely on considerations of symmetry and proportion. By the time of his 1930 paper he had studied the work of Bohr and Stoner and

---

[10] Niels Bohr, *The Theory of Spectra and Atomic Constitution: Three Essays.* (Cambridge: Cambridge University Press, 1922), 69.

[11] Charles Janet, *Essais de classification hélicoïdale des éléments chimiques.* (Beauvais: Imprimerie Départementale de l'Oise. April 1928).

[12] Charles Janet, *La classification hélicoïdale des éléments chimiques.* (Beauvais: Imprimerie Départementale de l'Oise. November 1928).

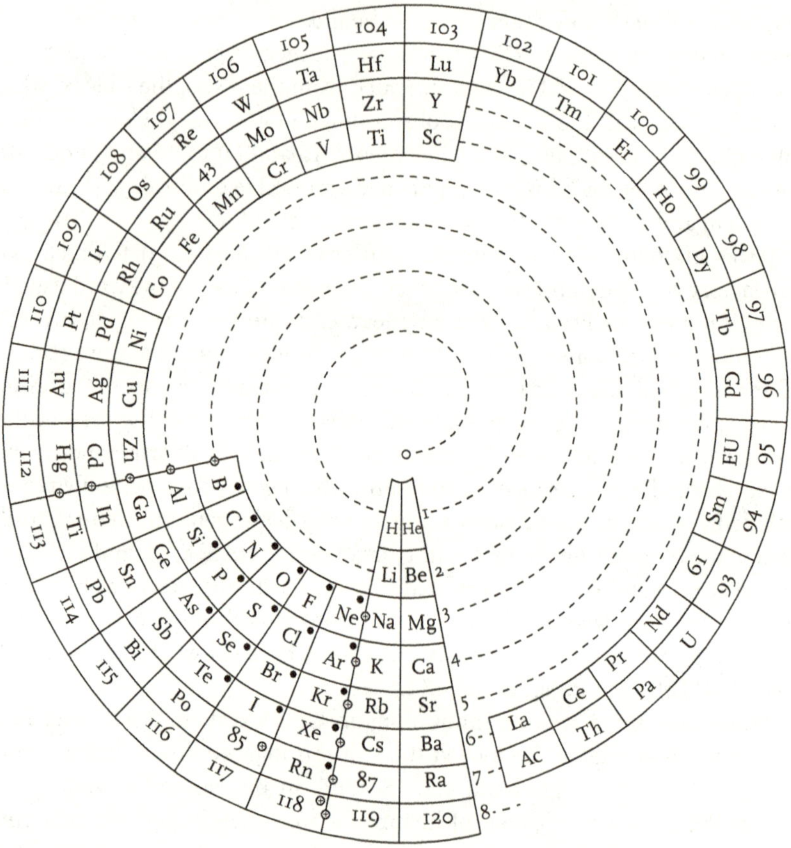

FIGURE 3.1 Periodic Spiral by Charles Janet, 1928.

realized that his table represented the electronic structure.[13] To make the correspondence perfect, he "corrected" the data for elements such as chromium, copper, palladium, and lanthanum, where the differentiating electron does not belong to the "right" block. For example, because lanthanum should be the first element in the f block, he assumed that it must have one f electron when in fact it has none.

Janet's logic in moving the s-block to the right-hand side had been the perfectly regular pattern of pairs of periods. In quantum terms, this replaced the second row of the conventional table, 2s+2p, by a third row consisting of 2p+3s. This brings his third and fourth rows into line with the general rule that each s orbital is filled before orbitals in subshells of the same or lower shells, for example 5s before 3d, 7s before 4f. He saw, and explicitly stated that each row corresponded to a value of the sum of the first two quantum numbers, n+ℓ.

[13] Charles Janet, *Concordance de l'arrangement quantique de base des électrons planétaires des atomes avec la classification scalariforme, hélicoïdale des elements chimiques.* (Beauvais: Imprimerie Départementale de l'Oise, 1930).

Generalizing this, he formulated what should be called the "Janet Rule" six years before Madelung.

Mendeleev usually referred to the horizontal series of his table as "rows," but the term "period" later became current. Strictly speaking a period can be any set that returns you to its starting point, for example Monday to Monday or Thursday to Thursday, carbon to silicon, or sulfur to selenium. People have argued about where rows should begin and end, but many solutions can claim to be periodic. By constructing a table in which each row corresponds to a new value of $n+\ell$, Janet was the first to provide the periods with an objective basis. This makes the irregularities less important. For example, whether lanthanum is counted as a member of the d block ($n=5$, $\ell=2$) or the f block ($n=4$, $\ell=3$), $n+\ell$ is the same: 7. The placing of hydrogen over the alkali metals goes back to Mendeleev and has been widely accepted (although some favor seeing it as a "halogen") despite its obviously being a non-metal, but the placing of a second non-metal, helium, at the head of the s block upsets chemists, although they are perfectly comfortable with the idea that the first row of the half metallic p block consists entirely of non-metals. They have not entirely got rid of the notion that chemical properties should dictate the placing of an element in the system, although since Bohr it is the electronic configuration that explains the chemical properties. The resemblance in behavior between helium and neon results from the fact that both have a full shell (unlike argon, krypton, and xenon), just as the similarity of hydrogen chemistry to that of carbon results from both being halfway to a full shell.

It cannot be said that professional chemists rejected Janet's system; very few of them even noticed it. He was unknown to them, his papers were privately printed and badly distributed, and his only English article, in *Chemical News*, was a travesty of whatever he had submitted in French. It was cut down to about 1100 words, the helix wound round nested cylinders was not shown, and the geometrical transformations were not properly explained.[14] His table was startling in his treatment of helium—it had eight rows instead of the seven of the medium-long form, and it broke with the convention of making all rows start with the alkali metals (including hydrogen). At least his *nappes* became established, translated as "blocks"—the first time, I believe, that the English term was used in this sense (p. 392).

Janet was not alone; Mazurs gives 90 references to other tables or spirals based on electronic configuration. The helices by Georg Schaltenbrand and by C. J. Monroe and W. D. Turner—both of the Missouri School of Mines—were similar to Janet's.[15] Other notable early examples are the right-step tables of O. M. Corbino, Roy Gardner, and Arrigo Mazzucchelli—all from 1930, which Mazurs combined in a colored version as a foldout at the end of his 1957

[14] Charles Janet, "The helicoidal classification of the elements," *Chem. News*, 138 (1929), 372 and 388.

[15] Georg Scaltenbrand, "Darstellung des periodischen System der Elemente durch eine räumliche Spirale," *Z. Anorg. Allgem. Chem.*, 112 (1920), 221. C. J. Monroe and W. D. Turner, "A new periodic table of the elements," *J. Chem. Ed.*, 3 (1926), 1058.

book.[16] L. M. Simmons, professor of chemistry at Scots College, Sydney, was a widely cited author. In 1947 he published Janet's April 1928 table, with hydrogen and helium above fluorine and neon. Basing himself on Janet's mangled English paper, he rather ungraciously wrote (p. 590): "Janet's insistence on coiling the chart, together with the paucity of explanation, cause Janet's table to be regarded as only touching the fringe of the present arrangement." The following year, he adopted Janet's definitive version of November 1928 but made no further mention of its originator.[17]

The other great historian of the periodic system, Johannes van Spronsen, concurred with Mazurs in preferring representations based on electronic configuration and he did himself publish such a system in 1951 (referenced in the endpapers of his 1969 book), though he later went back on the placing of helium (p. 257).[18] It is difficult to assess his debt to Mazurs; in a footnote to chapter 6 (p. 147) he wrote "After this chapter had been drafted, the work of Mazurs [1957] appeared." The chapter was presumably edited between 1957 and 1969.

Meanwhile, despite its irregularity and its gaps and interruptions, the standard medium-long table quietly pushed all others aside and became the ubiquitous chart on laboratory walls. Once IUPAC had adopted the Arabic numerals 1 to 18 for its groups it became almost official, although this numeration destroyed the relation to highest oxidation state, which had made Mendeleev's Roman numerals so useful. For the most part it the IUPAC version respects the blocks, except in the case of helium, but it has adopted a messy compromise over membership of the f block, which includes both lanthanum/actinium and lutecium/lawrencium; this makes it 15 elements wide, though there is only room for 14 elements in its seven orbitals.

The central feature of the periodic system is the noble gas group with, on one side, the halogens and the oxygen group, with their hunger for electrons and, on the other side, the alkali metals and the alkaline earths, with their ready shedding of electrons. Beyond the oxygen group there is the fairly homogeneous trivalent and largely non-metallic nitrogen group, but beyond the alkaline earths is a diverse set of trivalent metals, the first members of the p block, the d block and the f block, respectively. By putting the noble gases on the edge of the table, the standard medium long table—like Mendeleev's short table—obscures this central feature.

---

[16] O. M. Corbino. "Diagramma rappresentativo degli stati quantici e della formazione degli elementi del sistema periodico," *Rivista nuovo cimento*, 5 (1928), lvii. R. Gardner, "A table of electronic configurations of the elements," *Nature*, 125 (1930), 146. A. Mazzucchelli. "Una rappresentazione del diagramma degli elementi secondo Corbino," *Gazz. Chim. Ital.* 60 (1930), 531.

[17] L.M. Simmons, "A modification of the periodic table." *J. Chem. Ed.*, 24 (1947), 588.
L.M. Simmons, "The display of electronic configuration by a periodic table." *J. Chem. Ed.*, 25 (1948), 658.

[18] Johannes van Spronsen, *The Periodic System of Chemical Elements.* (Amsterdam: Elsevier. 1969).

Rather than abandon this unsatisfactory table, efforts have focused on alleviating its dullness by illustrating each element with a little picture, either—where possible—of the basic substance or of one of its uses or, beyond plutonium, of the person or place it is named after. Alternatively, it is crammed with information, not just on atomic number and atomic weight but also on isotopes, electronic configuration, density, melting and boiling point, and so on—data that are more conveniently found in a reference book or on the internet.

Despite the preference for tables, there has been continued production of spiral designs. Perhaps the simplest idea of all is that of Chancourtois, winding the whole sequence of elements round and round as a helix to bring together related elements. This was essentially the form proposed by M. Courtines (1925), with the s and p blocks wound together on a cylinder and the d and f groups added on at the side.[19] The version of D. F. Stedman (1947) was topologically the same, but more elegantly wound round a paraboloid with bulges on the side.[20] Its projection on to a plane gave a circular spiral with two limbs on the side. This is essentially the same as the spiral chart of Theodor Benfey (1964), which is often reproduced.[21] The simplest presentation of this system is the ribbon design of the physicist George Gamow (1961).[22] Roy Alexander, an amateur, obtained a patent for an almost identical arrangement in 1971.

Another genealogy goes back to John D. Clark, rocket-fuel expert and science fiction writer. His "arena" design was a flat spiral with an oval outline—a central rectangle with a semi-circle at each end, reminiscent of the Roman Circus Maximus. On the left, with a circular outline, are the combined s and p blocks, while the d block occupies the rest, with the f block occupying a single compartment. Arrows join Mendeleev's "a" groups (s and p blocks) in the circle with his "b" groups (d block) of elements in the right-hand half of the oval.[23] He called it a "chart," which is a more convenient term than "representation" to cover both tables and flat spirals, though it doesn't stretch to 3-D models.

An attractive color version of Clark's "arena," with elements of the f-block squeezed in, was used as a two-page spread in a special issue of *Life* magazine, with Glenn Seaborg and William Meggers cited as co-authors.[24] Besides Clark's arrows, it used colors to indicate the relationship between "a" and "b" groups. Color appears to have been first used in this way by Andreas von Antropoff in a table which is a virtual helix, with the noble gases on either

19 M. Courtines, "A model of periodic table." *J. Chem. Ed.*, 2 (1925), 107.

20 D.F. Stedman, "A periodic arrangement of the elements." *Canadian J. Res.*, 25B (1947), 199.

21 Theodore Benfey, "Spiral periodic chart." *Chemistry* (January 1970), 27.

22 George Gamow *The Atom and Its Nucleus* (Englewood Cliffs, NJ: Prentice-Hall, 1961), 10

23 John D. Clark. "A new periodic chart." *J. Chem. Ed.*, 10 (1933), 675.

24 "Prepared by *Life*, with assistance from Drs. William F. Meggers, Glenn T. Seaborg and John D. Clark," "Table of the Elements," *Life* (May 16, 1949), 82; credits 35.

FIGURE 3.2 Longman's mural, 1951.

side.[25] In 1950 Clark published an improved version, with the two outer coils enlarged to accommodate the f block, but this was aesthetically less pleasing and was never widely adopted.[26]

The *Life* version was used by Edgar Longman (1917-2005), a graphics artist and wartime poster designer, as the basis for a huge mural in the 1951 Science Exhibition of the Festival of Britain. With his artist's eye he added to its beauty by giving it an elliptical outline and a dynamic tilt (see Fig. 3.2). It was destroyed along with many other of the Festival's displays by order of the incoming Conservative Government later that year, but the wall on which it was painted is still there, covered with tiles.

I kept the memory of Longman's design alive, getting it published in an article by John Emsley, from which it was reproduced by Martin Kemp, finally finding its place in the autobiography of that great amateur of the periodic system, Oliver Sacks.[27] Seeing its perennial attraction, I used it as the basis for my

25 Andreas von Atrpopoff, "Eine neue Form des periodischen Systems der Elementen." *Zeitschrift für angewandte Chemie*, 39 (1926), 722.

26 John D. Clark. "A modern periodic chart of chemical elements." *Science* 111 (1950), 661.

27 John Emsley, "Mendeleev's dream table." *New Scientist* 105 (1985), 32. Martin Kemp, "Mendeleev's matrix." *Nature* 393 (1998), 527. Oliver Sacks, *Uncle Tungsten*. (London: Picador, 2001), 187.

own *Chemical Galaxy*, with progressively widening coils.[28] In the second edition (2006) I used colors to indicate the links between Mendeleev's A and B groups and to distinguish between lanthanoids with common oxidation states of 2, 3 and 4.

Two unusual spirals by non-chemists appeared in the 1990s. Jeff Moran is a software developer and former town supervisor of Woodstock, NY. He produced a rather spiky spiral design, reminiscent of that of Stedman and Benfey, in which the elements are inscribed in hexagons like those of Pozzi. This has been on the internet since 1991, but it achieved wider attention in 2006. Its main interest is that it is linked to a great deal of software giving information on the elements. In 1995, Melinda Green produced her Periodic Fractal, with echoes of the Mandelbrot Set. She describes herself as a "designer and developer of user interfaces, exotic and mundane," and she is creator of Buddhabrot—an image of the Buddha, an adaptation of the Mandelbrot Set.

Mazurs and van Spronsen both gave equal attention to tables and spirals, and both referred in the titles of their books to the "Periodic System." Amateurs have continued to favor spirals, but the chemists have held fast to their medium-long table. The narrowing of focus is evident in the titles of the books published in the 21st century, for example those by Dennis Rouvray and Bruce King, and by Eric Scerri.[29] The only spirals shown by the former are those of Chancourtois and Courtines and Fernando Dufour's Elementree and another anonymous model with stacked circles. Scerri shows only the spirals of Chancourtois and Hinrichs and Crookes' "pretzel." The message is clear: spirals are of interest as curios or historic relics.

For a short time, it seemed that Janet's table might replace the medium-long form, though chemists have so far shown no interest in his spirals and lemniscates. Gary Katz (2001) argued persuasively for it and presented detailed ideas about how it might be introduced into chemical education.[30] In his book, mentioned above, Scerri came near to endorsing it as the best form for the periodic table, having previously championed his own version of the Bayley-Bohr table. However, even before the book had been published, he changed his allegiance to what proved to be one of the three tables that Janet considered in April and rejected in November 1928. Scerri's rationale for this was that extra atomic number triads could be created between H, F, and Cl, and between He, Ne, and Ar. However, triads are a consequence of the structure of the system and cannot at the same time be its cause. These would be the only exceptions to the rule that the elements in the first row of each block never form part of atomic number triads (precisely because that is how the system is structured).

---

28 Philip Stewart. "A new image of the periodic table." *Educ. Chem.* 41 (2004), 156.

29 Dennis Rouvray and Bruce King. *The Periodic Table: Into the 21st Century.* (Baldock, England: Research Studies Press, 2004).

Eric Scerri. *The Periodic Table: Its Story and Its Significance.* (New York: Oxford University Press, 2007).

30 Gary Katz. The periodic table: an eight-period table for the 21st century. *Chem. Educ.* 6 (2001), 324.

Scerri then shifted this table around, so that the rows ran from the halogens (including hydrogen) to the oxygen group. In electronic terms, the first two rows thus read:

1s1 – 1s2 – 2s1 – 2s2 [gap] 2p1 – 2p2 – 2p3 – 2p4
2p5 – 2p6 – 3s1 – 3s2 [gap] 3p1 – 3p2 – 3p3 – 3p4

This was justified on the grounds of symmetry, with the first two rows standing as twin towers above the next four, from which the gaps are eliminated by moving the f block into a "footnote." This six-row table disagrees with the IUPAC numbering of both the columns and the rows. It is hard to imagine anything more confusing for chemical education. It is difficult to know the chronology of these changes of mind, first announced on internet sites.[31]

With the withdrawal of support from the most widely read writer on the periodic system, it seems that the opportunity opened by Katz's article has been lost. The medium-long table is again unchallenged, and its dominance is increased by its ubiquity on the internet and by the appearance of many decorated versions, with pretty pictures in the place of serious chemical and physical information. It seems that Janet's beautiful and electronically informed designs will have to continue their long wait for recognition.

However, there are two new developments. The first comes again from Scerri. In 2011 he shifted his allegiance to the "Stowe-Janet-Scerri Periodic Table"—an inverted tower of layers with increasing values of the sum of the first two quantum numbers n+l, with He and He at the base, under Li and Be.[32] Timothy Stowe's "Physicist's Chart of the Chemical Elements" was published in 1989 by Instruments Research and Industry Inc. It cannot strictly be called a table, and it is very difficult to read the sequence of elements, as the layers represent the values of n. Scerri's reordering brings it closer to being a table, but it clearly would be simpler just to adopt Janet's own version.

The second—and astonishing—development is the discovery by Valery Tsimmerman—an American engineer—that if the blocks, as delimited by Janet, are drawn using one square per orbital, their perimeters are all equal, which means they can be represented as equally spaced slices of a regular tetrahedron, parallel to two opposite edges. This was announced on the internet in 2007 and has not yet been widely enough noticed. In 2014 he arranged the production of this tetrahedron etched into a block of glass, the Adomah Periodic Cube—a beautiful and fascinating object.[33] Its visual impact, if

[31] Eric Scerri. "Presenting the left-step periodic table," *Educat. Chem.* 42 (2005), 135. Eric Scerri, "The role of triads in the evolution of the periodic table: past and present." *J. Chem. Ed.*, 85 (2008), 585.

[32] So far published only on an internet site. See http://www.meta-synthesis.com/webbook/35_pt/pt_database.php?PT_id=414 Also http://www.meta-synthesis.com/webbook/35_pt/pt_database.php?PT_id=759

[33] Valery Tsimmerman, "Derivation of Mathematical Expression of Mendeleev's Periodic Law and its Implications." http://www.perfectperiodictable.com/pres2016

seen by enough chemists, might be what is needed for them finally to accept Janet's table and his spirals and lemniscates as the best representations of the periodic system.

The number of rows in each Janet block is the number of values of the principal quantum number n, and the number of columns is twice the number of values of the magnetic quantum number m_l which can only be 1, 3, 5, or 7. The fact that the number of values of n is 8 for the s block, 6 for p, 4 for d, and 2 for f is the consequence of the relationship between the energy levels of the different forms of orbital. Slight irregularities begin to appear between 4s and 3d, they increase between 5s and 4d and between 4f and 5d, and they reach their greatest number between 5f and 6d, where 6 out of 14 actinoids have one or two electrons in d orbitals instead of the expected f.

Even without using orbital squares, the overall shape of the Janet table expresses the interplay between these two series of numbers, 1, 3, 5, 7 and 8, 6, 4, 2, which determines the overall structure of the periodic system. It is very satisfying to note that they are the first eight natural numbers. Janet intuitively grasped this symmetry without needing to know any quantum mechanics. Some chemists argue that, because of the slight fuzziness of the blocks, they are approximations and not suitable for geometrical generalizations like Tsimmerman's. It must be conceded that they are slightly idealized, but that does not invalidate the overall pattern. Their chemical significance was recognized long before their electronic basis was known, for example in Mendeleev's a and b series and in the collective term "rare earth elements." The elements that break the pattern do not disobey the $n+\ell$ rule. In general, the Nth element in a d or f sequence has N d or f electrons. Even in the most irregular sequences—4d and 5f—there is clearly a pattern to their irregularity.

To produce his Adomah periodic table, not only did Tsimmerman use one square per orbital, he also rotated the Janet table anticlockwise through 90° then inverted it top to bottom. As a result the rows became columns and the columns became rows. This is confusing for people who are used to the usual arrangement—which means most chemists—and it also goes against the usual order of reading the atomic numbers from left to right and from top to bottom. Tsimmerman also shifted the blocks, so that each column (or Janet row) corresponds to a single value of the principal quantum number, n. This makes what seems to me a useful distinction between periods and rows, which are generally confused. It is a good innovation, which Mazurs (1974, p. 134) suggested, but using one square per element, which makes the table inconveniently wide. I have reversed Tsimmerman's rotation and inversion to produce "Janet Rajeuni"—Janet rejuvenated (Fig. 3.2). An advantage of this is that the width to height ratio of the table is reduced to 2.2, which is very low for a long-form table. I have also designed an acrylic version of the Adomah Cube, enlarged for greater legibility and with the conventional rows and columns restored.

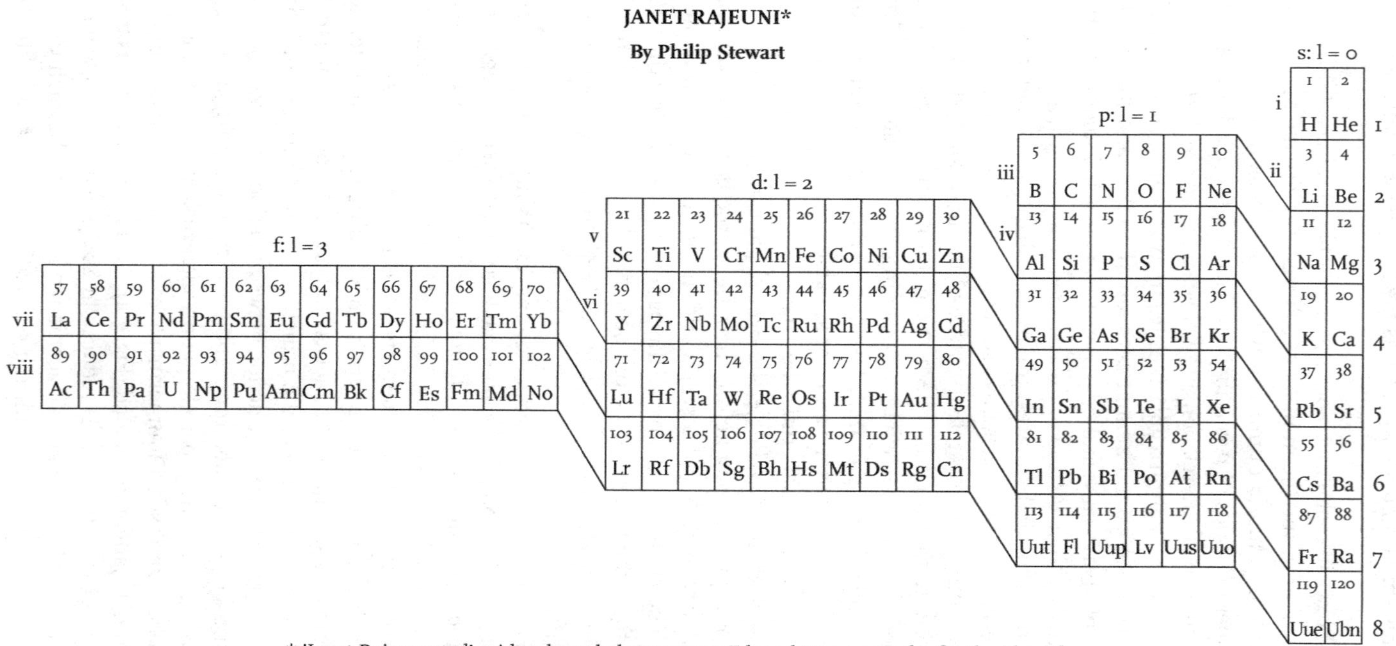

* 'Janet Rejuvenated', with acknowledgements to Edward Mazurs (1965) for the idea of shifting the blocks so that each row represents one value of n, the principal quantum number (Arabic numerals), and to Valery Tsimmerman for the idea of using one square per orbital. Periods are indicated by Roman numerals (i: n+l = 1; ii: n+l = 2; iii: n+l = 3; iv: n+l = 4; v: n+l = 5; vi: n+l = 6; vii: n+l = 7; viii: n+l = 8; l being the azimuthal quantum number).

FIGURE 3.3 Janet Rajeuni, by Philip Stewart.

Conclusion

The resistance of professional chemists to new ideas on the periodic system seems to be part of their culture. One explanation is no doubt the very familiarity of the medium-long form, on the basis of which they received their chemical education. Another factor may be that chemistry is potentially dangerous. Too wide a departure from the beaten path may lead to an explosion or a poisoning, which discourages big innovations. Also, many chemists work in industry, and unsuccessful innovation may have big financial costs. Such risks do not generally arise for mathematicians, astronomers, biologists, or even physicists. Whatever the reasons, chemists seem to be a rather cautious and conservative bunch.

Chemistry amateurs, on the other hand, are its lovers—the literal meaning of the word. A new idea on the periodic system can inflame their passions. Without discipline, this may lead them to fanatical devotion to an unviable idea. In the case of Chancourtois, Sedgewick, Janet, Longman, and Tsimmerman, however, it brought remarkable breakthroughs. Some chemists are grateful for these amateurs, but others rely on their professional prestige to repel interlopers. Science would be well served if they listened to each other.

CHAPTER 4

The Periodic System

A Mathematical Approach

GUILLERMO RESTREPO
Department of Computer Science, Leipzig University, Germany
Laboratorio de Química Teórica
Universidad de Pamplona, Colombia

THE PERIODIC TABLE, DESPITE its near 150 years, is still a vital scientific construct. Two instances of this vitality are the recent formulation of a periodic table of protein complexes (Ahnert et al. 2015) and the announcement of four new chemical elements (Van Noorden 2016). "Interestingly, there is no formal definition of 'Periodic Table'," claims Karol[1] (2017) in his chapter of the current volume. And even worse, the related concepts that come into play when referring to the periodic table (such as periodic law, chemical element, periodic system, and some others) overlap, leading to confusion.

In this chapter we explore the meaning of the periodic table and of some of its related terms. In so doing we highlight a few common mistakes that arise from confusion of those terms and from misinterpretation of others. By exploring the periodic table, we analyze its mathematics and discuss a recent comment by Hoffmann (2015): "No one in my experience tries to prove [the periodic table] wrong, they just want to find some underlying reason why it is right." We claim that if the periodic table were "wrong," its structure would be variable; however the test of the time, including similarity studies, show that it is rather invariable.

An approach to the structure of the periodic system we follow in this chapter is through similarity. In so doing we review seven works addressing the similarity of chemical elements accounting for different number of elements and using different properties, either chemical or physical ones.

1 Incidentally Paul J. Karol is the IUPAC committee chair for the project that analyzed the evidence of existence for the new elements.

Element and Its Relation to Basic and Simple Substance

The concept of "chemical element" has raised the interest of several scholars such as Paneth (1962) and is still a matter of discussion given the double meaning it has (see, e.g., Scerri 2007, Earley 2009, Ruthenberg 2009, Ghibaudi et al. 2013, van Brakel 2014, Restrepo & Harré 2015), which is confusing, leading to misconceptions. The two meanings of the concept of chemical element are basic and simple substance.

According to Paneth (1962), a *basic substance* belongs to the transcendental world and it is devoid of qualities, and therefore is not perceptible to our senses. It is the "indestructible substance present in compounds and simple substances"[2] (Paneth 1962), the "elementness" of some substances. It is, for example, the "boron-ness" of all substances whose chemical analysis yields boron as a constitutive part of them. In turn, a *simple substance* is *a* realization of an uncombined basic substance[3] (Fig. 4.1); that is, simple substances are allotropic forms of a basic substance. Thus, carbon, tellurium, and boron are basic substances whose respective realizations as simple substances are charcoal, diamond, fullerenes, and so on; amorphous and crystalline tellurium; and amorphous-, α-rhombohedral-, β-rhombohedral-, and so on, boron. Hence, *each box or position for an element in a periodic table represents a basic substance,* which is represented by a symbol.[4]

In the realm of properties of the elements, basic substances do not possess properties, but simple substances do. Hence, it is a mistake to state that carbon is black, for carbon, as a simple substance, is devoid of color; what is actually black is the charcoal realization of the basic substance carbon. Likewise, claiming that the density of carbon is 2.267 g/cm^3 is an error as the particular simple substance of carbon exhibiting such density needs to be declared. In fact, the different allotropic forms of carbon have different densities. Thus, *it is a mistake to write down properties for chemical elements on the periodic table,* unless the respective box of the element can be expanded into its different simple substances (allotropes). By the same token, Paneth (1962) criticizes approaching similarity of chemical elements through similarities of their simple substances. Paneth (1962, 152) asserts: "we need only recall how little resemblance there is between gaseous nitrogen and pure antimony"; and stresses the importance of compounds for similarity.

---

2 We do not agree with van Brakel's (2014) interpretation of basic substance as IUPAC's "'chemical element' [...]A species of atoms; all atoms with the same number of protons in the atomic nucleus." As Paneth (1962) stated and as van Brakel (2014) also mentions, "the concept of basic substance as such does not in itself contain any idea of atomism."

3 Restrepo & Harré (2015) have shown a mereological (part/whole) relationship between basic and simple substance, where simple substances are the whole and basic substances the parts.

4 There is a connection between the association of symbols to elements and to variables in the mathematical way of thinking postulated by Weyl (1940). The very assignment of the label Fe to all iron realizations, even under different conditions, requires abstraction and generalization, which are worth studying from a mathematical and philosophical view point.

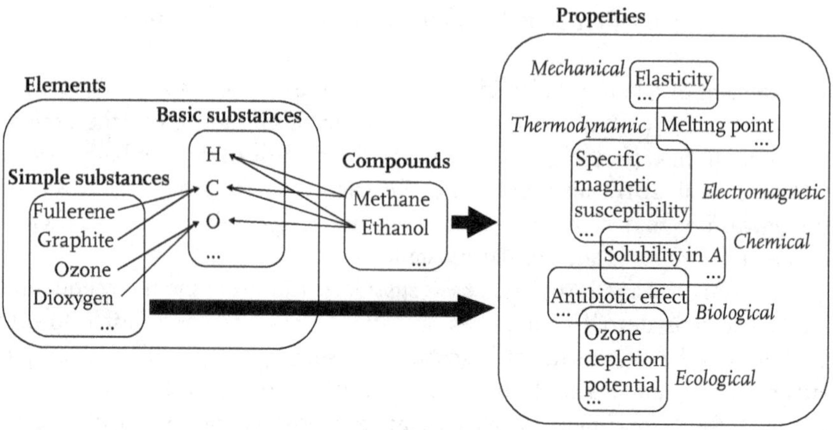

FIGURE 4.1 Basic and simple substances and their relationships with elements, compounds and material properties.

Compounds have played an important role in providing knowledge about chemical elements as basic substances; hence, there is something in each compound encoding information on its transcendental elements. That is why Paneth calls attention to the fallacious claim that "the properties of sulphur are just those properties which it exhibits to us in the special case when it is not combined with any other basic substance"[5] (Paneth 1962, 150). Our point here is that *a holistic representation of chemical elements as basic substances needs to consider their different simple substances as well as their compounds.*[6]

Periodic System, Table, and Law: How Different Are They?

If X represents the set of chemical elements (basic substances) and S the structure of similarity classes of chemical elements, then (X, S) is called a *periodic system*. As with any mathematical structure (Potter 2004), S is defined as $S = (X, R)$, with R a relation for the elements of X. What is the nature of such a relation? Is that relation unique or are there multiple relations? What are the causes of that relation? Those are questions that have driven research and interest on the periodic system since its discovery and which still need further

[5] This is a typical mistake when teaching the periodic system. Most, if not all, chemistry textbooks show electronic affinities and ionization energies as properties of chemical elements, disregarding valuable information on chemical reactivity and of compounds.

[6] Paneth has asserted "behind the 'simple substances' and 'compounds' endowed with constant properties, there stand the transcendental 'basic substances'" (Paneth 1962). This opens the question on the number of simple substances and compounds needed for such a description. Including them all is impossible, as the number of synthesized substances grows exponentially. Though, even if such a number were fixed, would all those substances be needed? Are some substances more important than others? Are binary compounds more important for representing elements than proteins?

research. For example, the nature of R can be analyzed by looking at Mendeleev's seminal statement (Mendeleev 1869, Jensen 2005, 16): "If one arranges the elements in vertical columns according to increasing atomic weight, such that the horizontal rows contain analogous elements, also arranged according to increasing atomic weight, one obtains the following table,"[7] where, remarkably, Mendeleev refers to arrangements and analogies. Thus, a periodic system is what results from taking chemical elements and ordering and classifying them in an entangled manner (Fig. 4.2).[8] Note that we refer to *a* periodic system rather than to *the* periodic system, for depending on the properties of the elements used for their characterization, different S may show up, therefore different periodic systems may also. Mendeleev ordered the elements by selecting as leading property the atomic weight (attaching a property to basic substances).[9] The similarity was based on different properties of simple substances and of compounds, such as oxides, hydrides, halides, and so on (Jensen 2005).

A mapping of the periodic system onto another space, normally a bi-dimensional space, is called a *periodic table*. As there may be different periodic systems and different spaces on which to map, the periodic table is not unique. Another widespread term when talking about chemical elements is *periodic law*, understood as the oscillating variation of some properties of chemical elements (simple substances) as a function of the atomic number Z ($f(Z)$).[10] Formally, it can be expressed as $(Z, f(Z))$. In this sense, there would be *a periodic law for each property of simple substances*, therefore no unique periodic table

7 Current depictions of the table have interchanged Mendeleev's columns and rows and the "arranging" criteria has been changed to the atomic number.

8 These ideas and their formal definitions are the subject of a paper with Wilmer Leal that is currently in preparation. Schwarz (2004) is one of the few researchers who has mentioned the importance of ordering *and* classification for the periodic system. Ordering was so important to Mendeleev that he criticized Chancourtuois and Newlands for their approaches to the periodic system lacking that relation (Mendeleev 1899, Jensen 2005 paper 11). Often claims about periodic systems of other entities different to chemical elements are found in the scientific literature, e.g., hydrocarbons, diatomic and triatomic molecules, fullerenes and protein complexes to name but a few. Do actually they meet the conditions of similarity and ordering of a periodic system?

9 Mendeleev believed in the transcendental character of chemical elements as Paneth (1962) later did and championed the concept of basic substance. Mendeleev regarded chemical elements as devoid of properties; however he relied on the atomic weight as *the* invariant property. Perhaps Mendeleev drew a distinction between properties of one sort, for example, density or melting point, for they may result from the direct measurement upon substances; this contrasts with the atomic weight, which comes from comparison of, however, measured properties, e.g., of vapor densities of gases containing the common element whose atomic weight is to be determined. A historical account of ways to determine atomic weights is found in Scerri (2007). Hence, Mendeleev could have regarded atomic weight as a more complex property than others such as density or melting point. By doing so he perhaps thought he was not hurting the transcendental character of chemical elements as basic substances.

10 Originally, Mendeleev mooted it using atomic weight but it is currently stated through atomic number.

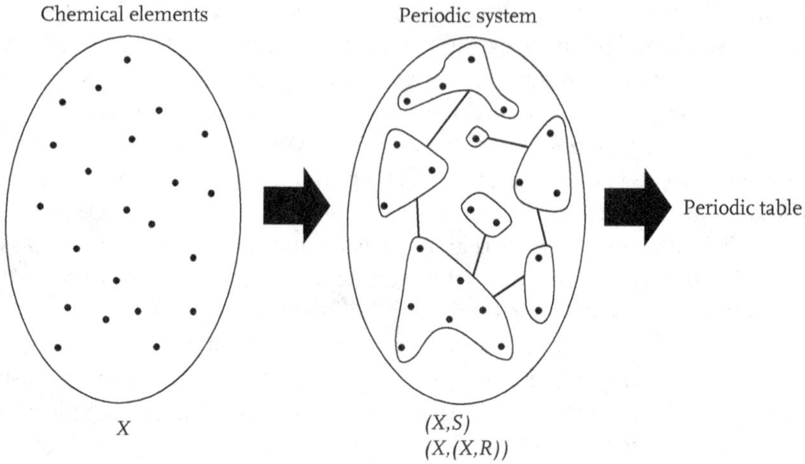

FIGURE 4.2 Relationship between chemical elements (X), periodic system, and periodic table, where S and R stand for structure and relation upon X, respectively.

would exist.[11] Moreover, contrary to general information spread by chemistry textbooks, even for the most common allotropes, not all of their properties are oscillating. In this respect Mendeleev's insight went too far when claiming that the law was applicable to all properties.[12]

Part of the success and the reason for the popularization of the periodic system lies in the use Mendeleev made of it for estimating properties of simple substances and compounds. Perhaps a better description of the prediction process would be relying not on the law but rather on the system,[13] which would avoid the problems caused by a lack of periodicity for certain properties. What is important and worth studying is the periodic *system*—its structure and the different functions (not necessarily periodic) that can be found between the system and a desired property.

[11] It is very unlikely that the range of $f(Z)$, for property f (say density), perfectly coincides with that of g (Z), for property g (say thermal conductivity); which is the requirement for unicity of the law. On the other hand, as the law refers to simple substances and each element can be represented by several allotropes, if all allotropes are regarded for all elements, the number of possible periodic laws is given by $\prod a_i$, with a_i being the number of allotropic forms for element i. Just by considering five allotropes for C, five for P, three for O, three for Se, seven for B, four for Sb, and five for Fe, for example, the number of periodic laws is 31,500.

[12] Mendeleev stated "not only in the forms of the compounds that we observe a regular dependency when the elements are arranged according to the magnitude of the atomic weights, but also in their other chemical and physical properties" (Jensen 2005, 45).

[13] Which was actually as Mendeleev derived algorithms for property estimations, i.e. not focusing on the "periodic law" but on the relations of classes and on the internal ordering of classes. See the translation into English of the original papers by Mendeleev in Jensen (2005).

From Similarities to the Periodic System

Mendeleev was interested in the mathematics underlying the periodic law, stressing that periods of elements cannot be treated as periods of continuous functions[14] but require a modification of continuous mathematics to cope with their discrete nature (Jensen 2005, 170; Jensen 2005, 221). Mendeleev wrote: "In my opinion, the reason one has so far been unable to represent the law using an analytical function is because the law relates to a field too little explored to allow for mathematical elaboration" (Jensen 2005, 221). He adds, "The reason for the absence of any explanation concerning the nature of the periodic law resides entirely in the fact that not a single rigorous, abstract expression of the law has been discovered" (Jensen 2005, 221). Indeed, such an abstraction is still absent, despite the different approaches from group theory, information theory, and statistics (King & Rouvray 2006). However, a different approach, based on classification and relations among classes was suggested in 2000 by Villaveces, who wrote: "It seems that the set of chemical substances is a topological space wherein relationships such as belonging to a class, neighbourhood or hierarchy of classes, etc. are more important than differential relationships"[15] (Villaveces 2000, 22). And he added: "This seems to be the mathematical structure underlying the periodic table and which would explain that predictions were possible when considering the whole set of chemical elements along with their equivalence classes" (Villaveces 2000, 22). This hypothetical structure of the periodic system is what we call *Villaveces' conjecture* and it has been the driving force of the similarity studies in which we have been involved and which are here analyzed along with other approaches.

Every similarity study looks for a classification of a set X of elements. The question that arises is whether it is possible to know the maximum number of classifications that may result given the number of elements in X. In other words, how many classifications are possible? A first approach is counting all possible subsets of X (power set of X); this is given by 2^n, with n the number of elements in X. Thus, for 118 elements, about 10^{35} subsets are possible. But in this counting there are subsets that may share elements, such as {Ga, Tc, Fr} and {Tc, Fr, Sb}. If one wants to avoid overlapping subsets and wants the union of subsets to build up the whole set, one is interested in counting the partitions[16] of X; which is given by the Bell numbers that grow faster than 2^n. A further refinement of this counting results when one has a rough idea of the number of classes of a particular size one is interested in—for example, the

[14] "neither the trigonometrical functions [...], nor the pendulum oscillations [...], nor the cubical curves [...], can [...] represent periods of chemical elements" (Jensen 2005, 170).

[15] Translated from Spanish by the author.

[16] For example, a partition of {Ga, Tc, Fr, Sb} is {{Ga, Tc}, {Fr, Sb}}. This corresponds to a crisp classification, where each element only belongs to a single class. However, there are fuzzy classifications where each element belongs to more than one class. See for example Rayner-Canham's classifications in this volume.

number of classes of four elements ($k = 4$) in a set of 10 elements. This is given by the Stirling numbers of the second kind $S(n,k)$ (Comtet 1974). If we assume that the number of classes of chemical elements is about the number of columns in the conventional medium-long form table (Scerri 2007) plus the lanthanoids and actinoids, the possible number of classifications is $S(118, 20) \approx 10^{135}$, but if one is just interested in metals and non-metals, then $S(118, 2) \approx 10^{35}$ and if semimetals are included $S(118, 3) \approx 10^{55}$. This gives an account of the colossal number of classifications of chemical elements that may result. Thus, if the periodic system were a random structure, each classification of chemical elements based on different properties would likely lead to a different classification. However, classifications of chemical elements show, again and again, that even if the properties are varied, the results change very little, which indicates a kind of underlying and invariable structure for the periodic system. In this chapter, we analyze seven similarity studies of chemical elements.

For similarity studies three kinds of input are needed: things to classify, properties characterizing them, and a mechanism to assess (dis)similarity. In formal terms, what is needed is a set of elements, their properties, and a (dis)similarity function. For chemical elements, the first question is the number of elements to analyze, which depends on the sort and number of properties characterizing them, for elements with little information are normally discarded.

Schummer (1998) has noted that properties characterizing materials are classified according to the context in which the material is embedded. Hence, there are (1) mechanical forces leading to *mechanical properties* like elasticity; (2) thermodynamic conditions like temperature associated with *thermodynamic properties* like melting point; (3) electromagnetic fields related to *electromagnetic properties* like specific magnetic susceptibility; (4) contexts of other chemical substances leading to *chemical properties* like the capacity for oxidation or the solubility in a certain liquid; (5) contexts of biological organisms associated to *biological properties* like antibiotic effects; and (6) contexts of ecological systems related to *ecological properties* like ozone depletion potential. Properties of type (1)–(3) are called *physical properties* and are characterized by "excluding the chemical factor" that is by working with inert container materials and atmospheres (Schummer 1998).

In the current chapter we discuss seven similarity studies of chemical elements, whose main features are shown in Table 4.1. The relationship of these approaches, besides similarity of chemical elements, is their phenomenological perspective; that is, mainly using experimental properties of bulk substances. There are some other non-phenomenological approaches based on theoretical descriptions of isotopes, atoms, and their electronic densities (Robert & Carbó-Dorca 1998, 2000, Khramov et al. 2006). All approaches of Table 4.1, except Leal et al (2012), characterize chemical substances by physical properties, while Leal et al (2012) is only based on chemical properties. Further we consider the most salient features of these two groups of approaches.

TABLE 4.1 Phenomenological similarity studies of the periodic system

AUTHOR(S) (PUBLICATION YEAR)	ZHOU ET AL. (2000)	SNEATH (2000)	KARAKASIDIS & GEORGIOU (2004)	RESTREPO ET AL. (2004)	RESTREPO ET AL. (2006A)	CHEN (2010)	LEAL ET AL. (2012)
Elements/ Atomic Numbers Z	H-Sn/Z=1-50	H-La, Hf-Bi/Z=1-57, 72-83	H-Nd, Sm-Dy, Y-Lu, Hf-Po, Rn, Ra-Ac/Z=1-60, 62-66, 70-71, 72-84, 86, 88-89	H-La, Hf-Rn/ Z=1-57, 72-86	H-La, Hf-Rn/ Z=1-57, 72-86	H-Xe/Z=1-54	H, Li-F, Na-Cl, K-At, Ra-Es/Z=1, 3-9, 11-17, 19-85, 88-99
Number of Elements	50	69	83	72	72	54	94
Number of Properties	7	54	2	31	128	10	4,700
Kinds of Properties	Physical	Physical and chemical	Physical	Physical and chemical	Physical and chemical	Physical and chemical	Chemical
Properties of Elements (simple substances)	X	X	X	X	X	X	
Properties of Compounds		X		X	X	X	X
(Dis)similarity Function	Fuzzy similarity	Gower's coefficient	Fuzzy similarity	Hamming, Euclidean, Gower and Cosine functions	Hamming, Euclidean, Gower and Cosine functions	Kohonen network	Symmetric difference

NOTES: Dedicated to Professor José L. Villaveces.

These six works are characterized by not stating the simple substances used.[17] The properties used vary in number, from 2 to 128, and in kind. In Karakasidis & Georgiou (2004) the only two properties used are first ionization potential and atomic number; while Chen used atomic mass, minimum oxidation state, atomic radius, electronegativity, state of matter, melting point, boiling point, heat of atomization, and ionization potential. But in Sneath and Restrepo chemical properties are added to the study. Sneath[18] used binary compounds of O, S, F, Cl, Br, and I; plus methyl-, ethyl-, and phenyl-compounds and complexes with donor oxygen, sulfur and nitrogen groups; and Restrepo used properties from geochemical origin.[19] In the majority of these studies atomic number and atomic weight are disregarded to assess to which extent the classification of chemical elements depends on these properties of great historical importance. The results indicate that the classifications are, in general, not affected by these properties.

Similarity results are depicted in Figure 4.3 and show a mixture of vertical (columns in the periodic table) and horizontal (rows) similarities but especially notorious are the horizontal similarities of Karakasidis & Georgiou (2004), which coincide with the periods of the table. This result is mainly given by the use of the first ionization potential to characterize the elements (the other property used was the atomic number, which monotonically increases throughout the structure of the table), a property that oscillates with the periods and it is in fact one of the properties used in the textbooks to show periodicity. Several of the other works include this property but its effect is combined with the effects of the other properties leading to the patchwork of vertical and horizontal similarities of the other tables in Figure 4.3.

Vertical similarities of elements close to the noble gases are in general invariant to the properties and substances selected for the different studies. In contrast, similarities of transition metals are more sensitive to properties and substances.[20] Supporting this, Sneath discusses the high degree of uncertainty in the classification of some transition elements like Cr, Tl, Pb, Pd and Co. In Zhou et al. (2000) Fe, Co, Ni, Ru, Rh and Pd show up as a class of similar elements, which are part of the so-called platinum metals with well reported chemical similarities (Greenwood & Earnshaw 2002).

---

[17] For example, which allotrope was chosen: α-rhombohedral- or β-rhombohedral-boron?

[18] Peter H. A. Sneath (1923–2011) was a leading biologist with a strong interest in chemistry; he conducted the work analyzed here (which is a fundamental piece for all subsequent similarity studies of the chemical elements) in 2000. He, along with Robert R. Sokal (1926–2012), was a pioneer of numerical taxonomy.

[19] These properties come from the work by Railsback (2003), which is further analyzed in his chapter in the current volume.

[20] This was already noted by Mendeleev: "The large periods [. . .] beginning with the alkali metals, which give the most powerful alkalies, end with the halogens, which correspond to the most powerful acids, and contain intermediate elements with less distinctly marked chemical properties" (Jensen 2005, 266).

Some other similarities are found, such as *diagonal relationships*, in other words, the similarity in chemical properties between an element and that to the lower right of it in medium-long periodic table—as in the similarity between Be and Al (Fig. 4.3). It is also found in all studies that H is not similar to any other element; Sc, Y, and La show up as a similarity class in all cases, except for Zhou et al. (2000), who did not consider La, and for Karakasidis & Georgiou (2004) who used properties mapping periods. There is evidence of the *singularity principle*, the principle that the chemistry of the second-period elements is often different to the later members of their respective groups (Rayner-Canham 2000), as exemplified by B, C, N, O, and F, which are in most of the cases not similar to the elements of their groups (columns).

A question treated in Sneath (2000) and Restrepo et al. (2006a) was about the *minimum number of properties needed to characterize the elements*. Sneath ran a principal component analysis study to reduce the 54-dimension space of properties to a new space of components resulting from linear combinations of the 54 properties used. The components attaining a high percentage of the variance of the original properties are regarded as the minimum number of properties needed to characterize the elements. He found that the 10 first most relevant combinations (principal components) accounted for about 80% of the total variance, indicating that at least 10 dimensions are needed to represent the elements with the kind of properties used. A principal component analysis study was also carried out in Restrepo et al. (2006a) and it was found that about 90% of the variance was attained by the first 31 principal components, which shows, as Sneath has already discovered, the difficulty in finding few components gathering most of the information of the chemical elements. This keeps raising the question of the minimum number of properties to describe chemical elements.

Sneath also clustered properties and found that most of them hold more than 98% resemblance, where chemical and physical properties are strikingly very similar. Not so similar properties are the first ionization energy and the boiling point, which makes one consider the discriminatory power of the properties and their mutual relationships. Results by Karakasidis & Georgiou (2004) indicate that part of the atypical character of the first ionization energy is depicting horizontal similarities, while the other properties show other trends. In any case, a detailed study of the distribution of all properties is needed to explore their similarities.

Besides the aforementioned results, Restrepo et al.'s work of 2004 introduced a novel approach to similarity: *chemotopology*, which takes the classifications to build up topological basis that are further used to explore similarities of different subsets of elements. These similarities are analyzed through different topological properties: closures, interiors, exteriors, derived sets, and boundaries (Restrepo et al. 2006b). In general, the *interior* of a set *A*, in terms of similarity, contains those elements that are more similar to the elements of *A* and may be regarded as representatives of *A*. The *exterior* of *A* gathers those elements that are different from elements in *A*. The *closure* of *A* contains the

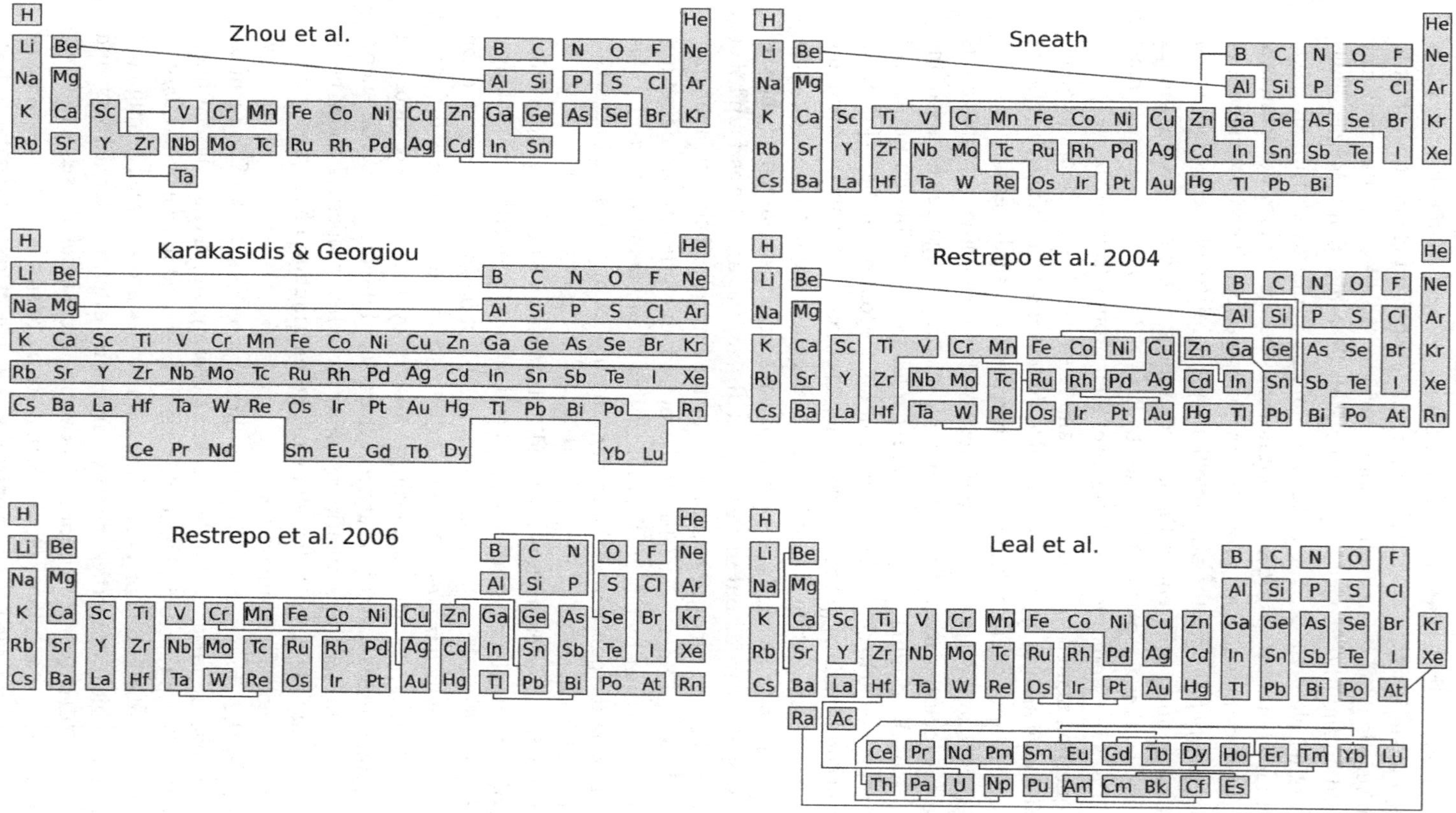

FIGURE 4.3 Classification results of the phenomenological similarity studies, where boxes indicate similarity classes.

elements that are similar to the elements of *A*; they do not have to be only similar to elements of *A* (as interior points are) but can hold resemblances with elements which are not included in *A*. The *derived set* of *A* gathers those elements that are similar to elements in *A*, not because of their own similarity, but because of the similarity of their neighbors. The *boundary* of *A* is given by those elements that are similar to the elements of *A* and also to the elements that are different to the elements of *A*. A review of chemotopology and its applications is found in Restrepo & Mesa (2011).

Chemotopology, applied to classification results of Restrepo et al. (2004, 2006a) shows that alkali metals, alkaline earth metals, {Sc, Y, La}, halogens, and noble gases are classes of elements whose interiors and closures are the same sets; for example, the interior and closure of {Sc, Y, La} is {Sc, Y, La}, which indicates that there is no other element similar to the elements of each class, a result strengthened by the finding of empty boundaries for those classes. The derived set shows (Restrepo et al. 2004) that every alkali metal is at least similar to another alkali metal; the same behavior is found for {Sc, Y, La}. But, in the same study, alkaline earth metals, halogens, and noble gases do not follow this similarity pattern, for Ba, F, and He turn out to be not part of the derived set of their respective groups.

Different topological properties, resulting from non-local and non-vertical similarities on the table, were found for most vertical classes (groups) of transition metals. Thus, for example, it is observed in Restrepo et al. 2004 that to the interior of {Fe, Ru, Os} only belongs Os, that to its closure belong, besides {Fe, Ru, Os}, Co, In, Ta, and W; that its boundary is made by the elements of the interior, except Os, indicating the lower resemblance of Os to Fe and Ru and the non-vertical similarities of Fe and Ru.

In Chen (2010) three main classes (metals, semimetals, and non-metals) are found, however without clear criteria for their selection and for the assignment of each element to each class.[21] Besides similarities found in previous works, two new classes were found: {Nb, Mo, Tc} and {Li, Mg}, {Li, Mg} being a case of diagonal relationship. A novelty introduced in Chen (2010) is the estimation of the classification of some few elements: Cs, Au, Hg, Pb, and Rn.[22]

Approach from Chemical Properties

A fundamental piece of information normally overlooked in chemistry textbooks is that Mendeleev's studies on chemical elements rested mainly on oxides, hydroxides, hydrides, and halides, whose resemblances were employed

21 It is presumed that the classification was based on the nearness of the neurons of the Kohonen network containing elements (characterized by two numbers representing two dimensions); however, sometimes the horizontal nearness is given more importance than the vertical one.

22 Unfortunately, the author did not mention the particular results for all these elements, but expressed that they were placed "in excellent locations"; it is only mentioned that Rn was placed in the same neuron as Xe.

to come up with similarities on chemical elements.[23] Mendeleev indicated the need to look at compounds rather than to properties of chemical elements, e.g. he claimed that "if CO_2 and SO_2 are two gases which closely resemble each other both in their physical and chemical properties, the reason of this must be looked for not in an analogy of sulphur and carbon, but in that identity of the type of combination, RX_4, which both oxides assume" (Jensen 2005, 263). And Mendeleev adds, "The elements, which are most chemically analogous, are characterized by the fact of their giving compounds of similar form RX_n" (Jensen 2005, 264).

In contrast to the popular neglect of compounds and the overemphasis of simple substances and their physical properties for studies on the periodic system, some authors have highlighted the important role of compounds and have mentioned the related problem of quantifying qualitative parts of chemistry for similarity studies (Schwarz 2007). The questions that arise are what can be done by having information of compounds and what kind of information is needed? Following Mendeleev, it is found that valences are important properties for chemical elements, which are obtained by stoichiometric decomposition of compounds in chemical analysis (Schwarz 2004).

Thus, considering historical aspects of the periodic system, especially its reliance on chemical compounds, valences, and stoichiometries of compounds, rather than on properties of isolated elements (simple substances); we ran a novel study based only on compounds. The study (Leal et al. 2012) is influenced by Schummer's (1998) idea that chemistry is rooted on the relational character of substances (Bernal & Daza 2010), which combined with concepts from category theory (Bernal et al. 2015) and social network analysis (Restrepo 2017) leads to the idea that a chemical substance (not necessarily an element) is not only characterized by properties measured upon the isolated substance, but, most important, by the other substances with which the substance in question is related, for example by chemical reactivity.

Schummer (1998) leads to the claim that (1) two substances belong to the same substance class if they are chemically similar and (2) they are similar if each of them react under the same conditions to form product substances of a common substance class. As (2) is difficult to meet, for it is not easy to find information about chemical reactions under the same reaction conditions, Leal et al. (2012) relaxed (1) and (2) as: (1´) two chemical elements belong to the same class if they are chemically similar and (2´) two chemical elements are chemically similar if each of them forms binary compounds of the same class. Hence, the general concept of chemical reaction is simplified to the study of reaction products, namely compounds (binary compounds) through

[23] Mendeleev thought that chemical elements were characterized by their interaction with others (as noted in Gordin's chapter of this book); it is an idea with philosophical basis. Mendeleev wrote: "Alone, you are simply nature's slave. Your individual is only zoological, animal, and all your humanness and all that you revere, all of this is from others, with others, and not for you alone, not personal but general."

the inspection of their formulas, where no attention is paid to the kind of reactions and the conditions leading to the binary compounds.

At this point, two chemical elements A and B are similar if they form binary compounds of the sort AC and BC, C being another chemical element. Hence, for example Na and K are similar as they form with O substances that later on react with water in a similar fashion. There are cases however, where the diversity of combinations is important for the discussion of similarity. For example, B, F, and Cl form binary compounds with H but it would be too crude to claim that B, F, and Cl are similar because of that fact. If the diversity of combinations is included, one finds that F and Cl are more similar to each other than to B, for B and H form by far more compounds (boranes) than F and Cl with H.

Leal et al. (2012) considered two chemical elements A and B as similar if they form binary compounds A_wC_x and B_yC_z, C being another chemical element, with similar values of w and y, on the one hand and of x and z on the other. The more combinations with different chemical elements and with similar stoichiometry, the more similar two chemical elements are. Leal et al. (2012) used 4,700 binary compounds accounting for 94 chemical elements for which at least one binary compound has been reported in the literature.[24] It is noteworthy that, so far, there is no other similarity study accounting for more elements than those considered in this study, which, for the first time, include actinoids.[25] Lanthanoids are only considered in Leal et al. (2012) and in Karakasidis & Georgiou (2004), but in the latter case with very little properties.

By exploring the formulae of the binary compounds, a neighborhood for each element *A* was defined gathering those elements with which *A* appears together in the formula along with information about the stoichiometry of the combination. Then the similarity among elements was assessed by contrasting their neighborhoods through their symmetric difference, which accounts for the differences between couples of neighborhoods.[26] These similarities led to the classification shown in Figure 4.3, where it is seen, again, that several groups of elements such as alkali, alkaline earth metals and halogens form similarity classes.

In Sneath (2000), Restrepo et al. (2004, 2006a), and Leal et al. (2012) the technical approach to similarity was through hierarchical cluster analysis (Everitt et al. 2011), which combines similarity functions and grouping methodologies to come up with a classification that is depicted as a *dendrogram*. A dendrogram is a hierarchy of classes, and there are different methods for selecting the most salient classes of the dendrogram (Everitt et al. 2011). In the

24 Restrepo (2017) explains how the procedure leads to a network and how the search for similarities corresponds to the reduction of the network by equivalences, using elements of social network analysis.

25 A more complete study of binary compounds is currently carried out by our research group accounting for more than 15,000 compounds.

26 In similarity studies it is customary to quantify distances among objects as the inverse of their similarity. Hence, distance values d may be transformed into similarities s by $s = 1\text{-}d$, with d in $[0, 1]$ (Deza & Deza 2006).

aforementioned works a compromise between the number of classes and their populations was sought so that highly populated classes are found. Such an approach was discussed in Uribe et al. (2005) and looks for classes of up to *n* elements. In Leal et al. (2012) *n* was set up to 4, therefore groups with five elements like alkali and alkaline earth metals were split, for example, {Li, Na} and {K, Rb, Cs}, and {Be, Sr, Ba} and {Mg, Ca}.

Similarities were found to be stronger for alkali metals than for alkaline earth metals; the dendrogram (Leal et al. 2012) shows that {Li, Na} and {K, Rb, Cs} are merged, while {Be, Sr, Ba} and {Mg, Ca} are not merged into a class of five elements but into the class merging {Be, Sr, Ba}, {Cd, Zn, Hg}, and {Mg, Ca}—indicating resemblance among alkaline earth metals but also of those metals with the above-mentioned transition metals. V and Zn groups are other examples of columns of the periodic table showing up as similarity classes.

There are some fragments of groups of the periodic table such as {Al, Ga, In, Tl}[27], {Ge, Sn, Pb}, {Sc, Y}, {Mo, W}, {Rh, Ir}, and {Cu, Ag} that form similarity classes. The presence of fragments rather than of complete groups indicates that not every element in a column of the periodic table combines in a similar fashion as the other elements of the column; that is the case of H, B, C, N, O, Cr, Ti, Mn, and Au (to name but a few), in their respective groups. In fact, there are 18 elements (single-classes) whose chemistry is totally different from that of members of their columns. Some of these single classes are instances of the singularity principle, such as B, C, N, and O; these have appeared in our previous similarity works (Restrepo et al. 2004, 2006a).

Zr and Hf turned out to be similar to Th and U, which are actinoids. This coincides with recent results on similarities of actinoids with 6th-row transition metals (Liu et al. 2017). Another example is the resemblance between Tc and Re and Pa and Np.

The well-known strong similarity among lanthanoids, which contrast with the reported relaxed similarity among actinoids, was observed by (see Liu et al. 2017). An interesting dissimilar actinoid is Pu, whose singularity has been reported by Schwarz (Liu et al. 2017).

Scerri (2007) has pointed out the discussion about the element at the beginning of the third row of the transition elements, which in some tables is La and in others Lu. Schwarz & Rich (2010) have asserted that Lu cannot be considered an *f*-block element (lanthanoid), for it does not fill *f* orbitals (as they are already filled), and has suggested that Lu should be regarded as a *d* element (transition metal). According to our results, La appears between two clusters, one of 11 lanthanoids and another of transition metals, namely {Y, Sc}. Lu is part of the clusters of 11 lanthanoids and the small four-element cluster containing it is {Er, Ho, Lu, Gd}. These results show that Lu is more similar to lanthanoids than to transition metals, while La shares similarities with lantha-

[27] If the dendrogram (Leal et al. 2012) is explored, it is found that {Al, Ga, In, Tl} is similar to Au and to {Cu, Ag}, rather than to B.

noids and with transition metals; therefore La must be the element located at the beginning of the third row of transition metals.

Scerri (2007) has also discussed the elements at the beginning of the fourth row of the transition elements, being Ac in some tables and Lr in others. Unfortunately no relevant data was available for Lr, therefore it is not possible to discuss its resemblance to other elements. Ac is found in a single class, which in turn is similar to a class of elements with a very small number of binary compounds—Ra, Kr, Xe and At.

Applying the chemotopological approach shows that the closure of each one of the following sets *X* turned out to be exactly the same sets *X*: halogens, alkali metals, alkaline earth metals, lanthanoids, V group, Cr group, VIII group (old IUPAC numbering), Cu group, Zn group, B group, C group, pnictogens, and chalcogens, as well as {B, C, N, O}, among others. This indicates that none of these classes has similarity with any other element outside of the respective class.

It is noteworthy that the classes {Fe, Co, Ni, Pd}, {Ru, Os, Pt} and {Rh, Ir} comprise the group VIII.[28] This reminds us that the similarity in the chemistry of these "platinum metals" was what led Mendeleev to group them together (Mendeleev 1889), as noted in his claim that "Only among these metals are compounds of the type RO_4 or R_2O_8 formed (which is why they are designated as the eighth group)" (Jensen 2005, 49). Our results show that it is valid to group them together, for they combine in a similar fashion (similar stoichiometry) with similar elements and constitute a set of elements with similar properties as those of the neighboring groups of the noble gases. That is, the interior and closure of platinum metals are the same elements; the boundary is empty and, interestingly, the derived set is the same set (leaving aside Pt), which shows that it is the least similar element to the "platinum metals."

Again, as in the previous works, the topological properties of the groups close to the noble gases show that their vertical similarities are the rule, that is, there are no similarities with side elements. Aside from the interesting result of the platinum metals, the topological properties of other transition metals are striking because they now turn out to be elements similar to their neighbors—that is,there are no longer similarities of, for example, Fe, Co, and In or Zn with Ga, Sn, and Pb (Restrepo et al. 2004).

Conclusions and Outlook

We think similarity studies of the chemical elements should avoid cases like this one: suppose only density and melting point are employed for the study, suppose the density for graphite and for α-rhombohedral boron are at hand as well as the melting point for diamond and for β-tetragonal boron. If the

28 In the old IUPAC group numbering or VIIIB in the CAS numbering and the VIII group of Mendeleev's periodic table (Mendeleev 1889).

similarity study is carried out, it would produce conclusions biased toward the kind of simple substances selected. Why are all the allotropes of C and B not selected? If they were selected, how should we treat, in mathematical terms, all those property values of a single basic substance? Is it right to characterize elements with a mixture of properties of different simple substances?

The same questions apply to compounds. Are we sure we are using properties of the same compounds? None of the studies discussed have made this point clear and it is very likely that they have mixed properties of different simple substances and compounds associated with the same chemical element (basic substance). Presumably all studies have retrieved substance properties from literature and databases where the statement about the clear identification of the simple substances is not normally made. Hence, before classifying elements and selecting properties for that task, *a clear statement about the kind of simple substances used for the classification is needed.*

How is it possible that the "mixture" of properties of allotropes and compounds of recent studies leads to similar classifications as those found in early 19th-century periodic systems? Is it an indication of a current compensation of errors serendipitously matching the right historical results? Or is it showing that the periodic system since its origins has mixed up properties of simple substances to come up with a general periodic system? We are tempted to suggest that the latter is the case: that the periodic system is a general depiction of ordering and similarity of chemical elements accounting for a large and diverse set of properties of different realizations of elements and compounds.[29]

We discussed the error of assigning properties of simple substances to the elements in the periodic table. It was mentioned that this can only be done if the boxes for elements in the table can be expanded into the different simple substances of the elements. This multidimensionality of each box could be shown with dynamic periodic tables in the Internet, which would be linked and fed by authoritative repositories of chemical information. Such a periodic table could also include a module for data analysis bringing up-to-date knowledge on the chemical elements based on experimental facts gathered in the repositories.[30]

Even if the current chapter is mainly dedicated to similarity approaches to the periodic system, we have shown that *it is a mistake to regard the system as a structure where only similarities are important.* By going back to the first publication by Mendeleev on the subject, we have found that order and similarity are the two key concepts for the system and they have been so since its discovery. Therefore, we agree with Schwarz's (2007) when claim that similarity is not enough to build up a periodic table, but we do not agree with his statement that

[29] By reading Mendeleev's papers on the periodic system and how he came up with it, the diversity of factual chemical knowledge that was included to devise relationships among elements is clear. The invitation is to devise mathematical algorithms able to cope with large amounts of chemical information to build up a more general periodic table avoiding biases related to properties and substances.

[30] This is an idea Eugenio J. Llanos has extensively discussed with this author.

the periodic table is needed to know the similarities of the elements. We have shown that similarity studies give a nice account of the resemblances among elements, in fact better than that depicted by periodic tables, for periodic tables cannot show all similarities (e.g. similarities among transition metals and actinoids) in just a depiction. This is a technical restriction resulting from the bi-dimensional character of most periodic tables, which combined with the ordering of elements make it impossible to show all similarities at once.[31] We then suggested that *if similarities are the aim, a better representation of bi-dimensional periodic tables is a dendrogram* or a structure of similarity classes depicting levels of similarity as the similarity landscape reported in Restrepo (2017).

Contrary to Mendeleev's claim, not all properties of chemical elements, as simple substances, are periodic (i.e., oscillating regarding elements ordered by atomic number). Hence, if periodicity was key to deriving the periodic system, how can it be explained that all the discussed similarity studies produce, in general, classes of elements that match to a high degree with the 19th-century periodic system? It seems that the oscillating character of the properties and in general the sort of distribution of property values is not required to come up with the periodic system. But then, what? Villaveces' conjecture sheds light on a possible answer: *it is the topological structure of the space of properties of simple substances and compounds, which when mapped onto basic substances, show the structure we know as periodic system of chemical elements.*

Regarding the minimum number of independent properties needed to characterize chemical elements, principal component analysis, as attempted by Sneath (2000) and Restrepo et al. (2006a), may give an answer. Schwarz (2007) claims that for about 100 elements described by 100 properties, between 5 and 10 principal properties (components) would be needed. This attending to $F \approx P^{1/n}$, with $n = 2\text{-}3$, where F is the number of components and P the number of properties. In fact, Sneath's (2000) work showed that 10 components are needed, while Restrepo et al. (2006a) needed fewer than 31. This is still an open question requiring, besides statistical analysis of the properties, a clear definition of what the intention of the classification is, which results in appropriate properties being selected for the classification.

The similarity results we have reported show that vertical similarities (columns in the periodic table) of elements close to the noble gases are in general invariant to the properties and substances selected for the different studies. In contrast, similarities of transition metals are more sensitive to properties and substances. This broad distinction of the similarity, ergo topological relationships, of chemical elements make us ponder, again, the properties needed to describe the elements. Perhaps attention needs to be focused on the transition metals, since finding properties able to characterize them will presumably not affect the similarities of groups around the noble gases. It is well-known, for

[31] The best depiction, if any, of these relationships constitutes an interesting optimisation problem. Note that the reference to the "best" is in mathematical terms, rather than in aesthetic ones.

example, how the similarities of transition metals change using only relational properties of elements into binary compounds (Leal et al. 2012) where a large amount of vertical relations are found, which contrasts with several horizontal similarity relations found in the other studies.

Scerri has posed the question whether classes of chemical elements constitute actual natural kinds. The results here discussed suggest that a positive answer can be given at least for alkali metals {Li, Na, K, Rb, Cs},[32] *pseudo*-alkali earth metals {Mg, Ca, Sr, Ba},[33] halogens {F, Cl, Br, I}, and noble gases {He, Ne, Ar, Kr, Xe, Rn}. Another promising set—at least in chemical terms—is made by platinum metals.

From a mathematical point of view, the stability of classes around noble gases and the relative lack of stability of transition metals is striking if one bears in mind the large number of classification possibilities given by the Stirling numbers of the second kind. Even if Pd is sometimes similar to Rh and Pt, some others to Ag and Cu, and still others to Ni, Co, and Fe; it has not shown up in the class {Pd, He, Gd, H, F}, or any other class that would suspiciously draw the attention of a chemist for its lack of chemical sense. What this shows, we claim, is that there is a chemical background underlying the similarity of the chemical elements and that *the classification of chemical elements is not a random result.*

In pedagogical terms the results from Leal et al. (2012) show that most of the similarity structure of the periodic system can be recovered by using only chemical properties, which contribute to the discussion on how to teach chemistry, whether from a microscopic and atomic level or from a macroscopic level of bulk substances (Nelson 2002). The work of Leal et al. (2012) favors the latter and it would be interesting to include this kind of approach in general chemistry courses by pointing out representative series of chemical reactions.

The results are mostly invariant to the inclusion or not of *atomic number and atomic weight*, which leads us to conclude that these properties, apart from their historical importance, *are not important for the similarity structure of the periodic system.* However, there is no doubt that their importance lays on the ordering structure of the system; the question that arises is whether only those properties are of importance for the ordering or whether there are others yielding the same structure.

The periodicity of the periodic system—those particular points in the ordering of elements where the order is "twisted" to bring closer similar elements—is not treated further.[34] Similarity studies do not bring complete information about that, for they disregard the order component of the system. However, the studies do indicate the elements that are similar, which, once ordering is regarded, become the indicators for the twisting. Hence, Schwarz's (2007)

---

[32] Further information on other elements like Fr is needed.

[33] We call them *pseudo*, for Be and Ra are not included.

[34] The study by Karakasidis & Georgiou (2004) shows that classes correspond to periods, but this study depends strongly on the first ionization energy and also lacks the ordering structure of the system.

question on the amount of periodicity encoded by similarity approaches is fully answered once the similarity classes are endowed with order relationships.[35]

Schwarz (2007) raised the question of the explanation of the similarities of the chemical elements. Similarity approaches bring classifications and now it is the turn of theoreticians to bring explanations. But this also raises the philosophical question of what an explanation *is*. Is an explanation of the periodic system the framing of some physical properties of chemical elements into the realm of quantum chemistry? Or is this just a change of language and of scientific tradition (discipline)? Sneath's results show that almost all properties are very similar, including chemical and physical ones, but only ionization energy and the boiling point differ. Why this happens requires further study and may contribute to the discussion of the reducibility of chemistry to physics (Hettema 2013). Are chemical properties actually condensed in physical properties of the elements? Or could it be taken the other way round—that the physical properties can be obtained by exploring chemical properties?[36]

Chen (2010) has mentioned that there is no information about the relationships between classes. This was also noted by Mendeleev, who claimed that this is where the importance of the periodic system lies and where research on the system must go (Jensen 2005, 59). Surprisingly, this is still something to be carried out; perhaps there is a super-structure for the periodic system where the objects of study are no longer elements but classes and now the similarities and ordering are for classes instead of for elements.

Only one similarity study has gone beyond the characterization of the similarities and its mathematics. This is Chen's (2010) approach, in which prediction of classifications of elements was done. Besides improving the methodological approaches to similarity and working on the selection of properties characterizing simple substances, similarity approaches should follow Chen's steps and venture in the realm of estimations. For example, a classification of chemical elements based on pure chemical relationships à la Leal et al. (2012) combined with Chen's (2010) ideas of training a network to predict classifications of elements is an interesting line of work to pursue. Moreover, classifications can be used to develop models of properties of elements and also of compounds.

Scerri (2007) has mentioned the resemblance between atom clusters of elements and chemical elements. It would be interesting to include properties of these simple substances as contributing to the description of elements as basic substances. Another dimension on properties to be explored is their values

35 Schwarz (2007) has shown how the twisting points on the order can be predicted by quantum chemical approaches for the bonded atoms, not for isolated ones, which makes the approach of Leal et al. (2012) closer to Schwarz's.

36 This position is not so strange, because a lot of the early motivation to quantum concepts came from chemistry, especially from the periodic system (Scerri 2007). But most importantly, as many of the quantum chemical calculations, if not all, somehow seek to bring insight on results of wet-lab chemistry.

under different conditions, for example, electrical conductivity is affected by density, which is affected by temperature and pressure (Hensel et al. 2015). If properties under different conditions are included, what would similarities look like? Likewise, Schwarz (2011) and Rayner-Canham (2011) have suggested using Pourbaix diagrams to run chemotopological studies. This is something to be carried out for further exploration of this chemical account of chemical properties.

If the structure of the periodic system is finally formalized as an entanglement of ordering and similarity, as suggested by Villaveces' conjecture, approaches such as those derived by Klein to estimate properties of substances based on their ordering relations could be applied to the chemical elements (Panda et al. 2013).

The studies here reported have shed light on the similarity structure of the periodic system, which have shown that Villaveces' conjecture points in the right direction; but the same conjecture states that hierarchy of classes is important for the structure of the system. Thus, further research in that direction is needed to finally accept or reject the conjecture as a whole.

Acknowledgments

I am indebted to Andrés Bernal, Ray Heffferlin, Douglas J. Klein, Michael Laing, Wilmer Leal, Eugenio J. Llanos, Héber Mesa, Geoff Rayner-Canham, Eric Scerri, Joachim Schummer, W. H. Eugen Schwarz, Peter Sneath, Pieter Thyssen, and José L. Villaveces for their valuable discussions and ideas. I thank the Alexander von Humboldt Foundation/Stiftung and the Universidad de Pamplona for the financial support to carry out this research.

References

Ahnert, S.E., Marsh, J.A., Hernández, H., Robinson, C.V. & Teichmann, S.A.: 2015. "Principles of assembly reveal a periodic table of protein complexes." *Science*, 350, aaa2245.

Bernal, A. & Daza, E.E.: 2010. "On the epistemological and ontological status of chemical relations." *Hyle—International Journal for Philosophy of Chemistry*, 16, 80–103.

Bernal, A., Llanos, E.J., Leal, W., & Restrepo, G.: 2015. "Similarity in chemical reaction networks: categories, concepts and closures." In: S.C. Basak, G. Restrepo & J.L. Villaveces (eds.), *Advances in Mathematical Chemistry and Applications*. Bentham: Sharjah, 24–54.

Chen D.Z.: 2010. "A new method for studying the periodic system based on a Kohonen neural network." *Journal of Chemical Education*, 87, 433–434.

Comtet, L.: 1974. *Advanced Combinatorics: The Art of Finite and Infinite Expansions*. D. Reidel Publishing Company: Dordrecht.

Deza, E. & Deza, M.-M.: 2006, *Dictionary of Distances*. Elsevier: Amsterdam.

Earley, J.E.: 2009." "How chemistry shifts horizons: element, substance, and the essential." *Foundations of Chemistry*, 11, 65–77.

Everitt, B.S., Landau, S., Leese, M. & Stahl, D.: 2011. *Cluster Analysis*. Wiley: Chichester.

Ghibaudi, E., Regis, A. & Roletto, E.: 2013. "What do chemists mean when they talk about elements?" *Journal of Chemical Education*, 90, 1626–1631.

Greenwood, N.N. & Earnshaw, A.: 2002. *Chemistry of the Elements*. Butterworth Heinemann: Oxford.

Hensel, F., Slocombe, D.R., & Edwards, P.P.: 2015. "On the occurrence of metallic character in the periodic table of chemical elements." *Philosophical Transactions of the Royal Society A*, 373, 20140477.

Hettema, H.: 2013. "Austere quantum mechanics as a reductive basis for chemistry." *Foundations of Chemistry*, 15, 311–326.

Hoffmann, R.: 2015. Personal communication.

Jensen, W.B.: 2005. *Mendeleev on the Periodic Law, Selected Writings, 1869–1905*. Dover: Mineola.

Karakasidis, T.E. & Georgiou, D.N.: 2004. "Partitioning elements of the periodic table via fuzzy clustering technique." *Soft Computing*, 8, 231–236.

Karol, P.J.: 2017. "Heavy, Superheavy...Quo Vadis?," In: E.R. Scerri & G. Restrepo (Eds.), *Mendeleev to Oganesson: A Multidisciplinary Perspective on the Periodic Table*.

Khramov, N.N., Bartenev, S.A., Markov, G.S., Romanovskiy, V.N., & Khlopin, V.G.: 2006. "Statistical modeling of chemical periodicity and prediction of the properties of the superheavy elements." In: B. King & D. Rouvray (eds.), *The Mathematics of the Periodic Table*. Nova: New York, 31–50.

King, B. & Rouvray, D. (Eds.): 2006. *The Mathematics of the Periodic Table*. Nova: New York.

Leal, W., Restrepo, G., & Bernal, A.: 2012."A network study of chemical elements: from binary compounds to chemical trends." *MATCH Communications in Mathematical and in Computer Chemistry*, 68, 417–442.

Liu, J-B., Chen, G.P., Huang, W., Clark, D.L., Schwarz, W.H.E. & Li, J.: 2017. "Bonding trends across the series of tricarbonato-actinyl anions $[(AnO_2)(CO_3)_3]^{4-}$ (An=U-Cm): The plutonium turn." *Dalton Transactions*, 46, 2542.

Mendeleev, D.: 1869. "Über die Beziehungen der Eigenschaften zu den Atomgewichten der Elemente." *Zeitschrift für Chemie*, 12, 405–406.

Mendeleev, D.: 1889. "The periodic law of the chemical elements." *Journal of the Chemical Society*, 55, 634–656.

Mendeleev, D.: 1899. "How I discovered the periodic system of elements." *Revue Générale de Chemie Pure et Appliquée*, 1, 211–214, 510–512.

Nelson, P.G.: 2002. "Teaching chemistry progressively: from substances, to atoms and molecules, to electrons and nuclei." *Chemistry Education Research and Practice*, 3, 215–228.

Panda, A., Vijayakumar, S., Klein, D.J., & Ryzhov, A.: 2013. "Network of secondary-substituted adamantane amines." *Journal of Physical Organic Chemistry*, 26, 917–926.

Paneth, F.A.: 1962. "The epistemological status of the chemical concept of element," *British Journal for the Philosophy of Science*, 13, 1–14 (Part I) and 144–160 (Part II).

The text of a public lecture given by Professor F.A. Paneth to the Gelehrte Gesellschaft of Königsberg in 1931. (German text published in Schriften der Königsberger Gelehrten Gesellschaft, Naturwissenschaftliche Klasse, 1931, Heft 4 [Max Niemeyer Verlag].) Translated by H.R. Paneth. A recent version of the paper was reprinted in *Foundations of Chemistry*, 5, 113–145 (2003).

Potter, M.: 2004. *Set Theory and Its Philosophy*. Oxford University Press: Oxford.

Railsback, L.B.: 2003. "An earth scientist's periodic table of the elements and their ions." *Geology*, 31, 737–740.

Rayner-Canham, G.: 2000. "Periodic patterns." *Journal of Chemical Education*, 77, 1053-1056.

Rayner-Canham, G.: 2011. "Relationships among the transition elements." *Foundations of Chemistry*, 13, 223–232.

Restrepo, G., Mesa, H., Llanos, E.J., & Villaveces, J.L.: 2004. "Topological study of the periodic system." *Journal of Chemical Information and Computer Sciences*, 44, 68–75.

Restrepo, G., Llanos, E.J., & Mesa, H.: 2006a. "Topological space of the chemical elements and its properties." *Journal of Mathematical Chemistry*, 39, 401–416.

Restrepo, G., Llanos, E.J., & Mesa, H.: 2006b. "On the topological sense of chemical sets." *Journal of Mathematical Chemistry*, 39, 363–376.

Restrepo, G. & Mesa, H.: 2011. "Chemotopology: beyond neighbourhoods," *Current Computer-Aided Drug Design*, 7, 90–9

Restrepo, G. & Harré, R.: 2015. "Mereology of quantitative structure-activity relationships models." *Hyle—International Journal for Philosophy of Chemistry*, 21, 19–38.

Restrepo, G.: 2017. "Building classes of similar chemical elements from binary compounds and their stoichiometries," In: *Elements Old and New: Discoveries, Developments, Challenges, and Environmental Implications*, Benvenuto, M. A.; Williamson, T., Eds.; ACS Symposium Series; American Chemical Society; Washington, DC, USA, 2017; Chapter 5, 95–110.

Robert, D. & Carbó-Dorca, R.: 1998. "On the extension of quantum similarity to atomic nuclei: Nuclear quantum similarity." *Journal of Mathematical Chemistry*, 23, 327–351.

Robert, D. & Carbó-Dorca, R.: 2000. "General trends in atomic and nuclear quantum similarity measures." *International Journal of Quantum Chemistry*, 77, 685–692.

Ruthenberg, K.: 2009. "Paneth, Kant, and the philosophy of chemistry." *Foundations of Chemistry*, 11, 79–91.

Scerri, E.R.: 2007. *The Periodic Table: Its Story and Its Significance*. Oxford University Press, New York.

Schummer, J.: 1998. "The chemical core of chemistry I: a conceptual approach." *Hyle—International Journal for Philosophy of Chemistry*, 4, 129–162.

Schwarz, W.H.E.: 2004. "Towards a physical explanation of the periodic table (PT) of chemical elements, achievements of the previous generations," In: E.J. Brändas & E.S. Kryachko (eds.), *Fundamental World of Quantum Chemistry*, III. 645–669.

Schwarz, W.H.E.: 2007. "Recommended questions on the road towards a scientific explanation of the periodic system of chemical elements with the help of the concepts of quantum physics." *Foundations of Chemistry*, 9, 139–188.

Schwarz, W.H.E. & Rich, R.L.: 2010. "Theoretical basis and correct explanation of the periodic system: review and update." *Journal of Chemical Education*, 87, 435–443.

Schwarz, W.H.E.: 2011. Personal communication.

Sneath, P.H.A.: 2000. "Numerical classification of the chemical elements and its relation to the periodic system." *Foundations of Chemistry*, 2, 237–263.

Uribe, E.A., Daza, M.C., & Restrepo, G.: 2005. "Chemotopological study of the fourth period mono-hydrides." *WSEAS Transactions on Information Science and Applications*, 2, 1085–1090.

van Brakel, J.: 2014. "Philosophy of science and philosophy of chemistry." *Hyle—International Journal for Philosophy of Chemistry*, 20, 11–57.

Van Noorden, R.: 2016. "Four chemical elements added to periodic table." *Nature*, doi:10.1038/nature.2016.19112.

Villaveces, J.L.: 2000. "Química y epistemología, una relación esquiva." *Revista Colombiana de Filosofía de la Ciencia*, 1, 9–26.

Weyl, H.: 1940. "The mathematical way of thinking." *Science*, 92, 437–446.

Zhou, X.-Z., Wei, K.-H., Chen, G.-Q., Fan, Z.-X., & Zhan, J.-J.: 2000. "Fuzzy cluster analysis of chemical elements." *Jisuanji Yu Yingyong Huaxue*, 17, 167–168.

CHAPTER 5

The "Chemical Mechanics" of the Periodic Table

ARNOUT CEULEMANS* AND PIETER THYSSEN**
*Department of Chemistry, Katholieke Universiteit Leuven, Belgium
**Institute of Philosophy, Katholieke Universiteit Leuven, Belgium

1. *Introduction*

In 1969, the centennial of Mendeleev's discovery of the periodic table was commemorated by an international conference devoted to the periodicity and symmetry of the elementary structure of matter. The conference was held in the Vatican and brought together a selected audience of first-rate atomic and nuclear scientists. In 1971, the proceedings were published in a joint publication [1] of the *Academy of Sciences* of Torino and the *National Academy* in Rome.

Among the many interesting contributions, the American cosmologist John Archibald Wheeler described a mind-boggling journey from "Mendeleev's atom to the collapsing star." According to Wheeler [2], Mendeleev was convinced that the atom is not "deathlike inactivity" but a dynamic reality and Mendeleev expressed his hope that the discovery of an orderly pattern would "hasten the advent of a true chemical mechanics."

This hope has certainly been met by Schrödinger's wave mechanics, which provides an accurate tool to simulate the properties of the elements. However, the overall structure and symmetry of the periodic table continues to defy understanding. The quest for an effective universal force law at the basis of the mechanics of multi-electron atoms forms the topic of this contribution.

2. *Bertrand's Theorem*

The search for central force laws should start with Bertrand's theorem in classical mechanics. In 1873, the French mathematician Joseph Louis Bertrand presented to the Paris Academy a short note [3, 4] on central force laws that give rise to stable orbits. For a proper understanding of the research question which Bertrand was addressing, we start from an everyday experiment. A mass attached to a string can easily be swept around in a perfectly circular orbit by simply pulling on the string. The only requirement is that the force should be fixed and directed toward the center of the orbit. If we want the mass to go faster, we simply have to pull harder. Newtonian mechanics tells us that there exists a simple relationship between the centripetal force, F, which we have to exert, and the speed of revolution, v:

$$F = \frac{mv^2}{r}. \tag{1}$$

Here m is the mass and r is the radius of gyration. So if the string is longer and the force stays the same, the mass will reach a higher speed. As this simple experiment shows, a fixed central attraction force can give rise to a closed circular orbit, provided the speed of revolution and the radius are in line with Equation (1). However, the resulting orbit is "fragile." Any change of F, v, or r leads to a loss of stability, which often results in chaotic behavior.

What Bertrand was looking for are attraction forces for which a body, launched arbitrarily with a speed less than a certain limit and pulled toward a given center, *necessarily* describes a closed curve about this center. In the solar system, the planets are attracted to the center of mass of the system by the gravitational force, and—happily for us—the solar system is robust and does not collapse but maintains stable orbits for ages. To guarantee the stability of such a system, we cannot invoke a fixed stable force, nor invisible ropes of a constant length; we need "force laws," which impose a functional form that relates the attraction force to the distance. Bertrand's communication to the Academy, which since then has been known as Bertrand's Theorem, proved that there are only two force laws with this property:

1. Newton's law of gravitation where the force is inversely proportional to the square of the distance to the center,
2. A central force which is directly proportional to the distance to the first power.

Both these laws are treated by Newton in his *Principia* [5]. The law of gravitation gives rise to the Kepler orbits, which are ellipses with the center of attraction in one of the focal points of the ellipse. Bertrand called this law "the one from Nature," since it describes the stability of the solar system. Later on, exactly the same distance dependence was found to be valid for the electrostatic attraction between point charges of opposite sign, as expressed in Coulomb's law. This attraction force gives rise to a potential which is inversely proportional to the distance $\sim 1/r$, as shown in Figure 5.1.

The second law, which Newton lapidary expressed as "the force is as the distance," is in fact a generalization of Hooke's law, which states that in a spring the elastic restoring force for small distortions from the equilibrium is simply proportional to the extent of elongation or contraction. It gives rise to

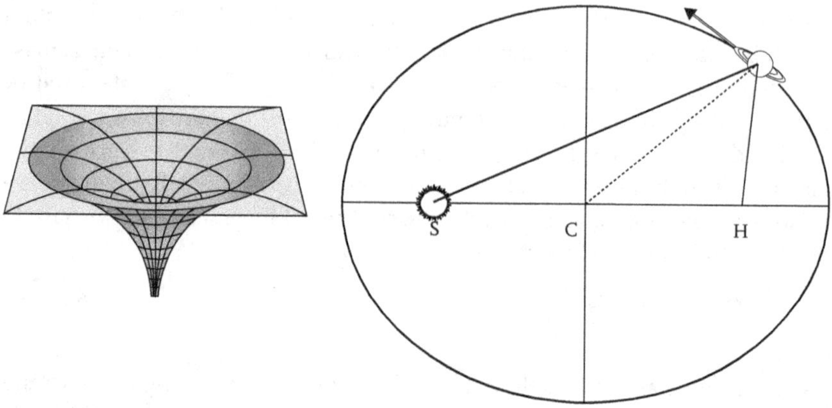

FIGURE 5.1 (Left) the attraction potential corresponding to an inverse square force law. (Right) The corresponding orbit is an ellipse with the attraction hole in one of the focal points.

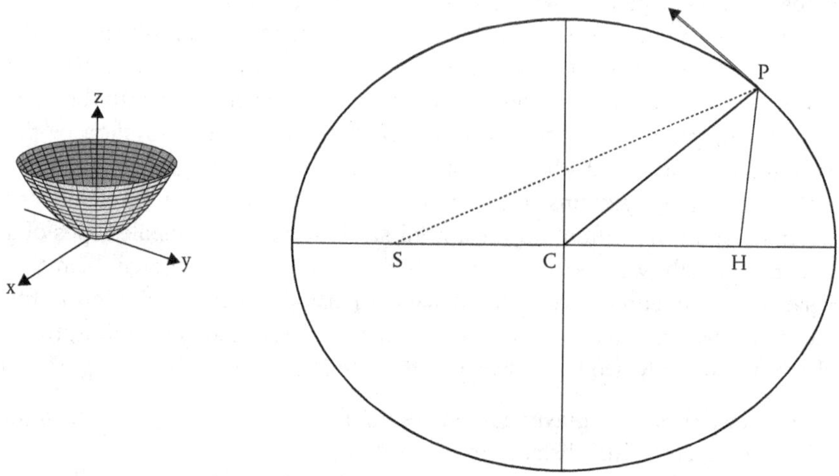

FIGURE 5.2 (Left) a parabolic attraction potential corresponding to a force proportional to the distance. (Right) The corresponding orbit is an ellipse with the attraction hole in the center.

the harmonic oscillator, the three-dimensional form of which is the spherical oscillator. The corresponding potential is a parabola, as shown in Figure 5.2.

In this case, Newton demonstrated that an elliptic orbit, with the center of attraction in the center of the ellipse, requires a centripetal force which is proportional to the distance. As a corollary he also noted the "inverse problem" that such a force must give rise to elliptic orbits, at least if the velocity is below the escape limit.

Hence both cases give rise to elliptical orbits, but with different symmetries: the Kepler orbit has a line of symmetry along the major axis of the ellipse, while the oscillator orbit has two lines of symmetry along the major and minor axes of the ellipse. In fact, Newton observed [6] that these two unique potentials are to a certain extent each other's dual. We briefly present Newton's result in the Appendix.

3. *Quantum Mechanics*

3.1 Stability and Conservation

The two cases distinguished by Bertrand in classical mechanics also are paradigmatic in quantum mechanics. The analogue of the planet gravitating toward the sun on Kepler orbits is of course Schrödinger's hydrogen atom, with gravity being replaced by the purely electrostatic Coulomb attraction, and the negatively charged electron orbiting the positively charged nucleus.

The analogue of the force-equals-distance law is the quantum mechanical spherical oscillator which is ubiquitous in physics, and which furnished the model potential on which the quark model of Gell-Mann was based. Not surprisingly then, the harmonic oscillator is often called "the mother of all quantum systems."

But what is so special about these two cases, which link the macrocosm of classical mechanics to the microcosm of quantum mechanics? Both cases stand out because in both cases the motions of the orbiting particle conserve quantities, that in turn explains the stability of the orbit. In the case of the Kepler system the conserved quantities are the angular momentum and the Runge-Lenz vector. Both are vectorial quantities and thus involve each three Cartesian components.

The angular momentum is a vector which is perpendicular to the orbit and thus conserves the plane of revolution. In order to destabilize this plane, an extra torque force should be exerted. This explains why the planetary orbits have not deviated much from a common ecliptic plane since the origin of the solar system. The special feature of the Kepler orbit, however, is the conservation of an extra vectorial quantity, known as the Runge-Lenz vector (or the Hermann-Bernoulli-Laplace-Hamilton-Runge-Lenz vector to be historically correct) [7]. This vector is perpendicular to the angular momentum and points along the major axis of the ellipse. Conservation of this vector implies that the in-plane orientation of the ellipse is maintained and that the precession of this orientation due to many-body effects is damped.[1]

In quantum mechanics both these quantities are replaced by three operators which commute with the Hamiltonian. The set of these six operators forms an algebra, which has been identified as the algebra of the SO(4) Lie group, standing for the special orthogonal group of rotations in four dimensions. The principal quantum number n is an eigenvalue of this algebra, and takes on the values:

$$n = 1, 2, 3, \ldots, \infty. \tag{2}$$

If the algebra is limited to the three angular momentum operators only, the group is reduced to the SO(3) group, which is the symmetry of a sphere in three dimensions and reflects the obvious spherical symmetry of the problem. It gives rise to the orbital angular momentum quantum number l, which stands for the angular dependence of the orbits. The l quantum number takes on the values from zero to a maximum of $n - 1$ in integer steps. For any given value of n, there are therefore n values of l:

$$l = 0, 1, 2, 3, \ldots, n-1. \qquad (3)$$
$$s, p, d, f, \ldots$$

Hence, in wave mechanics the trajectories of the electron no longer follow planetary orbits, but the connection is maintained through the conserved quantities.

Similar analogies can be drawn between the spherical oscillator in classical and quantum mechanics. In this case, conserved quantities can be gathered in a Lie algebra with eight operators, which generate the SU(3) group. (SU is an acronym for the special unitary group in three dimensions.) We will not consider this further in the present context. Both the SO(4) and SU(3) algebras are prime examples of Lie algebras. They explain why the corresponding problems can be solved exactly, and their solutions have given us the quantum numbers that describe the physical states of the hydrogen atom and quark matter. In fact, much of the success of quantum mechanics is due to the exact solvability of these simple model systems, and the quantum numbers coming out of these treatments play an undeniable key role in the whole of physics.

3.2 Force Laws and Quantum Numbers

But this is not all. There is still a deeper connection between the force laws and the quantization. This stems from a separate consideration of the radial and angular components of the momentum. In the hydrogen atom, the energy is inversely proportional to the square of an integer number n, which is the principal quantum number. It is given by:

$$n = 1 + n_r + l. \qquad (4)$$

Here n_r is the radial quantum number and l corresponds to the angular or orbital quantum number. The radial quantum number n_r counts the number of radial phase changes of the wavefunction when going from $r = 0$ to $r = \infty$. As an example: the 1*s* orbital is described by a monotonously decreasing exponential, having a cusp at the origin and approaching a zero asymptote toward infinity. This function has no radial nodes and thus $n_r(1s) = 0$. For the 2*s* orbital there is one sign change in the radial interval, and hence $n_r(2s) = 1$, and so on, yielding: $n_r(ns) = n - 1$.

The orbital quantum number l counts the number of angular nodes. These are recognized by the presence of nodal planes in the orbital graphs. The $2p_z$ orbital has a different sign on the northern and southern hemisphere, and hence becomes zero in the equatorial plane. Consequently one has: $l(2p_z) = 1$. An orbital such as d_{xz} obviously has two orthogonal nodal planes, one coinciding with the equatorial *xy*-plane, the other being the upright *yz*-plane. In more complicated cases such as d_{+2} or d_{z^2} it becomes less obvious to delineate these phase changes, and the application of the angular momentum operator is required. The contributions of n_r and l to the principal quantum number of hydrogen are listed in Table 5.1.

TABLE 5.1 Radial and angular quantum numbers, n_r and l, for the states of the hydrogen atom and the spherical oscillator

HYDROGEN				OSCILLATOR			
	n_r	l	n		n_r	l	n
1s	0	0	1	1s	0	0	0
2s	1	0	2	1p	0	1	1
2p	0	1	2	2s	1	0	2
3s	2	0	3	1d	0	2	2
3p	1	1	3	2p	1	1	3
3d	0	2	3	1f	0	3	3
4s	3	0	4	3s	2	0	4
4p	2	1	4	2d	1	2	4
4d	1	2	4	1g	0	4	4
4f	0	3	4				

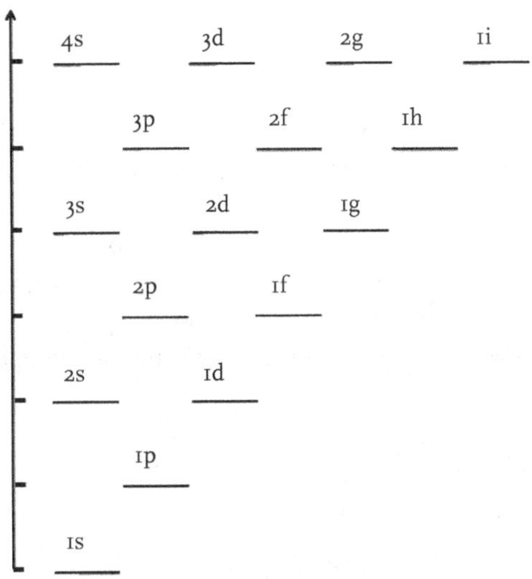

FIGURE 5.3 Energy spectrum of the spherical oscillator.

The fact that both quantum numbers appear with the same weight in Equation (4) can be understood when transposing the quantum result to the classical limit. It then means that the "frequency of revolution is the same as the frequency of excursions in the radial direction." This is self-evident for the ellipse in Figure 5.1 with the center of attraction in one of the focal points. One full turn around the perimeter will visit the perihelion point exactly once.

For the harmonic oscillator, the energy eigenvalues in units of $\hbar\omega$ are determined by a different expression:

$$E = 2n_r + l + \frac{3}{2}. \tag{5}$$

Here the term 3/2 is the zero-point energy of the oscillator. The sum $2n_r + l$ corresponds to the number of boson excitations. The states of the oscillator are also listed in Table 5.1, and the spectrum is shown in Figure 5.3 . Since now the radial frequency appears with a double weight, the same extrapolation will dictate that the radial frequency is twice as large as the angular one. Hence for the classical elliptic trajectory, when going around the center once, the shortest distance to the center will be visited twice, as is precisely the case for the elliptical orbit in Figure 5.2 with the attraction center in the origin.

4. *The Chemical Orbit*

This is precisely the point where Wheeler [2] steps in. For him, the chemical mechanics, which finds its expression in the periodic table, is not limited to the quantum mechanics of the hydrogen atom, but points to a more universal force law, at the basis of the internal symmetry of the periodic system. Crucial to this chemical mechanics, according to Wheeler, is the so-called Madelung rule, to which we turn next.

4.1 The Madelung Rule

The Madelung or $(n + l, n)$ rule was discovered in 1936 by the German physicist Erwin Madelung, and is generally defined as follows:

Definition 4.1. (The Madelung $(n + l, n)$ rule): With increasing nuclear charge Z, one-electron orbitals are filled according to increasing $N = n + l$, with n the principal quantum number and l the orbital quantum number. For fixed N, the orbitals are filled in order of increasing n.

Application of the $(n + l, n)$ rule gives rise to the following orbital sequence:

$$\underbrace{\overbrace{\{1s\}}^{n+l=1}}_{\text{dim}=2} \ll \underbrace{\overbrace{\{2s\}}^{n+l=2}}_{\text{dim}=2} \ll \underbrace{\overbrace{\{2p<3s\}}^{n+l=3}}_{\text{dim}=8} \ll \underbrace{\overbrace{\{3p<4s\}}^{n+l=4}}_{\text{dim}=8} \ll \underbrace{\overbrace{\{3d<4p<5s\}}^{n+l=5}}_{\text{dim}=18} \ll$$

$$\underbrace{\overbrace{\{4d<5p<6s\}}^{n+l=6}}_{\text{dim}=18} \ll \underbrace{\overbrace{\{4f<5d<6p<7s\}}^{n+l=7}}_{\text{dim}=32} \ll \underbrace{\overbrace{\{5f<6d<7p<8s\}}^{n+l=8}}_{\text{dim}=32} \ll \ldots,$$

where the orbitals have been grouped in sets of constant $N = n + l$ (see also Table 5.2). The dimensions of these sets correspond to a series of repeated "double squares" [8, p. 331]:

$$2-2-8-8-18-18-32-32-\ldots. \tag{6}$$

The hydrogenic dimensions appear exactly twice in the Madelung sequence—a phenomenon known as the "period doubling."

TABLE 5.2 Application of the Madelung rule according to which the orbitals in neutral atoms are filled in order of increasing $n + l$, and n for fixed $n + l$

$n + l$	$l = 3$	$l = 2$	$l = 1$	$l = 0$	Nn+lmax	$Z_i \to Z_f$	$X_i \to X_f$
1	—	—	—	$1s^2$	2	$1 \to 2$	H → He
2	—	—	—	$2s^2$	2	$3 \to 4$	Li → Be
3	—	—	$2p^6$	$3s^2$	8	$5 \to 12$	B → Mg
4	—	—	$3p^6$	$4s^2$	8	$13 \to 20$	Al → Ca
5	—	$3d^{10}$	$4p^6$	$5s^2$	18	$21 \to 38$	Sc → Sr
6	—	$4d^{10}$	$5p^6$	$6s^2$	18	$39 \to 56$	Y → Ba
7	$4f^{14}$	$5d^{10}$	$6p^6$	$7s^2$	32	$57 \to 88$	La → Ra
8	$5f^{14}$	$6d^{10}$	$7p^6$	$8s^2$	32	$89 \to 120$	Ac → 120

According to Goudsmit and Richards, the Madelung rule "is remarkably well obeyed throughout the periodic table" [9, p. 664]. Indeed, both the Madelung rule and the period doubling are considered to be of paramount importance to the periodic system. The Madelung rule flawlessly accounts for the overall structure of the periodic table, by predicting the onset of the different blocks in the periodic table, such as the start of the transition metal block (3*d*-block) after the 4*s*-block, or the inset of the lanthanide and actinide series (*f*-block elements) after the 6*s*- and 7*s*-elements respectively.

It can also be used as the basis for a novel representation of the periodic law, which is known as the "eight-period" or "left-step" periodic table (LSPT). The LSPT was first introduced by the amateur biologist Charles Janet in 1929, and offers a number of advantages as compared to the conventional format of the periodic law. First, the periods in the LSPT are characterized by a constant value of $N = n + l$, which suggests the possibility of elevating N to a new quantum number for the periodic system. Second, the period doubling is clearly highlighted in the LSPT. The pairing of the periods gives the LSPT its stepped profile, and leads to the set of cardinalities in Equation (6). Third, the ordering of the blocks, when read from right to left, follows the natural filling of the orbitals (*s*-*p*-*d*-*f*).

4.2 The Löwdin Challenge

Until now, however, the Madelung rule has remained an *empirical* (or *lexicographic*) rule. As a result, Allen and Knight [10, p. 83] have named it a "somewhat mysterious algorithm." In 1969, during the centennial anniversary of Mendeleev's discovery, Per-Olov Löwdin published "Some Comments on the Periodic System of the Elements" [8, p. 332], and noticed how remarkable it was that "in axiomatic quantum theory, the simple ($n + l$, n) energy rule has not yet been derived from first principles" (i.e., on the basis of the many-electron Schrödinger equation).

The search for such an *ab initio* derivation of the Madelung rule is now known as the "Löwdin challenge," and is considered [10, p. 83] the "oldest and largest standing problem in quantum chemistry." Many claims to a successful derivation have appeared in the scientific literature since Löwdin's plea, but most have been dismissed [11].

Consequently, both the validity and the utility of the Madelung rule have been called into question in recent years. One of the main voices in this debate is Eugen Schwarz [12–15], who deplores the importance that is generally given to the $(n + l, n)$ rule. According to Schwarz, the Löwdin challenge is impossible to meet since the Madelung rule is just an "approximate rule of thumb [...] at variance with too many facts" [14, p. 441]. Instead of praising the $n + l$ rule, we should be talking about the "$n + l$ blunder," dixit Schwarz. [12, p. 3412]

In order to back up this claim, Schwarz does raise a number of important issues which would benefit from closer scrutiny. But in our opinion, Schwarz dismisses the Madelung rule for the wrong reasons, and thereby fails to recognize the crucial role the $(n + l, n)$ rule has to play in the study of the periodic law.

4.3 Interpreting the Madelung Rule

Three interpretations of the Madelung rule can be distinguished in the scientific literature. The first two interpretations fail to hold in general—a point made most explicit by Schwarz. For Schwarz, this is sufficient to throw the Madelung rule into the dustbin of chemistry. But he thereby overlooks a possible third interpretation, which does apply universally. When Wheeler embarked on his quest for a "chemical mechanics," he probably had the second interpretation in mind (see section 4.4).

First interpretation. According to the first (and most common) interpretation, the Madelung rule is an Aufbau principle which provides the energy ordering of the different *nl* states:

$$E(1s) \ll E(2s) < E(2p) \ll E(3s) < E(3p) \ll E(4s) < E(3d) \ll \ldots \tag{7}$$

But this order is erroneously assumed to be fixed and universal. For example, according to the above energy sequence, the Aufbau process of a scandium atom proceeds as follows:

$$Sc^{3+}([Ar]) \rightarrow Sc^{2+}([Ar]4s^1) \rightarrow Sc^{+}([Ar]4s^2) \rightarrow Sc([Ar]3d^1 4s^2). \tag{8}$$

The 4*s* orbital is thus filled before the 3*d* orbital, as dictated by the Madelung order in Equation. (7). But this is contradicted by the empirical data which show that configurational reorganizations occur when adding electrons to a bare nucleus. As a result, the 3*d* orbital is initially lower in energy as compared to the 4*s* orbital, and is therefore filled first:

$$Sc^{3+}([Ar]) \rightarrow Sc^{2+}([Ar]3d^1) \rightarrow Sc^{+}([Ar]3d^1 4s^1) \rightarrow Sc([Ar]3d^1 4s^2). \tag{9}$$

This refutes the simplistic interpretation of the $(n + l, n)$ rule as a fixed energy ordering principle, which could somehow account for the entire process of building up atoms from scratch.

Second interpretation. According to the second interpretation, the Madelung rule merely predicts the final electronic configuration of neutral atoms. That is, the Madelung rule establishes in which *nl* spin orbital the differentiating

TABLE 5.3 Ground state electronic configurations for 19 elements with anomalous configurations that do not follow Madelung's rule

ELEMENT	PREDICTED MADELUNG GROUND STATE CONFIGURATION	EXPERIMENTALLY OBTAINED GROUND STATE CONFIGURATION
Cr	[Ar] $3d^4 4s^2$	[Ar] $3d^5 4s^1$
Cu	[Ar] $3d^9 4s^2$	[Ar] $3d^{10} 4s^1$
Nb	[Kr] $4d^3 5s^2$	[Kr] $4d^4 5s^1$
Mo	[Kr] $4d^4 5s^2$	[Kr] $4d^5 5s^1$
Ru	[Kr] $4d^6 5s^2$	[Kr] $4d^7 5s^1$
Rh	[Kr] $4d^7 5s^2$	[Kr] $4d^8 5s^1$
Pd	[Kr] $4d^8 5s^2$	[Kr] $4d^{10}$
Ag	[Kr] $4d^9 5s^2$	[Kr] $4d^{10} 5s^1$
La	[Xe] $4f^1 6s^2$	[Xe] $5d^1 6s^2$
Ce	[Xe] $4f^2 6s^2$	[Xe] $4f^1 5d^1 6s^2$
Gd	[Xe] $4f^8 6s^2$	[Xe] $4f^7 5d^1 6s^2$
Pt	[Xe] $4f^{14} 5d^8 6s^2$	[Xe] $4f^{14} 5d^9 6s^1$
Au	[Xe] $4f^{14} 5d^9 6s^2$	[Xe] $4f^{14} 5d^{10} 6s^1$
Ac	[Rn] $5f^1 7s^2$	[Rn] $6d^1 7s^2$
Th	[Rn] $5f^2 7s^2$	[Rn] $6d^2 7s^2$
Pa	[Rn] $5f^3 7s^2$	[Rn] $5f^2 6d^1 7s^2$
U	[Rn] $5f^4 7s^2$	[Rn] $5f^3 6d^1 7s^2$
Np	[Rn] $5f^5 7s^2$	[Rn] $5f^4 6d^1 7s^2$
Cm	[Rn] $5f^8 7s^2$	[Rn] $5f^7 6d^1 7s^2$

Data obtained from the National Institute of Standards and Technology (NIST) Atomic Spectra Database. Available online: http://physics.nist.gov/asd

electron should go. This clearly holds true for the scandium example above where the predicted Madelung-configuration and the experimentally obtained configuration for a neutral scandium atom are seen to agree.

However, even this weaker interpretation of the Madelung rule does not hold universally. Terry L. Meek and Leland C. Allen [16] have listed 19 elements with anomalous configurations that differ from those predicted by the Madelung rule (Table 5.3). Having said that, each of these 19 elements also has excited states near the ground state which *do* satisfy the Madelung rule. As a result, it is not clear how much weight should be attached to these exceptions.[2]

Third interpretation. According to the third and final interpretation, the Madelung rule foretells the onset of atomic subshell occupations in the Aufbau sequence of the periodic system. As noted above, it correctly predicts the start of the transition metals, lanthanide and actinide series. It thus accounts for the (at first sight ad hoc looking) layout of the different *s*-, *p*-, *d*- and *f*-blocks in Mendeleev's system, and this without exception. In doing so, it moreover discloses the period doubling as another characteristic feature of the periodic table.

4.4 The Chemical Orbit

In his 1971 contribution, Wheeler adhered to the second interpretation of the empirical Madelung rule, which leads to the conclusion that the energy of the outer electron is governed by the $n + l$ rule:

$$E = n + l = n_r + 2l + 1. \tag{10}$$

Interestingly, Wheeler adds in a footnote a further comment, prompted by a question of the Italian nuclear physicist and Nobel laureate Emilio Segrè.

Segrè asked if the formula often employed in spectroscopy for the correlation of atomic energy levels could also lead to this rule. The formula Segrè refers to is the empirical correction to the Rydberg expression for atomic spectra, and goes back to the early days of atomic spectroscopy. A comparison of the energy level diagrams in the alkali series to the hydrogen spectrum shows that the same pattern appears but energies are shifted. It was realized that the single outer electron in the alkali elements is moving about an atomic core, whose field shows marked deviations from the Coulomb field of a point charge. In order to cast the energies of the valence electrons in a Rydberg type formula, it is required to use effective quantum numbers which show a quantum defect with respect to the hydrogenic quantum numbers:

$$n^* = n - \Delta_{nl}. \tag{11}$$

The defect Δ_{nl} represents an empirical expression for the screening of the outer electrons by the core. Observations show that it is nearly independent of n, and is a rapidly decreasing function of l. This is consistent with the dominant role of the angular momentum in the screening properties. As an example, Table 5.4 lists values of Δ_{nl} for the rubidium atom [18, 19].

By approximating the defect as $\Delta_{nl} \sim a - bl$ the effective quantum number is rewritten as:

$$n^* = n - \Delta_{nl} = 1 - a + n_r + (1 + b)l. \tag{12}$$

So for $a = 0$ and $b = 1$ the empirical formula would converge to the Madelung expression.

From the perspective of Bertrand's Theorem two important aspects relate to the kinematic viewpoint of Wheeler. First, since the angular frequency is twice the radial frequency it can be inferred that the orbit of a classical particle in such a modified force field would involve two turns around the center before reaching the same closest radial distance again. Such an orbit was termed the "chemical orbit" or "double necklace" by Wheeler [2] and his collaborator

TABLE 5.4 Empirical quantum defect Δ_{nl} for the rubidium atom

s	p	d	f
3.13	2.64	1.35	0.016

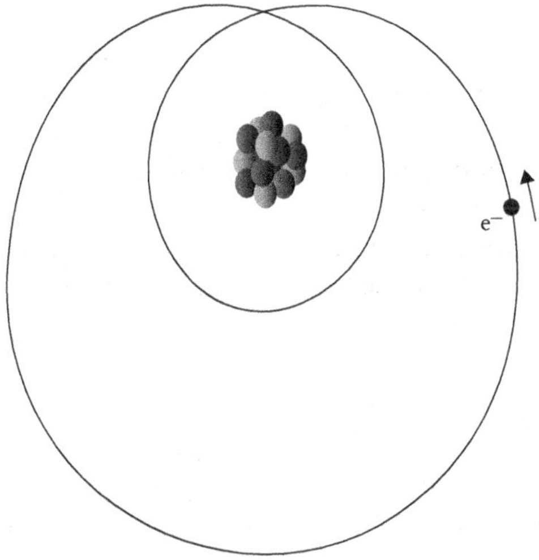

FIGURE 5.4 The chemical orbit as a double necklace.

Powers [20], as shown in Figure 5.4. Second, since Bertrand's Theorem does not yield other power laws than the two elliptic cases, it can be concluded that there is no simple universal force law that would stabilize classical chemical orbits, irrespective of the initial conditions.

5. *Contribution from Optics*

5.1 Maxwell's Fish-Eye Lens

The direct problem, originating from Wheeler's contribution, is how to infer a potential that would have the characteristics of the chemical orbit as its solution. An intriguing answer to this question was proposed by Demkov and Ostrovsky [21] who came forward with a potential solution, derived from a century old problem in optics, published in the *Cambridge and Dublin Mathematical Journal* in 1853. This journal was founded as the *Cambridge Mathematical Journal* in 1837 to support the revival of pure mathematics in Victorian England [22]. Not being patronized by an Academy or Society, its financial situation was often precarious and it relied heavily on the contributions of volunteers. In 1845 it was renamed the *Cambridge and Dublin Mathematical Journal* in an attempt to gain support from Trinity college in Dublin.

In 1853 as a rare instance of interference in editorial matters, the publisher MacMillan suggested attracting more readers by introducing a Problems Page. The solutions would appear in the following issue. One of these problems was to find a refraction law for a transparent medium such that all the rays proceeding from a given point would meet accurately at another point. The next issue then gave the solution which since then has been known as *Maxwell's fish-eye lens*. The problem and its solution were anonymous, as was quite often the case in those days since "it was the material presented which was important and it was a matter of accepted social form that the author should not draw attention to himself by self-advertisement" [22]. Anonymity was not entirely inspired by noble motives though, as it also enabled an attack to be mounted without the wider world being aware of the identity of the attacker.

Only later, when the collected works of Maxwell were published, was Maxwell's authorship established [23]. The solution considered a spherical lens, made of a transparent medium with a gradient index of refraction, provided by the function:

$$n(r) = n(0)\frac{a^2}{a^2 + r^2}. \tag{13}$$

Here $n(0)$ is the value of the refractive index in the center of the sphere. In a note the author explained that the possibility of the existence of a medium of this kind possessing such remarkable optical properties was suggested to him by the contemplation of the structure of the crystalline lens in fishes. The functioning of this lens can be illustrated through the analogy of the propagation of flexural waves in an elastic plate [24]. The desired gradient index profile in the plate was obtained by adjusting the thickness. In this way a mechanical analogue of the fish-eye lens could be built. Flexural waves emitted from a point-like pulse on one side of the lens travel through the medium, and are collimated again at precisely the opposite point on the sphere. Figure 5.5 presents the results of a simulation which proves to be in close correspondence with the experiments.

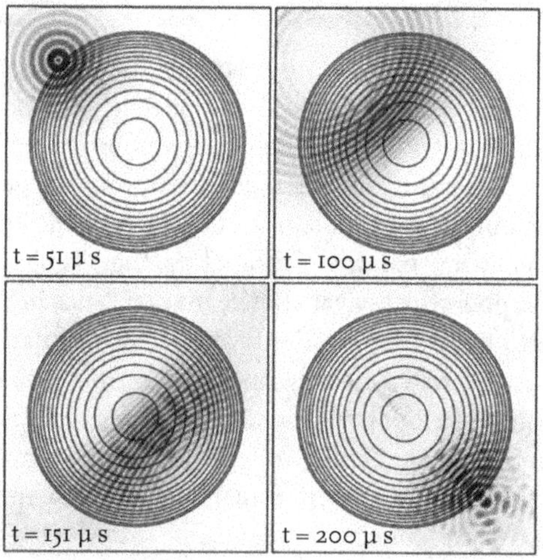

FIGURE 5.5 Illustration of the focusing property of a fish-eye lens by the flexural waves in an elastic plate. The contour lines indicate the spatial modulation of the thickness of the plate, corresponding to the refractive index in optics. The source is a Gaussian 60 kHz pulse at the top of the lens. The pulse propagates through the lens and is recovered at the bottom (from ref. [24]).

Wilhelm Lenz (the same Lenz as in the Runge-Lenz vector) later provided a generalization of this refraction law [25], which was expressed as follows:

$$n(r) = n(0)\frac{\left(\frac{r}{a}\right)^{\mu-1}}{1+\left(\frac{r}{a}\right)^{2\mu}} \tag{14}$$

For the index $\mu = 1$ this expression reduces to Maxwell's fish-eye lens, where the image point is exactly opposite the source. For a general value of the index, the image point is obtained at an angle of π/μ: for example, for $\mu = 2$, image points will be found at $\pi/2$, π, and $3\pi/2$. This equation has found important technological applications in the field of optics, but at present we are more interested in its potential relevance for electronic motions inside the atom. For this we have to cross the bridge that links optics to mechanics.

5.2 The Optical-Mechanical Analogy

In the history of physics, mechanics and optics have always been closely related, finally culminating in Schrödinger's wave mechanics in which the particle and wave nature came together. The optical-mechanical analogy, as expressed by Hamilton, refers to the isomorphism between the trajectories of a particle moving in a potential, and that of a light ray propagating through a medium [26]. It stems from analogous conservation laws and can be expressed as an equivalence between the momentum of a particle and the refraction index of a light ray:

$$p(\mathbf{x}) \Leftrightarrow n(\mathbf{x}). \tag{15}$$

In turn the momentum may be related to a potential energy by:

$$\frac{p^2}{2m} = E - V(\mathbf{x}). \tag{16}$$

Here E is a constant energy. In this way the refractive law for the generalized fish-eye lens may be turned into an attractive potential. A classical particle moving at zero energy is thus expected to describe characteristic orbits that would visit the focal points of the fish-eye lens. By combining Eqs. (14) and (16), with $E = 0$ this potential is easily obtained:

$$V_{\text{eff}}(\mathbf{r}) = -\frac{2v}{r^2 a^2[(r/a)^{-\mu} + (r/a)^{\mu}]^2}, \tag{17}$$

with v, μ and a constant parameters. For integer values of μ the trajectory of the particle will resemble a rose window with μ lobes. Hence for one turn it will exhibit μ radial oscillations, corresponding to the quantum rule $\mu\, n_r + l$. From this observation one can also immediately infer that the opposite ratio between radial and angular oscillations can simply be achieved by turning μ into a whole fraction. For $\mu = 1/2$ the trajectory will be characterized by half a radial oscillation for one full turn, or one radial oscillation for two angular oscillations, as required for the chemical orbit. This is the potential proposed by Demkov and Ostrovsky [21] to describe the multi-electronic atom. It should be clear though that this treatment does not yield a universal force law for the Madelung atom, as it does not describe bound states but refers to an isolated solution at zero energy.

6. *Conclusion*

Ultimately one is facing a paradox that on the one hand the Madelung rule hints at the existence of a universal force law that would regulate the mechanics in a multi-electronic atom, while on the other hand the unbreakable truth of Bertrand's theorem seems to exclude the existence of such a law.

The only escape from contradiction is to rethink the premises for Bertrand's law and the way they reflect on quantum mechanics. Along this vein, in our recent monograph we have explored the possibility of a more general pattern, not from the point of view of potentials, but from the point of view of the Lie algebras [11]. By imposing the Madelung rule in the manifold of bound states of the Coulomb atom we arrived at a non-linear Lie group structure. In essence this group still has the characteristics of the hydrogen SO(4) symmetry, but its structure parameters are no longer constants but functions of the operators representing the quantum numbers n and l+ 12 . This reflects the generalization of Kepler's law to a structure where n and l quantum numbers are combined.

7. *Appendix*

Book 1 of Newton's *Principia* is concerned with the motion of bodies. In proposition 7, corollary 3, Newton addresses the following problem: suppose a particle follows a given orbit, whether acted upon by a force F_C centered at a point C, or a force F_S acting toward a center S. What is then the ratio of the forces F_C and F_S? The corollary is then as follows:

> The force by which body P revolves in any orbit around the center of forces C is to the force by which the same body P can revolve in the same orbit and in the same periodic time around any other center of forces S as the solid $CP \times SP^2$ — contained under the distance of the body from the first center of forces C and the square of its distance from the second center of forces S—to the cube of the straight line CG, which is drawn from the first center of forces C to the tangent of the orbit PG and is parallel to the distance SP of the body from the second center of forces. For the forces in this orbit at any point of it are the same as in a circle of the same curvature. [27]

The ratio can thus be written as:

$$\frac{F_C}{F_S} = \frac{CP \times SP^2}{CG^3}. \tag{18}$$

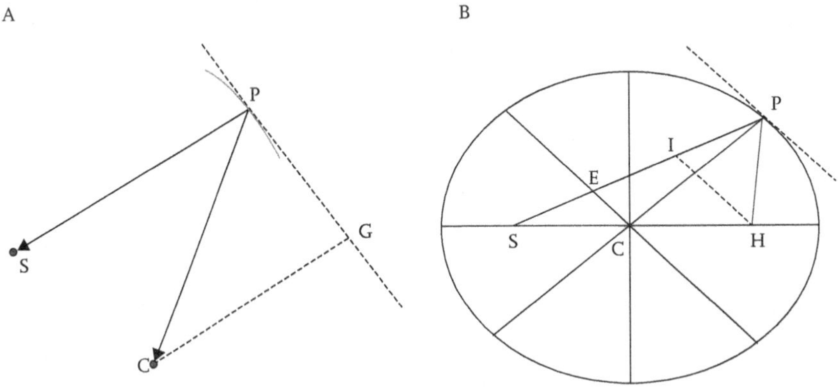

FIGURE 5.6 Comparison of attraction force to the center (*C*) and the focus (*S*) of an elliptical orbit. Part A illustrates Newton's proposition 7. Part B refers to proposition 11.

Figure 5.6A shows the configuration of these points. In proposition 10, Newton then proves that the centripetal force tending to the center *C* of an elliptic orbit is as the distance, as exemplified by Hooke's law. Proposition 11 is the famous proposition where Newton identifies the centripetal force toward the focus of an elliptic orbit as the inverse square law. This proposition is followed by a typical "idem aliter" clause, where the same result is proven in yet another way. The importance of this clause is that it really shows the duality between the two force laws considered in Bertrand's theorem.

In Figure 5.6B, the point *P* describes an elliptic orbit around the center *C* according to the force is as distance law. *S* and *H* denote the focus points of the ellips. The dashed line represents the tangent to the orbit in *P*. The lines *EC* and *IH* are parallel to the tangent through the center and the focus point, respectively. The force F_S that is oriented to the focus *S* and gives rise to the same orbit, is then related to the force F_C as:

$$\frac{F_C}{F_S} = \frac{CP \times SP^2}{EP^3} \tag{19}$$

Here we have applied Equation (18), noting that the distance *CG* in Figure 5.6A corresponds to *EP* in Figure 5.6B. Since $F_C \sim CP$, one thus has:

$$F_S \sim \frac{EP^3}{SP^2}. \tag{20}$$

Now, since the triangles ΔSEC and ΔSIH are proportional, one has:

$$\begin{aligned} SE &= \frac{SC}{SH} SI = \frac{1}{2} SI \\ &= EI. \end{aligned} \tag{21}$$

Also since the triangle ΔIPH is isosceles, one has:

$$IP = PH. \tag{22}$$

In an ellipse, the length a of the semi-major axis equals the average distance between a point on the ellips and the foci, hence:

$$a = \frac{SP + PH}{2} = \frac{EP + SE + PH}{2} = \frac{PE + EI + PI}{2} = EP. \tag{23}$$

Hence the distance EP is fixed, and thus Equation (20) reduces to:

$$F_S \sim \frac{1}{SP^2}. \tag{24}$$

Q.E.D.

References

[1] M. Verde, ed. *Atti del Convegno Mendeleeviano. Periodicità e Simmetrie Nella Struttura Elementaire Della Materia.* Accademia Delle Scienze di Torino, Turin and Rome, 1971.

[2] J.A. Wheeler. *From Mendeleev's Atom to the Collapsing Star*, 189–223. Vincenzo Bona, Turin, 1971.

[3] J. Bertrand. Théorème Relatif au Mouvement d'un Point Attiré vers un Centre Fixe. *Comptes rendus des séances de l'Academie des Sciences*, 77:849–853, 1873.

[4] F.C. Santos, V. Soares, and A.C. Tort. An English Translation of Bertrand's Theorem. *Latin American Journal of Physics Education*, 5(4):694–696, 2011.

[5] I. Newton. *The Principia, Mathematical Principles of Natural Philosophy.* Translated by I.B. Cohen and A. Whitman. University of California Press, Berkeley, 1999.

[6] A.K. Grant and J.L. Rosner. Classical Orbits in Power-Law Potentials. *American Journal of Physics*, 62(4):310–315, 1994.

[7] L. Motz. The Conservation Principles and Kepler's Laws of Planetary Motion. *American Journal of Physics*, 43(7):575–578, 1975.

[8] P.-O. Löwdin. Some Comments on the Periodic System of Elements. *International Journal of Quantum Chemistry*, S3:331–334, 1969.

[9] S.A. Goudsmit and P.I. Richards. The Order of Electron Shells in Ionized Atoms. *Proceedings of the National Academy of Sciences of the United States of America*, 51:664–671, 1964.

[10] L.C. Allen and E.T. Knight. The Löwdin Challenge: Origin of the $n + l$, n (Madelung) Rule for Filling the Orbital Configurations of the Periodic Table. *International Journal of Quantum Chemistry*, 90(1):80–88, 2002.

[11] P. Thyssen and A. Ceulemans. *Shattered Symmetry: Group Theory from the Eightfold Way to the Periodic Table.* Oxford University Press, New York, 2017.

[12] S.G. Wang and W.H.E. Schwarz. Icon of Chemistry: The Periodic System of Chemical Elements in the New Century. *Angewandte Chemie International Edition*, 48:3404–3415, 2009.

[13] W.H.E. Schwarz and S.G. Wang. Some Solved Problems of the Periodic System of Chemical Elements. *International Journal of Quantum Chemistry*, 110(8):1455–1465, 2010.

[14] W.H.E. Schwarz and R.L. Rich. Theoretical Basis and Correct Explanation of the Periodic System: Review and update. *Journal of Chemical Education*, 87(4):435–443, 2010.
[15] W.H.E. Schwarz. The Full Story of the Electron Configurations of the Transition Elements. *Journal of Chemical Education*, 87(4):444–448, 2010.
[16] T.L. Meek and L.C. Allen. Configuration Irregularities: Deviations from the Madelung Rule and Inversion of Orbital Energy Levels. *Chemical Physics Letters*, 362(5–6):362–364, 2002.
[17] Yu. N. Demkov and V.N. Ostrovsky. $n + l$ Filling Rule in the Periodic System and Focusing Potentials. *Soviet Physics JETP*, 35(1):66–69, 1972.
[18] W. Li, I. Mourachko, M.W. Noel, and T.F. Gallagher. Millimeter-Wave Spectroscopy of Cold Rb Rydberg Atoms in a Magneto-Optical Trap: Quantum Defects of the ns, np, and nd Series. *Physical Review A*, 67:052502, 2003.
[19] J. Han, Y. Jamil, D.V.L. Norum, P.J. Tanner, and T.F. Gallagher. Rb nf Quantum Defects from Millimeter-Wave Spectroscopy of Cold Rb Rydberg Atoms. *Physical Review A*, 74:054502, 2006.
[20] R.T. Powers. *Frequencies of Radial Oscillation and Revolution as Affected by Features of a Central Potential*, 235–242. Vincenzo Bona, Turin, 1971.
[21] Yu. N. Demkov and V.N. Ostrovsky. Internal Symmetry of the Maxwell "Fish-Eye" Problem and the Fock Group for the Hydrogen Atom. *Soviet Physics JETP*, 13(6):1083–1087, 1971.
[22] T. Crilly. The Cambridge Mathematical Journal and its Descendants: The Linchpin of a Research Community in the Early and Mid-Victorian Age. *Historia Mathematica*, 31:455–497, 2004.
[23] J. C. Maxwell. *The Scientific Papers of James Clerk Maxwell*. Dover, New York, 1965.
[24] G. Lefebvre, M. Dubois, R. Beauvais, Y. Achaoui, R. K. Ing, S. Guenneau, and P. Sebbah. Experiments on Maxwell's Fish-Eye Dynamics in Elastic Plates. *Applied Physics Letters*, 106:024101, 2015.
[25] W. Lenz. *Zur Theorie der optischen Abbildung*, 198–207. S. Hirzel, Leipzig, 1928.
[26] A.M. Bloch and A.G. Rojo. Optical Mechanical Analogy and Nonlinear Nonholonomic Constraints. *Physical Review E*, 93:023005, 2016.
[27] I. Newton. *o.c.* p. 457 (Point labels have been adapted to be in line with the subsequent application.).

Notes

1. If the major and minor axes of the ellipse did not retain their fixed orientation in the plane, the orbit would no longer be closed. The major axis would start to precess, and a "rosette" would be traced out over time. The orbit is then said to be "space-filling."

2. Demkov and Ostrovsky [17] have made an interesting and rather telling comparison in this regard between the hydrogenic (n, l) and Madelung $(n + l, n)$ order.

CHAPTER 6 | The Grand Periodic Function

JAN C. A. BOEYENS
Centre for the Advancement of Scholarship, University of Pretoria, South Africa

1. *Periodicity and Number Theory*

The discovery of material periodicity must rank as one of the major achievements of mankind. It reveals an ordered reality despite the gloomy pronouncements of quantum philosophers. Periodicity only appears in closed systems with well-defined boundary conditions [1]. This condition excludes an infinite Euclidean universe [2] and all forms of a chaotic multiverse.

Manifestations of cosmic order were observed and misinterpreted by the ancients as divine regulation of terrestrial events, dictated by celestial intervention. Analysis of observed patterns developed into the ancient sciences of astrology, alchemy and numerology, which appeared to magically predict the effects of the macrocosm on the microcosm.

The sciences of astronomy and chemistry have by now managed to outgrow the magic connotation, but number theory remains suspect as a scientific pursuit. The relationship between Fibonacci numbers and cosmic self-similarity is constantly being confused with spurious claims of religious and mystic codes, imagined to be revealed through the golden ratio in the architecture of the Great Pyramid and other structures such as the Temple of Luxor. The terminology which is shared by number theory and numerology, such as perfect number, magic number, tetrahedral number and many more, contributes to the confusion. It is not immediately obvious that number theory does not treat 3 as a sacred number, 13 as unlucky and 666 as an apocalyptic threat.

The relationship of physical systems to numbers is no more mysterious nor less potent than to differential calculus. Like a differential equation, number theory does not dictate, but only describes physical behavior. The way in which number theory describes the periodicity of matter, atomic structure, superconductivity, electronegativity, bond order, and covalent interaction was summarized in a recent volume [3]. The following brief summary of these results is augmented here by a discussion of atomic and molecular polarizabilities, as derived by number theory, and in all cases specified in relation to the grand periodic function that embodies self-similarity over all space-time.

2. *Periodic Properties*

2.1 Composition of Stable Isotopes

The periodicity of stable nuclides was first mooted by Harkins and the idea was developed in a number of papers, summarized in [4]. Harkins was also the first to postulate [5] the existence of neutrons in atomic nuclei, as N proton-electron pairs (pe). The stability of atomic nuclei was shown to depend on the ratio N/P $(P=A)$ and the atomic number Z. On plotting N/P *vs*. Z it was found that all points which represent atomic species lie on a set of curved festoons, converging with increased Z such that "no atomic species is known in which the ratio N/P is greater than 0.62."

This work was largely forgotten until independently rediscovered, almost a century later [6]. Using modern data it could be demonstrated rigorously [7] that the point of convergence occurs where $Z/N=(N-Z)/Z$, *i.e.*, $Z/N=\tau=0.61803\ldots$, the golden ratio. From this result it follows that the field of nuclide stability is limited by the convergence of Z/N from unity to τ, as defined by a series of Fibonacci fractions. Such convergence, being a central theme in number theory, is interpreted to imply that the observed periodicity could best be formulated as a function, based on Fibonacci sequences.

2.2 The Fibonacci Sequence

The well-known Fibonacci sequence:

$$\{n\}=1\quad 1\quad 2\quad 3\quad 5\quad 8\quad 13\quad 21\quad 34\quad 55\ldots$$

is generated by adding the first two numbers to produce the third, and extending the process to all higher terms. The remarkable convergence of the associated fractions, in

$$\lim_{n\to\infty}\frac{n}{n+1}=\tau$$

is not unique to the Fibonacci sequence, but occurs for any sequence generated by the same procedure, starting from an arbitrary pair of seed values, that is:

$$x\quad y\quad x+y\quad x+2y\quad 2x+3y\quad 3x+5y\quad 5x+8y\ldots \tag{1}$$

By way of illustration:

$$\text{For}\quad x=5,\quad y=11:\quad \frac{15+55}{25+88}=\frac{70}{113}=0.619\sim\tau$$

It is equally remarkable how *all* Pythagorian triangles are generated by sets of four consecutive terms in a general sequence such as (1) [8]. For the sequence

$$a\quad b\quad c\quad d$$

TABLE 6.1 The first Farey sequences

$\frac{0}{1}$										$\frac{1}{1}$
$\frac{0}{1}$					$\frac{1}{2}$					$\frac{1}{1}$
$\frac{0}{1}$			$\frac{1}{3}$		$\frac{1}{2}$		$\frac{2}{3}$			$\frac{1}{1}$
$\frac{0}{1}$		$\frac{1}{4}$	$\frac{1}{3}$		$\frac{1}{2}$		$\frac{2}{3}$	$\frac{3}{4}$		$\frac{1}{1}$
$\frac{0}{1}$	$\frac{1}{5}$	$\frac{1}{4}$	$\frac{1}{3}$	$\frac{2}{5}$	$\frac{1}{2}$	$\frac{3}{5}$	$\frac{2}{3}$	$\frac{3}{4}$	$\frac{4}{5}$	$\frac{1}{1}$
.	.	.	.	.	.	.	.	.	.	.
							$\frac{5}{8}$			

the three sides of the triangle are specified as:

$$\left.\begin{array}{rcl} s & = & ad \\ t & = & 2bc \\ h & = & b^2 + c^2 \end{array}\right\} \qquad (2)$$

By way of illustration:

$$1 \quad 1 \quad 2 \quad 3 \Rightarrow \Delta \quad 3 \quad 4 \quad 5$$

The prescription works equally well for rational fractions and all right-angled triangles, such as:

$$\frac{1}{4} \quad \frac{1}{2} \quad \frac{3}{4} \quad \frac{5}{4} \Rightarrow \Delta \quad \frac{5}{16} \quad \frac{3}{4} \quad \frac{13}{16}$$

2.2.1 The Farey Sequence

An important variation on the Fibonacci theme has developed as a strategy for the enumeration of rational fractions, known as the Farey sequence. It starts from the seed values of 0/1 and 1/1. New intermediate fractions are generated by separate addition of numerators and denominators, as shown for the first few steps in Table 6.1.

Progression along the sequence in bold print represents the normal Fibonacci convergence to

τ.

The well-documented [9] and remarkable correlation between a large variety of natural phenomena and Fibonacci systems argues convincingly for the formulation of a new number-based calculus to explore the self-similar nature and periodic properties of matter [10]. An especially powerful computational element is the Fibonacci logarithmic spiral that has been used [10] to analyze the packing of nucleons, the electronic structure of atoms, atomic radii, covalent bond order and the structure of the solar system.

2.3 The Periodicity of Matter

The distinction between stable and radioactive nuclides is not always a trivial one. Radioactive half-life is an obvious criterion to use and in most cases the distinction is beyond dispute. However, in a smaller number of borderline cases disagreement on limiting half-lives and error margins introduces a measure of uncertainty. Despite some lack of consensus there still is sufficient agreement on the number of stable nuclides at about 264, with minor differences in preference, which do not seriously affect the general argument.

Evidence of a periodic relationship between the stable nuclides is made abundantly clear by the updated version of the Harkins plot, $N-Z$ *vs* A, shown in Figure 6.1

The neutron excess increases in digital steps from -1 to 44, separating the 264 isotopes of 81 elements into 45 subgroups. The nuclear binding energy and solar abundance of the nuclides in each subgroup increase in both directions through a local maximum [7, 11]. The 44 subgroups, $(N-Z\geq 0)$, combine into 11 groups, comprising the modular subgroups $A(\mathrm{mod}4)=0\rightarrow 3$ in each group. The two sets of nuclei of even A consist of 81 members and each of the odd ones of 51. Each line of nuclides can be reconstructed by adding the equivalent of an α-particle to the previous member, starting from the seed compositions $n^{*}n(A=0)$, $n^{*}n^{*}(A=-2)$, $n^{*}(A=-1)$, $n(A=1)$, and allowing for β^{+} decay at the end of each subgroup. The two elements $Z=43$ and 61 do not occur in this scheme.

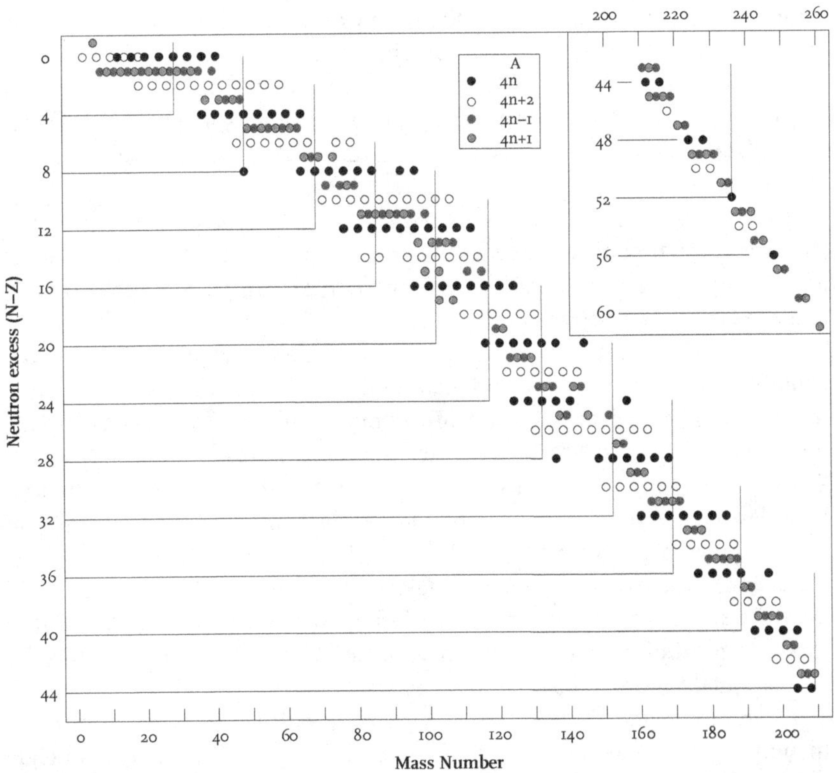

FIGURE 6.1 Variation of neutron excess with mass number.

Continuation, within the implied convergence limits that control the stability of stable nuclides, of the α-addition scheme beyond $A=209$, is shown on the inset in Figure 6.1. Convergence of $(N-Z)/Z \to \tau$, at ^{261}Fm terminates the series at 300 isotopes of 100 elements.[1] Noting that $300=24!=12\times 25=24\times 12\frac{1}{2}$ and $264=11\times 24$, an attractive possibility is to look for a periodic function in terms of the 11 modular quartets, rather than 44 subgroups. This decomposition is shown in Figure 6.1. The vertical hemlines that separate the 264 stable nuclides into 11 groups of 24 correspond convincingly well to the mean binding energy and abundance maxima in each segment [7], hence providing a convincing physical basis of the proposed periodicity.

2.3.1 Golden Excess

The demonstrated convergence of Z/N to the golden ratio at the point where $A=267$, $N=\tau A=165$ and $Z=\tau^2 A=102$ may be interpreted to imply that this ratio represents the most efficient packing of protons and neutrons in a heavy nucleus.[2]

In self-similarity with botanical growth the implied three-dimensional arrangement is interpreted to resemble Fibonacci phyllotaxis, in balance with nuclear stability considerations. The result is a characteristic golden proton excess, $g=Z-\tau N$, that occurs as a positively charged surface layer for each stable nuclide, at the point where the supply of neutrons runs out [7]. For the isotopes of any element the factor g approaches a characteristic maximum with increasing N (A). Beyond this limit, excess g is reduced in a natural way by β^+ emission or electron capture. In a plot of g *vs* A, with a scale factor of 1:8, the positions of stable nuclides occur on a 44×44 square grid, as shown in Figure 6.2.

A continuous limiting profile, drawn to outline the spread of nuclear sites, shows a number of distinct breaks at $A=41.5$, 51, 82.5, 133.5, and 165, that is, at $2A/267=\tau/2, \tau^2, \tau, 1$, and 2τ, reaches a maximum at $A\sim 102$ $(267\tau^2)$, and ends at $A=216(267/2\tau)$.

All of these special points are related to the golden ratio, noting that the triple $\tau^2, \tau, 1$ defines the fundamental Kepler triangle. In the same way that the ratio 267/102 relates the mass and atomic numbers of the 100th element, the ratio 216/82.5 characterizes the 81st, final stable element $(Z=83)$. Whereas 102/2=51, 83/2=41.5; whereas the neutron number for natural nuclides converges to $267\tau=165$, for stable nuclides it converges to $216\tau=133.5$; $41.5/133.5=\tau/2$.

As an alternative the five major breaks in the curve may be read to approximate the symmetric closure of 2, 2, 3, 2, 2 periodic groups in sequence. Either way, all aspects of the diagram appear to be conditioned by the golden ratio, the fundamental feature of both natural and stable nuclide periodicity.

It can be concluded categorically that the periodic function concerns more than the extranuclear electronic configuration of atoms. The reality of nuclide periodicity implies a mathematically closed function. From Figure 6.2, closed

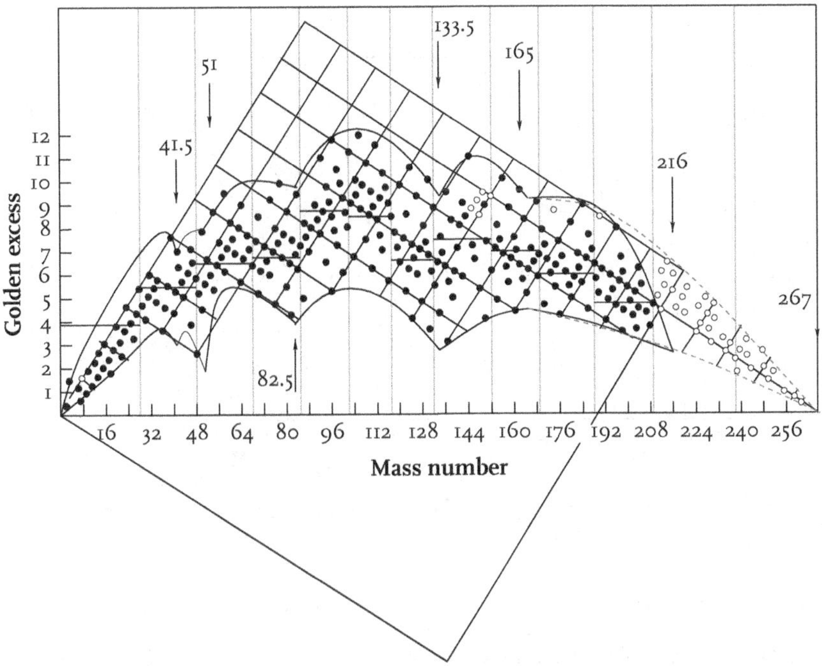

FIGURE 6.2 The golden excess, $Z-\tau N$, plotted for all stable nuclides as a function of A, with the conjectured continuation of α-unstable nuclides, represented by open circles.

intervals $A[0,267]$, $N[0,165]$, and $Z[0,102]$, over the natural elements, appear as viable candidates, with the subsets $A[0,216]$, $N[0,134]$, and $Z[0,83]$ for stable nuclides. All of these have been confirmed empirically [7]. Of these, the well-known chemical periodicity of the elements is the most informative on the ultimate origin of the function.

In a previous study [7] an absolute correlation between golden excess and the potency of nuclides to induce superconducting properties could be shown to relate to the accumulation of unpaired nuclear spins. The horizontal line across each period in Figure 6.2 identifies both high-spin nuclides and potential superconductors.

2.3.2 Elemental Periodicity

There are several equivalent ways of demonstrating how elemental periodicity derives from the golden ratio, using, *inter alia*, Fibonacci series, Farey sequences, Ford circles, and the logarithmic spiral. Visually, the most appealing way is to insert the hemlines of nuclide periodicity into a plot as shown in Figure 6.3.

Sampled along the Z-axis at $Z/N=\tau$ the intersecting hemlines generate a set of numbers that correlate precisely with empirically known periodic atomic numbers. At $Z=0.58$ the points of intersection correspond in detail with the periodic arrangement that matches the electronic spectrum of H as predicted by Schrödinger's equation, that is:

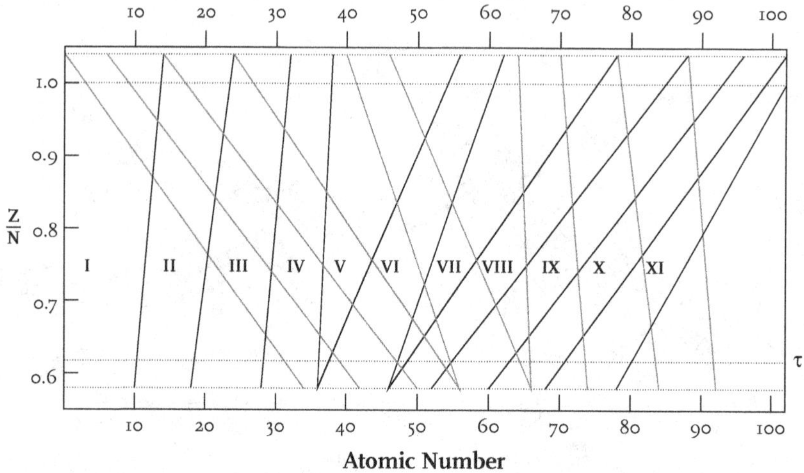

FIGURE 6.3 Graphical presentation of elemental periodicity as a function of space-time curvature.

$$1s < 2s < 2p < 3s < 3p < 3d < 4s < 4p < 4d < 4f, etc.$$

The exact inverse of this sequence is mapped out at $Z/N = 1.04$, *i.e.*:

$$4f < 3d < 2d < 1s < 4d < 3p < 2s\, etc.$$

In the same way the pattern at $Z/N = 1$ is inverse to that at τ.

These observations are in satisfying accord with the notion that space-time is curved to an extent that depends on the golden ratio [3] and that an inversion of energy levels occurs in atoms under high pressure [12]. As space-time curvature parallels gravitational compression, it is logical to associate the isolated Schrödinger H atom with zero curvature $(Z/N = 0.58)$ in hypothetical Euclidean space-time. Observed periodicity, by the same argument, reflects the natural world with general space-time curvature, quantified by the golden ratio. The ratio $Z/N = 1$ is commensurate with equilibrium nucleogenesis by α-particle fusion, conjectured to happen in massive stars, producing the 100 natural elements in the form of 300 isotopes. On dispersion into domains of lower curvature the heaviest nuclides decay radioactively, with only 81 stable elements surviving.

The ratio of $Z/N = 1.04$ implies infinite curvature, the inverse of flat Euclidean space-time, at the presumed singularity at the center of a black hole. The effect is total conversion of matter into anti-matter. The mirror image of the periodic hemlines, shown in bold on Figure 6.3, represents the complementary periodicity of antimatter. Specification of the overall periodicity requires a four-dimensional description of the involuted topology, by projective geometry, in elliptic space-time [11].

The layout, Figure 6.4 (top), of the elemental periodic system is readily reconstructed along a Fibonacci spiral. Using divergence angles along a double spiral of $4\pi/(2n-1)$ radians, known to predict the periodic variation of

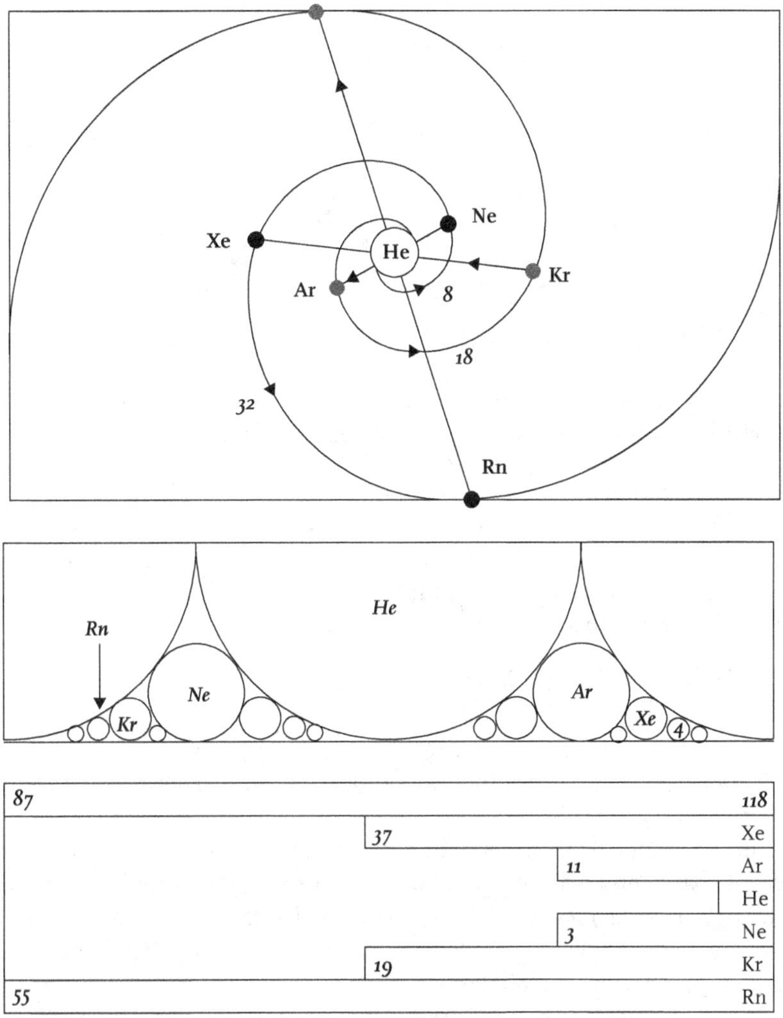

FIGURE 6.4 Periodic system of the elements mapped by Fibonacci spirals, Ford circles and Farey sequence.

atomic electron density [13], the generated points correlate with the periodic spread of the noble gases. The radial displacements of Ne, Kr, and Rn on one leg scale to values of $2n^2 = 8, 18, 32$, with Ne $\rightarrow$ Ar = 8 and Kr $\rightarrow$ Xe = 18. This result is equivalent to the system predicted by the Ford-circle representation (Fig. 6.4 middle) based on Farey sequence and shown at the bottom of Figure 6.4.

2.3.3 Magic Numbers

Periodicity of the stable nuclides as a function of neutron number is recognized on mapping the 11×24 periodicity in the form of Figure 6.5, in order to reveal the periodicity as a function of A, rather than the number of nuclides

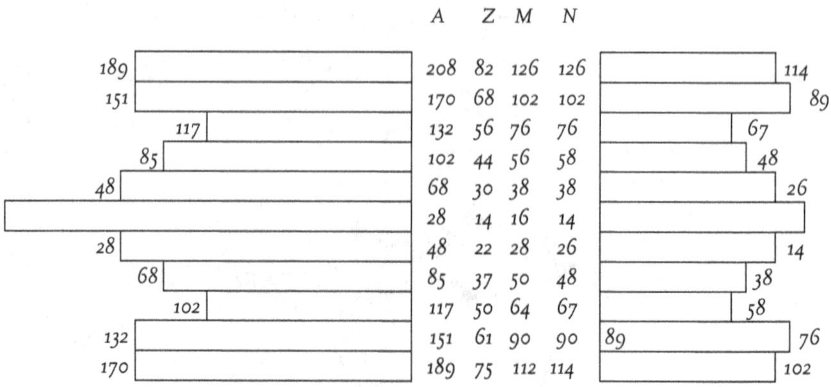

FIGURE 6.5 Diagram to demonstrate the relationship between the three nuclide periodic functions and the magic numbers of nuclear physics.

per periodic group. The result is shown in Figure 6.6 (top). The only adjustment required to produce a symmetrical layout was to replace the isotope ^{152}Sm by ^{151}Pm, which, by definition, could not be identified to close period 8. The Z/N ratio at the periodic points, from 14/14 to 68/102 is approximated well by the unimodular Farey sequence

$$\frac{1}{1} \quad \frac{5}{6} \quad \frac{4}{5} \quad \frac{7}{9} \quad \frac{10}{13} \quad \frac{3}{4} \quad \frac{5}{7} \quad \frac{7}{10} \quad \frac{2}{3}$$

that further converges to τ at 102/165, assuming period 10 closed at ^{189}Re. This convergence translates into the set of Ford circles shown in Figure 6.6 (middle and bottom).

These findings finally substantiate the initial assumption of 11×24 periodicity [6] based on Plichta's prime-number cross [14], as arising from standard number theory. Elemental periodicity is not correctly reflected by the atomic numbers that appear in this scheme, primarily because the hemlines of Figure 6.3 are slanted from the vertical.

On the other hand, neutron periodicity in symmetrical arrangement is consistent with mass-number periodicity, as shown on the right-hand side of Figure 6.5. The predicted neutron numbers (N) are in fair agreement with the empirical magic numbers (M) of nuclear physics [7, p. 235]. The reason is that the hemlines as a function of N generate a single invariant set of values at $Z/N=\tau$ and 1. This observation is interpreted to mean that neutron levels are not sensitive to compression. A neutron star may hence be viewed as one massive atom.

The puzzling near-periodicity of 8, 11, and 19 in Z, N and A, observed before [7, p. 236], is now fully accounted for.

Mass-number and neutron periodicities are analyzed graphically in Figure 6.6. The Z/N ratios observed at the position of the hemlines, which decrease in unimodular Farey steps are displayed as a set of kissing Ford circles. The convergence from unity to τ is thereby confirmed, not to be a continuous change, but occurs in Fibonacci steps that give rise to the periodicity.

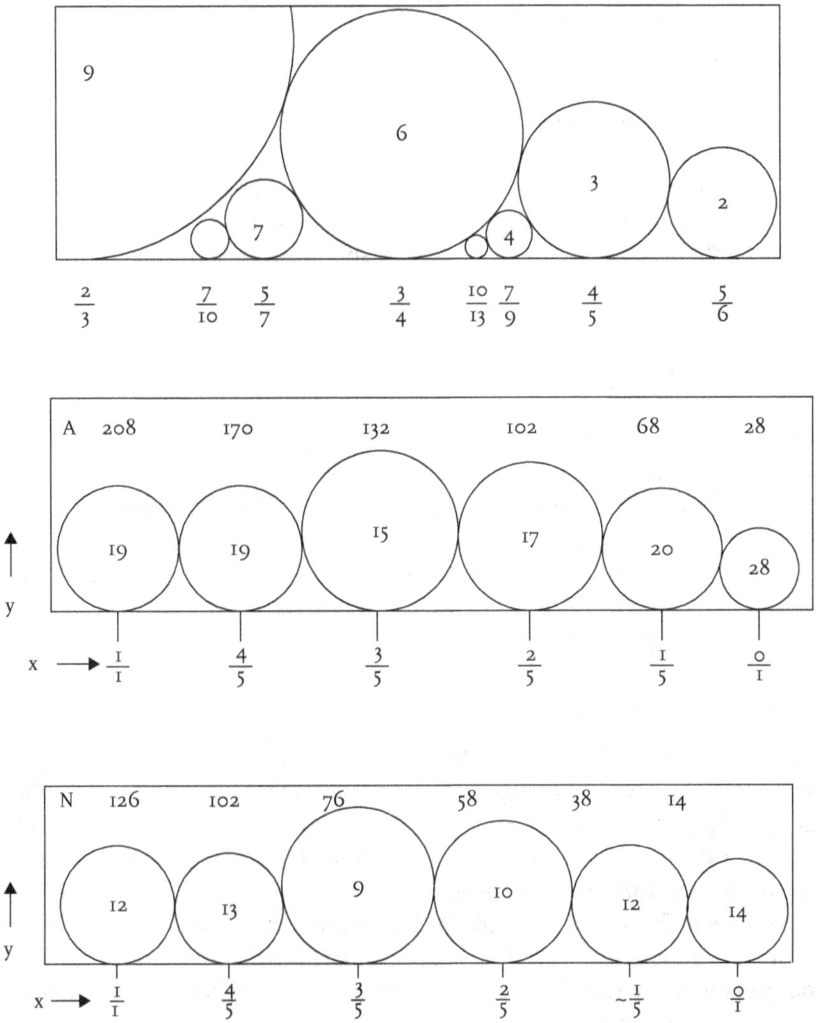

FIGURE 6.6 Ford-circle mapping of the periodic variation of the ratio Z/N, and the observed periodicity, as a function of A and N.

That this periodicity is a regular function of mass number is also shown in the middle diagram of Figure 6.6. This diagram is constructed by equating the mass-number range between hemlines, $\Delta A = 2k^2$. The quantity $1/\Delta A$, as before, then represents both the y-coordinate and the radius of a Ford circle. The x-coordinate in the form h/k follows by positioning the resulting circles in kissing mode.

The final solution, only half of which is represented in the diagram, shows that the 11 periods divide the field into segments of $\Delta(h/k) = 1/10$. This unanticipated beautiful result gives new meaning to the periodic function. The numbers 5 and 10 are well known to be closely associated with the golden ratio, already securely identified as the essential factor in the periodicity of atomic matter.

As an encore, neutron periodicity follows exactly the same pattern, which at the same time explains the origin of the famous magic numbers of nuclear physics.

The slightly subjective choice of the operational 264 stable nuclides now appears of less importance. Even the precise count of 24 per period is no longer decisive, but is not anticipated to deviate substantially from this in the final analysis.

3. *Chemical Properties*

The successful number-theory simulation of the periodic function implies that the same approach could be of benefit in the calculation of important atomic chemical properties, such as atomic size, chemical affinity (covalence), electronegativity and polarizability. Available methods for the calculation of these quantities traditionally involve highly complicated integrations, if not totally impossible. Previous results are briefly summarized and some new applications are discussed in more detail.

3.1 Atomic Structure

The optimization of atomic electron densities on Fibonacci spirals [13] has been mentioned with Figure 6.4. The simulation that involves no more than the specification of the $4\pi/(2n-1)$ divergence angles produces results, equivalent to extensive Thomas-Fermi computations, readily adjusted to match Hartree-Fock results for all atoms.

By an equally simple procedure the simulation is extended to produce atomic ionization radii, previously only obtained by computerized isotropic compression of Hartree-Fock densities [15]. The best-defined and most reliable atomic electronegativities derive directly from these radii [16].

3.2 Covalence Parameters

Although the periodic function strictly applies to atomic matter only the techniques that simulate atomic properties, model interatomic covalent interactions equally well [17]. The relevant periodic parameters are the characteristic atomic ionization radii, r_0. Together with a set of bond-order-dependent dimensionless constants, d_b', all homonuclear interatomic distances are modeled by the linear relationship, $d = d_b' r_0$ for each bond order.

Optimized values of d_b' are obtained from a Fibonacci spiral with divergence angles of $\pi/16$ from $d_0' = 1$ for bond order zero to $d_4' = \tau$ for bond order 4. Corresponding dissociation energies are obtained as $D = K r_0^2 \tau^n$, where the exponent n varies regularly within periodic groups and with period number.

For heteronuclear interaction the formulae involve $R_o = \sqrt{r_o(1) \cdot r_o(2)}$ to yield

$$d = d_b' R_o \quad and \quad D = K r_o^3(1)\tau^n / r_o(2), \quad r_o(1) > r_o(2).$$

Less detailed analyses [17] also enabled the derivation of harmonic stretching force constants and diatomic dipole moments by an extension of these methods. The logical inference is that all chemical phenomena that display periodic trends should be amenable to analysis by procedures based on number theory, the golden ratio, and Fibonacci spirals. By way of demonstration direct optimization of polarizabilities, which are notoriously difficult to model for both atoms and molecules, is described next.

3.3 Polarizability

Polarizability is measured in the units of volume and therefore conveniently optimized in the form of an atomic polarization radius, such that

$$\alpha = \frac{4}{3}\pi r_\alpha^3 \quad .$$

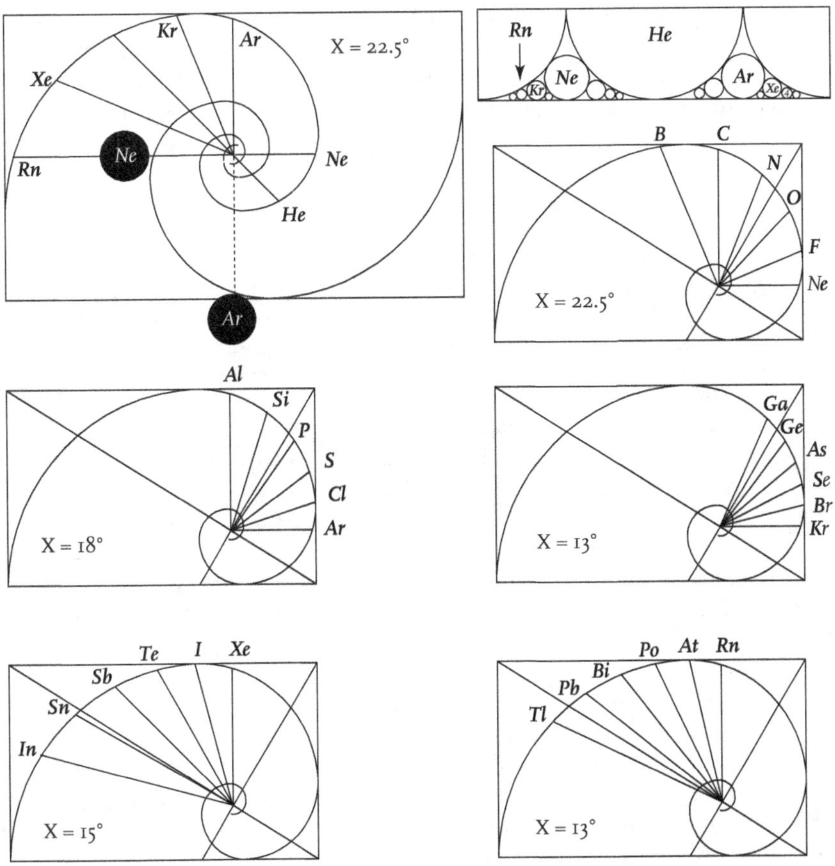

FIGURE 6.7 Polarization radii (Å) and atomic polarizabilities (Å^3) for p-block elements.

As shown at the top-left of Figure 6.7 an optimization based on divergence angles in multiples of 22.5° results in an arrangement of noble-gas atoms along a set of Fibonacci spirals with the same symmetry as the Ford-circle mapping of the periodic table.

The radial distances at which the atoms are placed in this arrangement are in agreement with the best recommended values of their polarizabilities [18].

Calculated polarizabilities are scaled in terms of the "exact" value for α(He) = 0.205 in the commonly used units of 10^{-24} cm^{-3} ($Å^3$). The optimized values, shown in Table 6.2, are next used as starting parameters in the optimization of the associated p-block polarizabilities, as in Figure 6.7. The results are collated in Table 6.2.

The calculated polarizabilities are as reliable as the recommended values, if not more so, and hence appropriate as input in the estimation of molecular polarizabilities. Even the naive approach that assumes each chemical bond to constrain the atomic sum of polarizabilities by a characteristic amount, gives surprisingly good results.

Starting from atomic polarizabilities of α(H)=0.667 and α(C)=1.76 $Å^3$ with bonding factors of α(C–C) = −0.33 and α(C–H) = −0.46 $Å^3$ the results for some alkanes are shown in Table 6.3.

The bond factor will most likely turn out to be related to bond dissociation energy, which, as shown before, is a function of the golden ratio. It may therefore eventually transpire that molecular polarizabilities of homologous series can be simulated directly by Fibonacci optimization. This conjecture is supported by the observed periodicity in the spectroscopic excitation of diatomic

TABLE 6.2 Polarization radii (Å) and atomic polarizabilities for p-block elements ($Å^3$)

	r_α	α	α (REC)		r_α	α	α (REC)
He	0.366	0.205					
Ne	0.46	0.41	0.40	As	1.02	4.45	4.31
F	0.51	0.56	0.56	Ge	1.11	5.73	6.07
O	0.58	0.82	0.80	Ga	1.24	7.99	8.12
N	0.65	1.15	1.10	Xe	0.99	4.06	4.04
C	0.76	1.84	1.76	I	1.05	4.85	4.7
B	0.86	2.66	3.03	Te	1.12	5.88	5.5
Ar	0.74	1.70	1.64	Sb	1.19	7.06	6.6
Cl	0.81	2.23	2.18	Sn	1.31	9.42	7.7
S	0.89	2.95	2.90	In	1.46	13.04	10.2
P	0.98	3.94	3.63	Rn	1.09	5.42	5.3
Si	1.10	5.58	5.38	At	1.14	6.21	6.0
Al	1.21	7.42	6.8	Po	1.20	7.24	6.8
Kr	0.85	2.57	2.48	Bi	1.27	8.58	7.4
Br	0.90	3.05	3.05	Pb	1.38	11.00	6.8
Se	0.96	3.71	3.77	Tl	1.52	14.71	7.6

TABLE 6.3 Calculated and measured [18] polarizabilities (Å^3) of some alkanes

MOLECULE	ATOMS	BONDS	α (MOL.)	α (MEAS.)
CH_4	4.428	1.84	2.59	2.593
C_2H_6	7.522	3.09	4.43	4.43
C_5H_{12}	16.804	6.84	9.96	9.99
C_8H_{18}	26.086	10.59	15.50	15.9
$C_{11}H_{24}$	35.368	14.34	21.03	21.03
C_3H_6	9.282	3.75	5.51	5.66
C_5H_{10}	15.47	6.25	9.22	9.15

molecular series, focused on hypothetical rare-gas molecules [19]. Fibonacci-spiral optimization of these trends should be feasible in principle.

4. *Cosmic Self-similarity*

The prescient pioneering work of Nagaoka and Bohr on the self-similar relationship between atomic and astronomical structures is vidicated by the periodicity of matter as revealed by number theory. The same logarithmic-spiral optimization of periodic atomic and chemical structures also accounts for structures in the solar system [20] and of spiral galaxies. It has also been recognized in patterns that develop in lava flows on Mars (Fig. 6.8). Andy Ryan [21] spotted the spirals using the HIRISE (High Resolution Imaging Science Experiment) instrument on board the Mars Reconnaissance Orbiter. Similar spirals are reported to occur in lava flows on the Hawaiian Islands.

Equally interesting is the alignment of earthquakes in North America [22]. Events recorded over a 24-hour period shown in Figure 6.9 are seen to be distributed along a spiral focussed on the caldera of the super volcano at Yellowstone. The monitoring was done on the USGS (US Geological Survey) web site (http://earthquake.usgs.gov/) by Michael Chobanian. From this site, the USGS reports earthquakes of magnitude >1.0 for the continental United States, Alaska, Hawaii, and Puerto Rico.

In all of these instances the golden ratio features as a vital factor. The second common factor is the topology of space-time. The implied link between the two factors argues for projective cosmology, at variance with the big bang.[3]

In the same way that the periodic function is closed at $Z_0 = Z_{102}$, as in Figure 6.5, space-time is closed at $\infty = -\infty$, in the projective point at infinity [11]. The consequences of this conclusion cut both ways. Schrödinger's equation, despite its seductive rationalization of elemental periodicity is no more than a useful approximation. It is incomplete in its failure to recognize the

FIGURE 6.8 Spiral patterns in Martian lava flows.

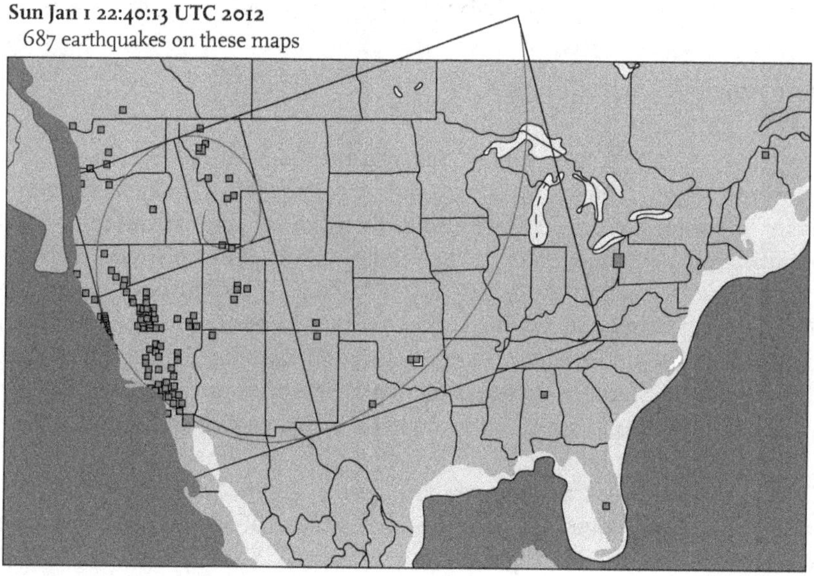

FIGURE 6.9 A spiral of 687 earthquakes. Spiral enhanced.

four-dimensional nature of space-time and hence its failure to account for electron spin. Its failure to recognize the inherent nonlinearity of curved space disqualifies it as a model of chemical behavior [10].

On the flip side it militates against cosmological expansion into infinite Euclidean space. On both scores it emphasizes the need of a new scientific paradigm that incorporates nonlinear effects, analyzed in four dimensions. Nowhere is this need more urgent than in theoretical chemistry which is dominated by the popular pursuit of *Quantum Chemistry*, demonstrably based on a set of discredited assumptions [25, 26].

5. *Conclusion*

Commenting on the discovery of atomic number, Bohr stated [27]:

> This interpretation of the atomic number may be said to signify an important step towards the solution of a problem which for a long time has been one of the boldest dreams of natural science, namely to build up an understanding of the regularities of nature upon consideration of pure numbers.

Recognition of the grand periodic function based on atomic mass number has now widened the scope to the point where phenomena such as nuclear structure, superconductivity, cold fusion, covalence, atomic structure, electronegativity [10], and polarizability, phenomena that are conspicuously resistant to quantum-mechanical analysis, are recognized as functions of the golden ratio and elementary number theory, just like biological growth structures, geological patterns and astronomical structures.

Periodicity of the elements, a subset of the grand periodic function, emerges as a closed function that varies with the curvature of space-time. It is inferred from this observation that the topology of the cosmos, consistent with the appearance of both matter and antimatter, must be projective in four dimensions and intrinsically nonlinear [10]. Traditional quantum theory does not reflect these features, which explains why it fails to account for four-dimensional action (spin) and appears to be at variance with the theory of relativity.

The ubiquitous symmetry, known as self-similarity, neatly accounts for the growing number of apparently unrelated regularities of Nature that derive from a single number pattern, based on the golden ratio. The unavoidable conclusion is that this pattern is generated by the topology of space-time. In this sense discovery of the periodic function is a further step in Bohr's conjectural scheme.

l'Envoi

Humans experience the four-dimensional Riemannian space-time manifold in three-dimensional Euclidean tangent space. Constantly unaware of space-time curvature there is little appreciation of objects and events that unfold naturally in their four-dimensional environment in a way that appears anomalous in tangent space. A mysterious force seems to regulate the growth pattern of all structures, great and small, with self-similar symmetry, universally related by an equally mysterious mathematical parameter, known as the golden ratio. In the same way that a flat-earther ponders the mode in which an approaching ship on the flat ocean appears to change its shape in a mysterious way, observers in tangent space encounter growth patterns that defy comprehension.

The obvious explanation is that the golden ratio is a measure of space-time curvature. Whatever develops in curved space-time follows this curvature, most commonly expressed as a golden logarithmic spiral which becomes embedded in the structure of atoms, molecules, nanoparticles, flowers, skeletons, lava flows, hurricanes, planetary and solar systems, and galaxies. The only unanswered question is why the curvature matches the golden ratio.

The relationship between periodicity and the golden ratio derives from the notion that the stability of atomic nuclei varies with pressure, which on a cosmic scale depends on space-time curvature. Nuclei, generated under conditions of high pressure and curvature, on release into intergalactic space-time decay radioactively to an extent limited by the local golden-ratio curvature. The field of nuclide stability as reflected by the periodic function is then clearly conditioned by the golden ratio. In projective space-time the involuted periodicity of antimatter is complementary to the periodicity of matter.

It is no accident that the grand periodic function and all periodic phenomena are based on number theory, the golden ratio, logarithmic spirals, and self-similarity. The richest variety of periodic trends has been identified in chemistry and that is where a new number-based calculus is expected to develop.

References

[1] A. Deitmar, *A First Course in Harmonic Analysis*, 2nd ed., Springer, New York, 2005.

[2] E. Schrödinger, "The proper vibrations of the expanding universe," Physica 6 (1939), 899–912.

[3] J.C.A. Boeyens and P. Comba (eds.), "Electronic structure and number theory," Struct. Bond. 148 (2013).

[4] W.D. Harkins, "The periodic system of atomic nuclei and the principle of regularity and continuity of series," Phys. Rev. 38 (1931) 1270–1288.

[5] W.D. Harkins, "The stability of atoms as related to the positive and negative electrons in their nuclei, and the hydrogen, helium, H_3, H_2 theory of atomic structures," J. Am. Chem. Soc. 42 (1920) 1965–1997.

[6] J.C.A. Boeyens, "Periodicity of the stable isotopes," J. Radioanal. Nucl. Chem. 257 (2003) 33–43.

[7] J.C.A. Boeyens and D.C. Levendis, *Number Theory and the Periodicity of Matter*, Springer.com, 2008.

[8] C.W. Raine, "Pythagorean triangles from the Fibonacci series," Scripta Math. 14 (1948) 164–165.

[9] J. Kappraff, *Beyond Measure*, World Scientific, Singapore, 2002.

[10] J.C.A. Boeyens, *The Chemistry of Matter Waves*, Springer, Dordrecht, 2013.

[11] J.C.A. Boeyens, *Chemical Cosmology*, Springer.com, 2010.

[12] S. Goldman and C. Joslin, "Spectroscopic properties of an isotropically compressed hydrogen atom," J. Phys. Chem. 96 (1992) 6021–6027.

[13] J.C.A. Boeyens, "Calculation of atomic structure," Struct. Bond. 148 (2013) 71–91.

[14] P. Plichta, *God's Secret Formula*, Element, Boston, 1998.

[15] J.C.A. Boeyens, "Ionization radii of compressed atoms," J.C.S. Faraday Trans. 90 (1994) 3377–3381.

[16] J.C.A. Boeyens, "The periodic electronegativity table," Z. Naturforsch. 63b (2008) 199–209.
[17] J.C.A. Boeyens, "Covalent Interaction," Struct. Bond. 148 (2013) 93–135.
[18] D.R. Lide (ed.), *Handbook of Chemistry and Physics*, 86th ed., CRC Press, Boca Raton, 2005–2006.
[19] R. Hefferlin, J. Sackett, and J. Tatum, "Why do molecules echo atomic periodicity?," Int. J. Quan. Chem. DOI: 10.1002/qua.24469 (2013) 1–12.
[20] J.C.A. Boeyens, "Commensurability in the solar system," Physics Essays 22 (2009) 493–499.
[21] A. Ryan, "How lava created strange spirals on Mars," Pop. Mech., April 26, 2012.
[22] M. Chobanian, Personal communication, 2013.
[23] J.C.A. Boeyens, "Cosmology and Science," in A. Travenna and B. Soren (eds.), *Recent Advances in Cosmology*, Nova, New York, 2013, 129–157.
[24] A. Travenna and B. Soren (eds.), *Recent Advances in Cosmology*, Nova, New York, 2013.
[25] J.C.A. Boeyens and C.J.H. Schutte, "The Assumptions of Quantum Chemistry," in M.V. Putz (ed.) *Chemical Information and Computational Challenges*, Nova, New York, 2012,. 153–180.
[26] M.V. Putz (ed.) *Chemical Information and Computational Challenges*, Nova, New York, 2012.
[27] N. Bohr, *Atomic Theory and the Description of Nature*, Cambridge University Press, London, 1934.

Notes

1. Referred to as natural elements.
2. The limiting value of 102 corresponds to the appearance of 100 natural elements as $Z = 43$ and $Z = 61$ do not feature.
3. Not considered to be sensible cosmology [23, 24].

CHAPTER 7 | What Elements Belong in Group 3 of the Periodic Table?

ERIC R. SCERRI AND WILLIAM PARSONS
Department of Chemistry and Biochemistry, UCLA, USA

The Group 3 Debate

The question of precisely which elements should be placed in group 3 of the periodic table has been debated from time to time[1] with apparently no resolution. This question has also received a recent impetus from several science news articles following an article in *Nature Magazine*[2] in which the measurement of the ionization energy of the element lawrencium was reported for the first time.[3]

We believe that this question is of considerable importance for chemists and physicists as well as students of these subjects. It is our experience that students are typically puzzled by the fact that published periodic tables show variation in the way that group 3 is displayed. Instructors typically cannot answer questions that students may have on this matter. The aim of this chapter

[1] R.W. Clark, G.D. White, "The fly-leaf periodic table," *J. Chem. Educ.*, 85, 497, 2008; W.B. Jensen, "The positions of lanthanum (actinium) and lutetium (lawrencium) in the periodic table," 59, 634, 1982. L. Lavelle, "Lanthanum (La) and Actinium (Ac) should remain in the d-block," *J. Chem. Educ.*, 2008, 85, 1482–1483; E.R. Scerri, Which Elements Belong in Group 3?," *J. Chem Educ.*, 86, 1188–1188, 2009; W.B. Jensen, "Misapplying the Periodic Law," *J. Chem. Educ.*, 2008, 85, 1491; E. Scerri, *A Very Short Introduction to the Periodic Table*, Oxford University Press, Oxford, 2013.
[2] T.K. Sato et al., "Measurement of the First Ionization Potential of Lawrencium, Element 103" *Nature*, 520 (9 April), 209–212, 2015.
[3] D. Castelvecchi, "Exotic Atom Struggles to Find Its Place in the Periodic Table," *Nature*, April 8, 2015, http://www.nature.com/news/exotic-atom-struggles-to-find-its-place-in-the-periodic-table-1.17275; J. Kemsley, "Lawrencium Ionization Energy Measured," *Chem. & Eng. News*, 2015, 93 (13 April), 8; M. Gunther, "Lawrencium Experiment Could Shake Up the Periodic Table," *Chemistry World*, 9 April, 2015, http://www.rsc.org/chemistryworld/2015/04/lawrencium-experiment-could-shake-periodic-table; R. Gray, "Is the Periodic Table Wrong? Elements May Need to Be Reordered After Scientists Find Lawrencium Looks Out of Place," *Daily Mail*, 10 April, 2015. http://www.dailymail.co.uk/sciencetech/article-3033570/Is-periodic-table-WRONG-Elements-need-reordered-scientists-L awrencium-looks-place.html

is to make a clear-cut recommendation regarding the membership of group 3, which we believe should consist of the elements scandium, yttrium, lutetium, and lawrencium.

Although the arguments in favor of replacing lanthanum and actinium by lutetium and lawrencium are rather persuasive there is a popular and mistaken belief that IUPAC supports the traditional periodic table with lanthanum and actinium in group 3. This view has been disputed by Jeffrey Leigh in an interesting article in which he made it clear that IUPAC has not traditionally taken a view as to the correctness of any version of the periodic table and that there is no such thing as an officially approved IUPAC periodic table.[4]

We will briefly review the previous arguments that have been provided in favor of moving lutetium and lawrencium into group 3 of the periodic table in place of lanthanum and actinium. We will then reiterate what we take to be a categorical argument in favor of this placement and will discuss any remaining issues. When added to other arguments made over more than 50 years it becomes clear that the time may have arrived for IUPAC to make a ruling on this question.[5]

A Historical Survey of Evidence and Arguments in Favor of Placing Lu and Lr into Group 3

Since the advent of quantum mechanics and the precise determination of the electronic configurations of atoms there has been a general belief that such an approach should settle any remaining questions having to do with the details of the periodic table. Although we believe this premise to be incorrect we will begin by discussing evidence based on the electronic configurations of some of the atoms in question.

The early determination of the configurations of the elements ytterbium (70) and lutetium (71) seemed to indicate the following electronic configurations for their atoms:

Yb [Xe] $4f^{13}\ 5d^{1}\ 6s^{2}$

Lu [Xe] $4f^{14}\ 5d^{1}\ 6s^{2}$

As a consequence it was believed that lutetium should mark the end of the lanthanide series. Further, the discovery of lutetium had occurred at about the same time as the discovery of several other rare earth elements, which meant that it was generally regarded as simply being a rare earth element.[6]

---

[4] J. Leigh, "Periodic Tables and IUPAC" *Chem. Int.*, Vol. 31 No. 1, January-February 2009 http://www.iupac.org/publications/ci/2009/3101/1_leigh.html

[5] Since the writing of this article IUPAC has approved the formation of a task group to discuss the constitution of group 3 and to make recommendations to them. This task group is under the chairmanship of one of us (E.S.).

[6] C.H. Evans, *Episodes from the History of the Rare Earth Elements*, Springer, Berlin, 1996.

In 1937 Meggers and Scribner published an article in which they reported that, contrary to earlier observations, the configuration of ytterbium should be assigned as,[7]

Yb [Xe] f^{14} $6s^2$

However, the authors did not comment on any possible ramifications of placing the subsequent element lutetium in the periodic table. For the sake of clarity we will briefly discuss this consequence now.

If ytterbium possesses 14 f-electrons, rather than 13 as formerly believed, it can genuinely be thought of as the final rare earth element. Consequently, the next element, lutetium, can be regarded as a d-block element, thus placing it under scandium and yttrium in group 3. In the years following the correct configuration of ytterbium a few books and published periodic tables incorporated the newly assigned configuration but refrained from discussing any possible reassignment of the placement of lutetium.

The first statement that we have been able to find that these configurations provided grounds for regarding lutetium as a d-block rather than as an f-block element comes from the well-known, perhaps even classic, 1959 book on quantum mechanics by Landau and Lifshitz.[8] Here the authors stated categorically that,

> In books on chemistry, lutetium is also placed with the rare earth elements. This, however is incorrect, since the 4f shell is complete in lutetium...

It would appear that this simple notion was then resurfaced separately by a number of other authors, working in different subdisciplines, although none of these proposals seemed to have any impact on the way in which the periodic table was presented in textbooks or elsewhere—with the possible exception of the work of Luder.[9] Several of the authors who proposed that lutetium should replace lanthanum in group 3 were physicists, a factor that may have contributed to them being ignored by the chemical community.[10]

In 1982 a major development took place in the story of which elements to place in group 3. William Jensen published a widely cited article in *Journal of Chemical Education* in which he reviewed previous evidence and made perhaps the first concerted plea, to chemists, for periodic tables to be changed so that lutetium replaced lanthanum and lawrencium replaced actinium in group 3.[11]

---

7 W.F. Meggers, B.F. Scribner, "Arc and Spark Spectra of Ytterbium," *J. Res. Nat. Bur. Stand.*, 19, 651, 1937.

8 L.D. Landau, E.M. Lifshitz, *Quantum Mechanics*, Pergamon, London, 1959, p. 245 footnote.

9 W. F. Luder, "Electronic Configuration as the Basis of the Periodic Table," J. Chem. Educ., 20, 21, 1943.

10 D.C. Hamilton, M.A. Jensen, "Mechanism for Superconductivity in Lanthanum and Uranium," *Phys. Rev. Letters*, 11, 205, 1963; B.T. Matthias, W.H. Zacharisen, G.W. Webb, J.J. Englehardt, "Melting Point Anomalies," *Phys. Rev. Letters*, 18, 781, 1967.

11 W.B. Jensen, "The positions of lanthanum (actinium) and lutetium (lawrencium) in the periodic table," *Chem. Educ.*, 1982, 59, 634–636.

It would seem that this article has not convinced many authors since the majority of periodic tables have remained unchanged and the debate has continued.

Brief Analysis of Jensen's 1982 Article

We believe that the article by Jensen represents a major step toward the reassignment of lutetium and lawrencium to group 3, but that the proposal suffers from some limitations that have lessened its impact.

The first limitation is rather minor, but nevertheless worth mentioning in this context. While reporting the reassignment of the configuration of ytterbium Jensen wrote,

> More recent spectroscopic work, however, has revised the earlier electronic configurations (4). Only three of the rare earths in period 6 (La, Gd, and Lu) are now known to have the ground state $[Xe]4f^{x-1}5d^{1}6s^{2}$, all of the rest having the configuration $[Xe]4f^{x}6s^{2}$... Ytterbium and nobelium should now have configuration [Noble Gas]$(n-2)f^{14}(n-1)d^{1}ns^{2}$ resulting in a d rather than an f differentiating electron for both lutetium and lawrencium and making them equally good candidates for the first members of the d-block in periods 6 and 7... both lanthanum and actinium should be considered the first members of the f-block (rather than Lu and Lr), and lutetium and lawrencium (rather than La and Ac) should be considered the first member of the d-block in periods 6 and 7 and assigned to Group IIIB along with scandium and yttrium.

It appears that Jensen did not take the trouble to investigate just when the assignment had first been proposed since it was a matter of about 50 years rather than "More recent spectroscopic work." If anything, this makes it more puzzling that the relocation of lutetium was not seriously considered earlier.[12]

The second limitation is that we believe that Jensen has been too selective in the evidence he puts forward to support his case (Figure 7.1). We will show that if other forms of data are appealed to, the case becomes somewhat less compelling.

It should also be mentioned that Jensen was not personally responsible for assembling such data or in presenting it in the fashion shown in his article. As Jensen acknowledged, he was using data collected and plotted some years earlier by Christyakov.[13] In fact, these data are somewhat more comprehensive than what Jensen chose to display in his 1982 article; again one could say that he is introducing a further element of selectivity in his choice of what evidence to omit from Christyakov's article.

---

12 Some authors did place lutetium and element 103 in the same group as scandium and yttrium but for quite different reasons. One such example is Charles Janet; see P. Stewart, "Charles Janet: Unrecognized Genius of the Periodic System," *Found. Chem.*, 12, 5–15, 2010.

13 V.M. Christyakov, "*Secondary Periodicity of Biron" in Secondary d-Subgroups of the Short Periodic Table*," *Zh. Obshch. Khim.*, 38(2), 209, 1968.

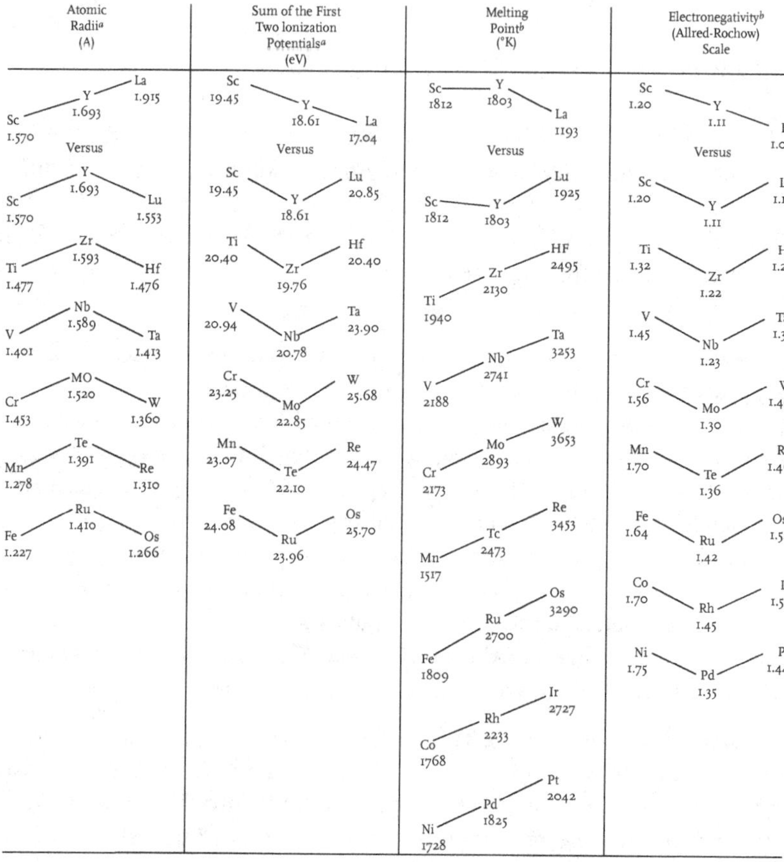

FIGURE 7.1 Jensen's re-presentation of Christyakov's data.

To investigate more thoroughly the kinds of trends that Christyakov and Jensen appeal to we set out to examine how other properties vary among elements in several transition metal groups. For example, it is not at all clear why Christyakov (and following him, Jensen), chose to focus only on the sum of the first two ionization energies of the elements in question rather than, say, the sum of the first three ionization energies. The latter quantity is in fact more relevant given the overwhelming tendency of actinides to form ions with a +3 charge.

Sum of First Three Ionization Energies

A casual inspection of Figure 7.2, in which the sum of the first three ionization energies is plotted, shows that lanthanum fits the general downward trend among transition group elements better than does lutetium. In the case of

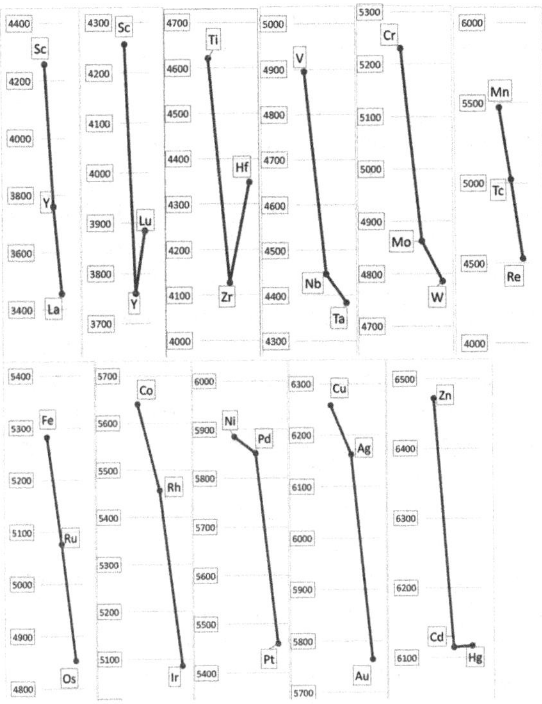

FIGURE 7.2 Sum of first three ionization energies in kJ/mol for first three elements in groups 3 to 12, inclusive.

lutetium there is a slight increase which appears to be anomalous since, apart from group 12, most of the other sets of transition metal groups show a decrease in the value for the final member of the group. This feature emphasizes the fact that appealing to specific chemical or physical data on the elements in question is inconclusive.

The data for group 3 metals in Figure 7.2 are presented in two ways, first with lanthanum and second with lutetium in place of lanthanum. Lanthanum seems to conform to the general downward trend better than lutetium, contrary to Jensen's argument. This underlies the inconclusive nature of arguments based on specific chemical and physical data.

A Conclusive Argument in Favor of Sc, Y, Lu, Lr

We have not yet presented a conclusive argument to settle the membership of group 3 in this chapter. As we have seen, arguments drawing on changes in electronic configuration suggest the need to replace lanthanum with lutetium but are, by no means conclusive. Similarly the chemical and physical data first presented by Christyakov, and represented more recently by Jensen, also support the replacement of lanthanum with lutetium—but these are also not conclusive. We have seen that in the case of ionization energy, focusing on the more relevant sum of the first three ionization energies leads to the conclusion that lanthanum should *not* be replaced.

Let us return to the need for a completely general and conclusive argument that does not depend on either individual data on the elements concerned or their electronic configurations. Such an argument[14] has been published; it depends on two very modest requirements. The first is to present the periodic table in a 32-column-long format rather than the more frequently displayed medium-long (18-column) format. The long form can be said to be a more correct representation given that it incorporates the f-block into the main body of the periodic table. In the 18-column table the f-block appears as a disconnected footnote, but this format has survived for pragmatic reasons. The 32-column table is rather wide, with the result that it becomes difficult to represent on wall charts or even on the printed page since the size of each element symbol must be rendered smaller than in the 18-column version. However, it is generally accepted that this is merely a pragmatic consideration.

The second modest requirement is that the elements be presented in such a fashion that they show a smooth increase in atomic number as one progresses through the periodic table, from left to right across each period. If these two recommendations are followed it becomes quite clear that group 3 should contain lutetium and lawrencium rather than lanthanum and actinium. As can be seen in Figure 7.3, this arrangement makes for a smooth and regular sequence in the atomic numbers of all the elements. On the other hand if one insists on retaining lanthanum and actinium in group 3, the sequence of increasing atomic number becomes highly anomalous as highlighted in Figure 7.4.

Finally, for the sake of completeness we should point out that there is another remote possibility that would still preserve the traditional placement of lanthanum and actinium in group 3 while still showing all elements with a regular increase in atomic number. This is a long-form periodic table, which is occasionally seen in textbooks and articles. However, it comes at a high price. As can be seen in Figure 7.5, this presentation of the periodic table requires that the d-block elements be split into two highly uneven parts; one part consists of only one group of d-block elements, while the remaining nine groups are shown together. Nowhere else in the periodic table does this occur, leading one to think that it is a rather unnatural presentation that can be safely ignored.

Finally, we have carried out some very recent work that suggests that the viability of the "split d-block option" displayed in Figure 7.5 is even more improbable than was previously believed. We now add a third requirement to the criteria that should be used in selecting an optimal periodic table. Up to now we have required that the periodic table be displayed in a 32-column format and that the elements should follow each other sequentially in terms of

[14] E. Scerri, *A Very Short Introduction to the Periodic Table*, Oxford University Press, Oxford, 2013, chapter 10.

H 1																															He 2
Li 3	Be 4																									B 5	C 6	N 7	O 8	F 9	Ne 10
Na 11	Mg 12																									Al 13	Si 14	P 15	S 16	Cl 17	Ar 18
K 19	Ca 20															Sc 21	Ti 22	V 23	Cr 24	Mn 25	Fe 26	Co 27	Ni 28	Cu 29	Zn 30	Ga 31	Ge 32	As 33	Se 34	Br 35	Kr 36
Rb 37	Sr 38															Y 39	Zr 40	Nb 41	Mo 42	Tc 43	Ru 44	Rh 45	Pd 46	Ag 47	Cd 48	In 49	Sn 50	Sb 51	Te 52	I 53	Xe 54
Cs 55	Ba 56	La 57	Ce 58	Pr 59	Nd 60	Pm 61	Sm 62	Eu 63	Gd 64	Tb 65	Dy 66	Ho 67	Er 68	Tm 69	Yb 70	*Lu* 71	Hf 72	Ta 73	W 74	Re 75	Os 76	Ir 77	Pt 78	Au 79	Hg 80	Tl 81	Pb 82	Bi 83	Po 84	At 85	Rn 86
Fr 87	Ra 88	Ac 89	Th 90	Pa 91	U 92	Np 93	Pu 94	Am 95	Cm 96	Bk 97	Cf 98	Es 99	Fm 100	Md 101	No 102	*Lr* 103	Rf 104	Db 105	Sg 106	Bh 107	Hs 108	Mt 109	Ds 110	Rg 111	Cn 112	Nh 113	Fl 114	Mc 115	Lv 116	Ts 117	Og 118

FIGURE 7.3 Long-form periodic table with lutetium and lawrencium in group 3. The increase in atomic number shows no anomalies whatsoever.

H 1																															He 2
Li 3	Be 4																									B 5	C 6	N 7	O 8	F 9	Ne 10
Na 11	Mg 12																									Al 13	Si 14	P 15	S 16	Cl 17	Ar 18
K 19	Ca 20															Sc 21	Ti 22	V 23	Cr 24	Mn 25	Fe 26	Co 27	Ni 28	Cu 29	Zn 30	Ga 31	Ge 32	As 33	Se 34	Br 35	Kr 36
Rb 37	Sr 38															Y 39	Zr 40	Nb 41	Mo 42	Tc 43	Ru 44	Rh 45	Pd 46	Ag 47	Cd 48	In 49	Sn 50	Sb 51	Te 52	I 53	Xe 54
Cs 55	Ba 56	Ce 58	Pr 59	Nd 60	Pm 61	Sm 62	Eu 63	Gd 64	Tb 65	Dy 66	Ho 67	Er 68	Tm 69	Yb 70	Lu 71	*La* *57*	Hf 72	Ta 73	W 74	Re 75	Os 76	Ir 77	Pt 78	Au 79	Hg 80	Tl 81	Pb 82	Bi 83	Po 84	At 85	Rn 86
Fr 87	Ra 88	Th 90	Pa 91	U 92	Np 93	Pu 94	Am 95	Cm 96	Bk 97	Cf 98	Es 99	Fm 100	Md 101	No 102	Lr 103	*Ac* *89*	Rf 104	Db 105	Sg 106	Bh 107	Hs 108	Mt 109	Ds 110	Rg 111	Cn 112	Nh 113	Fl 114	Mc 115	Lv 116	Ts 117	Og 118

FIGURE 7.4 Long form periodic table with lanthanum and actinium in group 3. Atomic number no longer increases in a regular fashion. This suggests that La and Ac are misplaced.

H 1																														He 2
Li 3	Be 4																								B 5	C 6	N 7	O 8	F 9	Ne 10
Na 11	Mg 12																								Al 13	Si 14	P 15	S 16	Cl 17	Ar 18
K 19	Ca 20	Sc 21														Ti 22	V 23	Cr 24	Mn 25	Fe 26	Co 27	Ni 28	Cu 29	Zn 30	Ga 31	Ge 32	As 33	Se 34	Br 35	Kr 36
Rb 37	Sr 38	Y 39														Zr 40	Nb 41	Mo 42	Tc 43	Ru 44	Rh 45	Pd 46	Ag 47	Cd 48	In 49	Sn 50	Sb 51	Te 52	I 53	Xe 54
Cs 55	Ba 56	La 58	Ce 59	Pr 60	Nd 61	Pm 62	Sm 63	Eu 64	Gd 65	Tb 66	Ho 67	Er 68	Tm 69	Yb 70	Lu 71	Hf 72	Ta 73	W 74	Re 75	Os 76	Ir 77	Pt 78	Au 79	Hg 80	Tl 81	Pb 82	Bi 83	Po 84	At 85	Rn 86
Fr 87	Ra 88	Ac 89	Pa 91	U 92	Np 93	Pu 94	Am 95	Cm 96	Bk 97	Cf 98	Es 99	Fm 100	Md 101	No 102	Lr 103	Rf 104	Db 105	Sg 106	Bh 107	Hs 108	Mt 109	Ds 110	Rg 111	Cn 112	Nh 113	Fl 114	Mc 115	Lv 116	Ts 117	Og 118

FIGURE 7.5 Alternative version of long-form periodic table in which the d-block is split into two highly uneven portions consisting of one and nine groups of elements.

increasing atomic number. As the reader may recall this allowed us to rule out the option presented in Figure 7.4, but not the one shown in Figure 7.5. We now propose to add the requirement that in progressing sequentially through the blocks of the periodic table one should adhere to both parts of the Madelung Rule, namely that the order of orbital filling should proceed with increasing values of $n + \ell$ and in cases of equal values of $n + \ell$ in an increasing order of the n quantum number.

If one accepts this proposal it becomes clear that the splitting of the s-block of the periodic table, which is routinely carried out in periodic tables that maintain helium in the noble gas group, is categorically different from the split d-block option of Figure 7.5. Simply stated, the familiar splitting of the s-block does not violate either part of the Madelung Rule. On the other hand, the split d-block periodic table entails a violation of the second part of Madelung's rule as will be explained below.

As can be seen in Figure 7.5, the split d-block table features the following two sequences of elements in the sixth and seventh rows, respectively. In the sixth row we encounter barium followed by lanthanum, which is followed in turn by cerium.

Focusing now on the blocks of the periodic table that are crossed on moving through these three elements we find the sequence of s-block, d-block, and then f-block. More specifically, the values of sum of the $n + \ell$ quantum numbers and n for these atoms appear as follows.

	Ba	**La**	**Ce**
differentiating electron	6s	5d	4f
$n + \ell$	6	7	7
n	6	5	4

Whereas the first part of Madelung's rule is obeyed on moving through the three elements, the movement from lanthanum to cerium shows a violation of the second part of the rule since for the same value of $n + \ell$, namely 7, the sequence should follow the order of increasing n whereas this case shows a decrease from 5 to 4.

In the seventh period a precisely analogous case can be made for the atoms of radium, actinium and thorium:

	Ra	**Ac**	**Th**
differentiating electron	7s	6d	5f
$n + \ell$	7	8	8
n	7	6	5

We acknowledge that the Madelung rule does not always give the correct order of occupation of orbitals for a particular atom, as, for example, in the case of scandium.[15] However, apart from about 20 anomalies, the Madelung rule does provide the order of orbital occupation as one moves through the periodic table. For example, in the case of potassium and calcium the 4s orbital is indeed preferentially occupied whereas in scandium it is the 3d orbitals that are preferentially occupied. The distinction between order of filling in a particular atom starting from a bare nucleus and order of filling as one progresses through the periodic table has been recently stressed in the literature.[16]

Moreover, for the purposes of selecting an optimal periodic table we prefer to consider block membership as a global property in which we focus on the predominant differentiating electron. We readily acknowledge the fact that the atoms of Mn, Zn, Tc, Cd, Pt, Hg, Lr, La, Gd, Ac, Th, and Cm are all anomalous in that they have a differentiating electron that is atypical of the block that they are situated in. These anomalies should not challenge our attempts to establish the overall structure of the periodic table in terms of sequences of blocks in the periodic table and as a result our recommendations for the membership of group 3 of the table.

Similarly, we draw attention to the fact that there are anomalous cases such as the thorium atom, which features no f-orbital electrons—yet nobody disputes that this element belongs in the f-block of the periodic table.

The hoped for categorical argument for claiming that group 3 should consist of Sc, Y, Lu, and Lr has therefore been strengthened, although admittedly not rendered completely categorical, by means of our further requirement for the adherence to both parts of the Madelung rule on progressing through the periodic table. We suggest that this further requirement makes the split d-block periodic table an even weaker possibility. In addition, we do not need to appeal to arguments involving the aesthetic undesirability, or asymmetry of splitting the d-block into two unequal portions consisting a one-group and a nine-group part.

To conclude, there is now highly persuasive evidence as well as a rather categorical argument to support the placement of lutetium and lawrencium into group 3 of the periodic table in place of lanthanum and actinium. We strongly recommend that IUPAC support this proposal and make an official ruling on the matter.

---

[15] S-G. Wang, W.H.E. Schwarz, "The Icon of Chemistry," *Ang. Chemie Int. Edn.* 48, 3404–3415, 2009; E.R. Scerri, "The Trouble with the Aufbau Principle," *Educ. Chem.*, November, 24–26, 2013.

[16] S. Salehzadeh, F. Maleki, The 4s and 3d subshells: Which one fills first in progressing through the periodic table and which one fills first in any particular atom? Found. Chem., 18, 57–65, 2016.

CHAPTER 8

The Periodic Table Retrieved from Density Functional Theory Based Concepts

The Electron Density, the Shape Function, and the Linear Response Function

PAUL GEERLINGS
General Chemistry, Free University of Brussels-VUB, Belgium

1. *The Periodic Table as Interpreted by Quantum Mechanics: The Orbital Paradigm.*

"The periodic table of the elements is one of the most powerful icons in science: a single document that captures the essence of chemistry in an elegant pattern."

This statement taken from Eric Scerri's marvelous book *The Periodic Table: Its Story and Its Significance* (Scerri 2007) grasps in one simple sentence the status that the periodic table has acquired in chemistry, but not only in chemistry: every person all over the world who took high school chemistry remembers for the rest of his/her life at least one thing from it (and from science courses in general): this mysterious table of the elements omnipresent in all books and documents, often decorating the classroom. Why? Although the answer is not easy it is probably because a whole discipline of science is condensed in a simple table or scheme offering at one glimpse the essence (and the beauty) of a part of science which for the rest of the lives of many students will remain unexplored—and even for which these students will create an aversion among others in view of the "polluting role" of chemistry. Nevertheless at some moment in their lives these students felt "something" that was remarkable and which they always remember, as witnessed by their comments when visiting their old school of university with their children. In the hierarchy of the sciences chemistry is often considered, mainly by physicists, as the "physics of the outer shell." It is sometimes said that chemistry, as also quoted by Scerri (2007), has no deep ideas, not a few fundamental laws like in physics or biology (such as those governing quantum mechanics, relativity and evolution), but the not-so-distant observer will disagree: *chemical periodicity as precisely reflected in the periodic table is in my view not only the most fundamental law of chemistry*, but is a law or *if you want* an *"organizing principle" with the same status as these famous laws in adjacent disciplines!* Chemical periodicity is at the heart of reducing an astonishing amount of experimental data (and nowadays theoretical data as well) to a limited number of patterns often with common origin, enabling one to understand and interpret the properties of the now more-than 50 million compounds registered in the databases of the Chemical Abstracts Services (Toussant, 2009).

Why then is this periodic table (PT) so powerful? Currently, high school and college freshman chemistry classes introduce the periodic table in a way which deviates from the historical development. Whereas Mendeleev and his predecessors "built" the table in the second half of the nineteenth century starting from experimental data on the elements and their compounds, lacking basic physics at the atomic level (emerging only at the beginning of the twentieth century with the advent of quantum mechanics), modern textbooks (see for example Atkins et al. 2013) often take the opposite approach—starting from the basic laws of quantum mechanics (often summarized at the freshman level in an approximate, sometimes sloppy, "for want of a better choice," quantum mechanical description of the hydrogen atom).

Modern textbooks build the PT starting from the extension of the hydrogen atom, conserving the orbital and orbital energy concepts. As a result, many high school students or freshman who are not chemistry majors spend the rest of their lives with the idea that the PT stems from the orbital paradigm, completely neglecting the history of science—the PT was originally presented, based on experimental facts, before the advent of quantum mechanics, even before the discovery of the electron!

Even when passing to sophomore or even more elaborate Physical Chemistry textbooks (Atkins and De Paula 2009) I, not myself a historian of chemistry but having taught general, physical, and quantum chemistry for about 30 years, still witness a reductionist approach, aiming to underpin the PT from "first principles." This is not a problem as long as its limitations are recognized, the first one being that the underpinning always concentrates on gas phase atoms whereas the elements in the periodic table are *substances*, not *atoms*! A second limitation which may be less evident but which I have already mentioned is the central role of the orbital concept yielding electronic configurations. Every (quantum) chemist is taught that orbitals are by definition one-electron atom functions, which thus can correspond to the true wavefunction only in one-electron systems as H, He^{+}, Li^{2+},....

Orbitals for many electron systems are in fact only mathematical constructs to express the astonishingly complicated wavefunction (the description of a simple carbon atom with six electrons already asks for a 24-variable wavefunction if spin is taken into account) (Levine 2014). Let me be clear that orbitals are unobservables despite the provocative title of the famous 1999 *Nature* paper by Zuo et al., "Direct observation of d-orbitals holes" discussed and refuted by, among others, Scerri (2001) (for a recent account see Mulder 2011).

The traditional Hartree-Fock approach in which the orbitals are gathered in a single Slater Determinant that yields an electronic configuration of the type $(1s)^2(2s)^2(2p)^2$ for carbon is only a first-order approach to describe the electronic structure of the gaseous atom of carbon in its ground state. (See, for example, Bransden and Joachain 2003). (For an extension based on relativistic Dirac Fock calculations and extending up to Z=172 see the 2010 account by Pyykkö).

Let us summarize the main concerns.

1. The PT is not a periodic table of gaseous atoms in its original construction and as it is "explained" nowadays via quantum mechanics (QM).
2. Even if the QM reductionist approach is followed (Ostrovsky, 2001) the orbital paradigm should be looked upon with caution. Moreover at the QM level some intricacies of the PT were never properly explained, as there was no conclusive evidence for either the length of the periods or the so-called Madelung n+ℓ rule for the ordering of the energy levels (Löwdin 1969, Allen and Knight 2001, Scerri 2006).

Let it be understood that, as stated by Scerri, "the application of quantum mechanics to the periodic table represents a reductionist approach with the inevitable result that the focus of interest shifts from *macroscopic elements to the atoms* of the element" (Scerri 2012; emphasis added). In my view there is no objection to such a strategy as long as the intricacies in the notion "element" are recognized and the reduction from chemistry to physics, in this case quantum mechanics is not considered "full and complete" (Pyykkö 2010) (for a discussion of the different notions associated with the word "element" see E. Schwarz's review in Wang and Schwarz 2009).

In this context, we concentrate in this chapter, within the limitations of the first concern (we will concentrate on gaseous atoms) on alternatives for the second concern, where we will drop the orbital concept and the whole orbital paradigm by focusing on a fundamental different carrier of information as nowadays frequently used in quantum chemistry: we will abandon the wavefunction approach and move on to the electron density as fundamental carrier of information (Hohenberg and Kohn 1964, Parr and Yang 1989).

In section 2 I will highlight the importance of the Hohenberg-Kohn theorem stating that sufficient information on the ground state of an atom or molecule is contained in its electron density, a function of merely 3 variables as opposed to the 4N variables present in an N-electron function. In section 3 I will briefly describe which role the density and related quantities derivable from it, can play to describe the change of systems (atoms, molecules,...) upon perturbation, as for example during a chemical reaction. In section 4 I will offer a solution to the problem that the density of all atoms shows highly similar behavior and extract the different information they *should* bear: information theory will be the key tool to retrieve periodicity from the density, or from an even simpler carrier of information, the shape function. In section 5 I will describe how, by introducing a perturbation, differences in the response of the density to that perturbation show a much more diversified answer from which the periodicity is easily and without any prejudice extracted.

2. *An Alternative: The Electron Density, the Basic Carrier of Information in Density Functional Theory*

In traditional quantum mechanics or quantum chemistry the wavefunction Ψ is the basic carrier of information depending on all spatial and spin variables (4N in total for a N electron system). This leads to a horribly complicated wavefunction.

$$\psi = \psi(\underline{x}_1, \underline{x}_2, \ldots \underline{x}_N) \tag{1}$$

where the spatial (x_i, y_i, z_i) and spin (σ_i) variables of electron i are gathered in a four-vector $\underline{x}_i$. In 1964 Walter Kohn and Pierre Hohenberg gave an astonishingly short and elegant proof that a much simpler function, the electron density function $\rho(\underline{r})$, which upon multiplications by an elementary volume dv gives the number of electrons contained in that volume, contains all essential information necessary to characterize the system's ground state (Hohenberg and Kohn 1964). The crux of the proof is that $\rho(\underline{r})$ not only directly yields N, the total number of electrons of the system by simple integration

$$\int \rho(\underline{r})\, d\underline{r} = N \tag{2}$$

Indeed, $\rho(\underline{r})$ also determines the external potential $v(\underline{r})$, that is, the potential felt by an electron due to the nuclei. To put it another way, if you are given the electron density of a ground-state system (which, as opposed to the wavefunction ψ is experimentally accessible), this density is compatible with a single constellation of nuclei (their number, nuclear charge and position) as pictorially represented in Figure 8.1.

Once N and $v(\underline{r})$ are known the atomic or molecular Hamiltonian is also known—for example in the atomic case (Levine 2014, Bransden and Joachain 2003) by

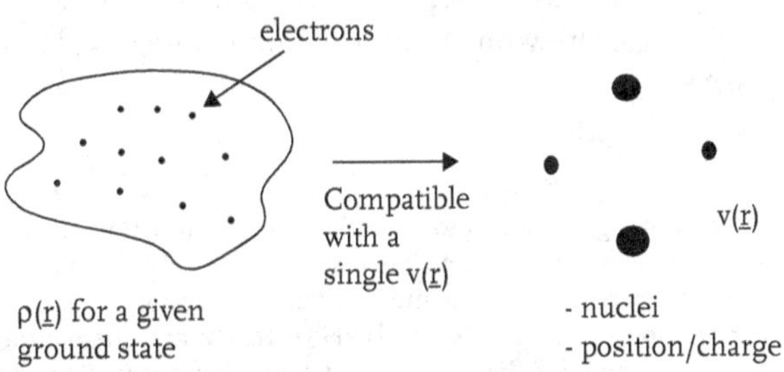

FIGURE 8.1 Pictorial Representation of the Hohenberg Kohn Theorem.

$$H_{op} = \sum_{i=1}^{N}\left(-\frac{\hbar^2}{2m}\nabla_i^2 + v(\underline{r}_i)\right) + \sum_{i<j}\sum \frac{e^2}{r_{ij}} \tag{3}$$

where the first summation contains the electronic kinetic energy operator and the interaction with the external potential of all electrons with $v(\underline{r}_i) = -Z/r_i$ with r_i the distance from electron i to the atomic nucleus with charge Z. The second term contains the electron-electron interaction energy (r_{ij} denoting the distance between electrons i and j). It is clear that knowing N, Z, and the positions of the nucleus this Hamiltonian can be written and, in principle, be solved (Scheme 1).

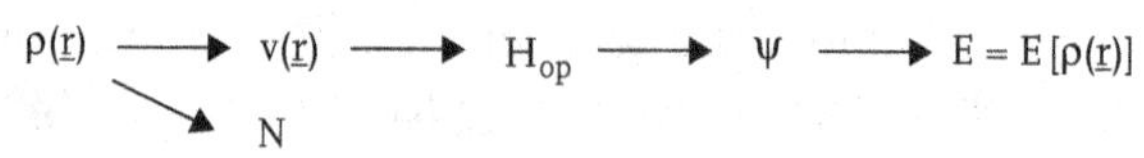

SCHEME 1 The E vs $\rho(\underline{r})$ functional

The atomic ground state is then, at least in principle, accessible. The Hohenberg-Kohn theorem paved the way for Density Functional Theory (DFT) where, making use of a variational theorem for E as a functional of ρ($E = E[\rho]$)[1] instead of the usual wavefunctional–based variational theorem, an optimal ρ and thus an optimal energy can be found. We will not address the intricacies of the technical implementation of this idea leading to the celebrated Kohn-Sham equations (Kohn and Sham 1965). They are the counterpart of the Hartree-Fock equations in wavefunction theory (Bransden and Joachain 2003, Hartree 1957). They however bear the advantage that their solutions contain part of the electron correlation energy as opposed to Hartree-Fock and that they can be obtained at much lower computational expense than in wavefunction theory (for a review by Kohn see Kohn 1999). These qualities paved the way for DFT as the present day workhorse for computational studies on medium and large systems (Koch and Holthausen 2001). Coming back to the essence of DFT, ρ as the key carrier of information, all this prompts us to think whether one should stick to a wavefunction (more precisely an approximate orbital-based one) interpretation/justification of even the gas phase atom table and whether a direct approach via the electron density is not possible/advisable.

It is at first sight remarkable that the direct relation between $\rho(\underline{r})$ and Ψ, shown in eqn. (4) reveals how much information is integrated out when passing from Ψ to $\rho(\underline{r})$.

1 A *functional* is a rule linking a function to a number. In wave function quantum mechanics, for example, E is a functional of the wave function Ψ written as $E = E[\psi]$ as to each (quadratically integrable) function Ψan energy (and thus a number) can be associated by calculating the expectation value of the Hamiltonian operator H_{op} as $\int \psi^* H_{op} \psi d\tau$. For a mathematical discourse see, for example, I.M. Gelfand and S.V. Fomin, *Calculations of Variations*, Englewood Cliffs, NJ: Prentice Hall, 1963.

$$\rho(\underline{r}) = N \int \Psi^*(\underline{x}, \underline{x}_2, \ldots, \underline{x}_N) \Psi(\underline{x}, \underline{x}_2, \ldots, \underline{x}_N) ds\, d\underline{x}_2 \ldots d\underline{x}_N \quad (4)$$

One passes from 4N to 3 variables by integrating over 4N-3 electron coordinates (McWeeny and Sutcliffe 1969).

This approach has been followed at various occasions in the literature, both by others and our own research group. In order to situate $\rho(\underline{r})$ and the quantities related to it, we first introduce in section 3 in a very concise and intuitive way a subdomain of DFT commonly called, after its founder R.G. Parr (Parr and Yang 1995), Conceptual DFT (Geerlings et al. 2003, Chermette 1999), the more appropriate name being Chemical (Reactivity) (Ayers et al. 2005) theory as evidently concepts (and certainly the more subtle concepts on its basics) are everywhere in DFT.

3. *An Introduction to Conceptual DFT*

Conceptual DFT quintessentially studies the functional E=E[N,v] (Parr and Yang 1989, Geerlings et al.2003, Ayers et al. 2005), or to put it in a qualitative, chemical way, it is concerned with the way the energy of a system (say an atom) changes when it is perturbed by a reagent leading to changes in N and $v(\underline{r})$. This means that, say up to first order, we consider the following differential:

$$dE = \left(\frac{\partial E}{\partial N}\right)_{v(\underline{r})} dN + \int \left(\frac{\delta E}{\delta v(\underline{r})}\right)_N \delta v(\underline{r}) d\underline{r} \quad (5)$$

The functional derivative[2] $\left(\delta E / \delta v(\underline{r})\right)_N$ indicates the sensitivity of the system when it is perturbed by a variation in $v(\underline{r})$, $\delta v(\underline{r})$ just as the partial derivative $\left(\partial E / \partial N\right)_v$, indicates the sensitivity of the system when its number of electrons N is changed by dN (e.g., in an electron transfer reaction).

In this way the two derivatives involved are commonly called *response functions* (Geerlings and De Proft 2008) because they contain the basic information of the magnitude of the perturbation in the system for a given $\delta v(\underline{r})$ and/or dN. They are quantities typical for each system and intrinsic (i.e., independent of the magnitude of the perturbation).

Parr and coworkers proved, in a groundbreaking paper (Parr et al. 1978), that $\left(\partial E / \partial N\right)_v$ can be identified with the negative electronegativity in the Mulliken sense (Mulliken 1934) (commonly written as the average of ionization energy I and electron affinity A)

$$\left(\frac{\partial E}{\partial N}\right)_v = -\chi_M = -\frac{1}{2}(I + A) \quad (6)$$

2 A functional derivative can be intuitively defined as follows. Consider a functional $E = E[f(\underline{r})]$. If the function $f(\underline{r})$ is varied infinitesimally to $f(\underline{r}) + \delta f(\underline{r})$ the first order change in E can be written as $dE = \int A(\underline{r}) \delta f(\underline{r}) d\underline{r}$ where $A(\underline{r})$ gives the sensitivity to E of the variation of f at position r: it is written as $\delta E / \delta f(\underline{r})$ and called the functional derivative of E with respect to $f(r)$.

First order perturbation theory (Parr and Yang 1989, Geerlings et al. 2003), on the other hand, yields $\rho(\underline{r})$, in other words, the electron density itself.

$$\left(\frac{\delta E}{\delta v(\underline{r})}\right)_N = \rho(\underline{r}) \tag{7}$$

As easily seen two fundamental quantities in DFT and/or chemistry are retrieved from the study of the E=E[N,v] functional up to first order. The same exercise can be done up to second order; the corresponding response functions have been identified as follows:

- $\left(\partial^2 E / \partial N^2\right)_v$ has been identified as the Chemical Hardness (Parr and Pearson 1983), introduced by Pearson (Pearson 1997) in the context of systematizing Lewis acid-base reactions, leading to the well know HSAB Principle stating that hard acids prefer to react with hard bases and soft acids with soft bases.
- The mixed derivative $\frac{\partial^2 E}{\partial N \delta v(\underline{r})} = \left(\frac{\partial \rho(\underline{r})}{\partial N}\right)_v$ (via eqn.7) has been identified as a generalization of Fukui's (Fukui et al. 1952) frontier orbital concepts in the study of regioselectivity. This so-called fukui function describes how sensitive the electron density at a given point $\underline{r}$ is to a change in the total number of electrons (Parr and Yang 1984).
- Finally the (at first sight) awkward quantity $\left(\delta^2 E / \delta v(\underline{r}) \delta v(\underline{r}')\right)_N$ gets more significance when again using eqn. (7), it turns into $\left(\delta \rho(\underline{r}) / \delta v(\underline{r}')\right)_N$,

that is, the change in density at a given point $\underline{r}$ when the system is perturbed at a different point $\underline{r}'$ which is of major concern in any chemical reaction (see section 5).

It should also be noted that several of the (at first) complicated third order derivatives were seen to bear chemical meaning (for a review see Geerlings and De Proft 2008) but we will not consider them further.

Coming back to our basic question, how to retrieve periodicity, or more generally the structure of the periodic table, from a DFT context, this section describes different options.

In the *first order derivatives* $\rho(\underline{r})$ itself is obviously a candidate; we will investigate this in section 4.

As we will see, extracting information from $\rho(\underline{r})$ itself is not always easy and so one might consider the idea that the information *carried by* $\rho(\underline{r})$ might become more tangible or visible when the system it describes is disturbed. Candidates to be considered are $f(\underline{r})$ and the so-called linear response function $\chi(\underline{r}, \underline{r}')$ defined as

$$\chi(\underline{r}, \underline{r}') = \left(\frac{\delta^2 E}{\delta v(\underline{r}) \delta v(\underline{r}')}\right)_N = \left(\frac{\delta \rho(\underline{r})}{\delta v(\underline{r}')}\right)_N \tag{8}$$

Based on our experience, among others with the technicalities of taking partial derivatives with respect to N (Yang et al. 2012), and on the simple argument that $\left({}^{\delta\rho(\underline{r})}\!/_{\delta v(\underline{r}')} \right)$ is obviously more sensitive when *local* perturbations are at stake (in $\left({}^{\partial\rho(\underline{r})}\!/_{\partial N} \right)_v$ only the total number of electrons is changing) we prefer in this chapter $\chi(\underline{r},\underline{r}')$ (in a simplified form, vide infra) as second descriptor as compared to the Fukui Function $f(\underline{r})$.

4. *The Electron Density and the Shape Function: Reading Their Information Content*

4.1. From the Density and Its Radial Density Distribution Function to a Piecewise Exponential Decay

At first sight, getting information from $\rho(\underline{r})$ is trivial. In view of the (here assumed to be omnipresent) spherical symmetry of the electron density function of atoms in their ground state, a plot of $\rho(r)$ (density at a given distance from r irrespective of the direction) versus r is an obvious choice.

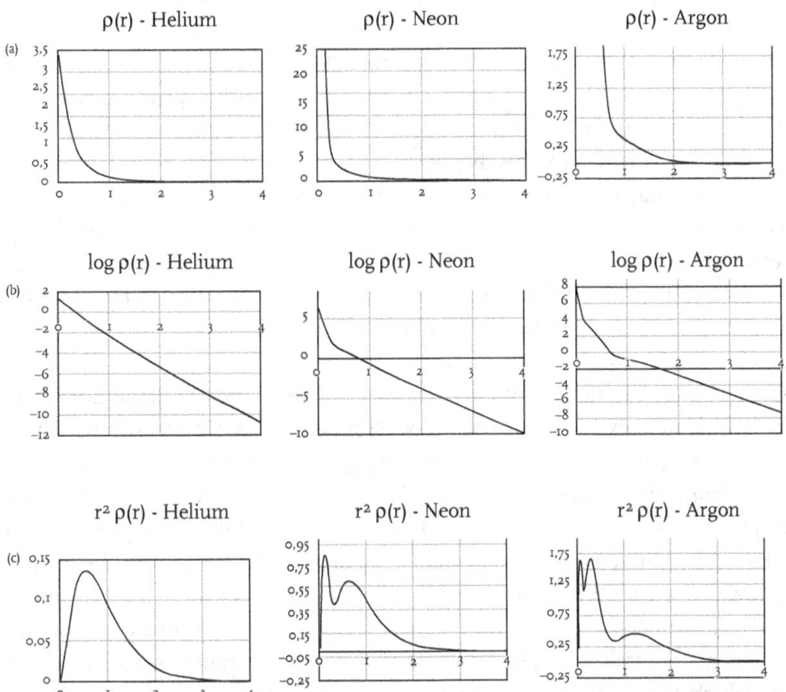

FIGURE 8.2 Electron Density (a), Logarithm of the Density (b) and Radial Density Distribution Function (c) versus r for He, Ne and Ar. (Values, in a.u., obtained at the PBE-DFT level with a aug-cc-pVT Z basis set.)

The result shown in Figure 8.2a for He, Ne, and Ar may be surprising; it is certainly disappointing. As stated by Weinstein, Politzer, and Srebrenik (1975), ρ(r) is a continuously decreasing function for all atoms in the ground state, a feature which was also seen in some previously published density plots (e.g., Sperber, 1971) but has been largely overlooked (or its significance not appreciated). Weinstein et al. commented in their 1975 paper that there appears to be a fairly widespread belief that the electronic density of a ground-state atom, when plotted against the radial distance from the nucleus, will show one or more relative maxima for $r > 0$. As in our Figure. 8.2a, these figures for the ground state of different atoms invariably show a monotonically decreasing decay of the density. For the heavier atoms there are some small "bumps and wiggles" but there is no distance from the nucleus for which a relative maximum exists. At first glance this computational result shows almost no structure at all and yields the idea that all atoms are "about equal" and that $\rho(\underline{r})$ keeps its information—which it should have according to the Hohenberg Kohn theorems—behind closed doors. [Note that a formal proof of this continuous decrease of $\rho(\underline{r})$ has not been given yet (Ayers and Parr 2003, Parr 2005) although computational evidence invariably points in this direction (Eickerling and Reiher 2008).]

Closer inspection of the density plots leads to the observation that for the heavier atoms the rate of decrease of ρ(r) as a function of r is not completely uniform. Wang and Parr (1975) proposed that these plots are continuous piecewise exponential functions of r with as many different regions as there are principal quantum numbers. Consequently, plotting log ρ(r) vs r would yield a series of intersecting lines as seen in the case of Ne and Ar (in Fig. 8.2b) with two transition regions indicating what is commonly expected as the passage from one shell to another (see below).

The main reason for this "Misconception Concerning the Electronic Density Distribution of an Atom" (the title of the Weinstein, Politzer, and Srebrenik paper (Weinstein et al. 1977) is most probably that the traditional way to disclose information (e.g., about the shell structure of atoms) has been to look at the so-called radial density distribution function. This function is well known, appearing even in general chemistry textbooks (Atkins et al. 2013) for the H-atom, as well as in most physical chemistry textbooks (Atkins and De Paula 2009). The idea behind the extension to many-electron atoms is simple (Bransden and Joachain 2003, Eickerling and Reiher 2008): consider the electron population derived from an (if necessary spherically averaged) density in a spherical shell with inner radius r and thickness dr. The resulting quantity is the product of the (averaged) density, ρ(r) and the volume of the spherical shell $4\pi r^2 dr$. The resulting function $D(r) = 4\pi r^2 \rho(r)$ is the many-electron analogue of the radial distribution function for the hydrogen atom. In Figure 8.2c the plots are shown for the same atoms depicted with their ρ(r) curve, revealing a much more structured pattern from which the shell structure of atoms can be retrieved—at least for simple atoms. Note that in 1953 Bartell and Brockway experimentally resolved the K, L, and M shells of argon from electron diffraction experiments (Bartell & Brockway, 1953). Several later theoretical investigations discovered that the correct number of maxima can only be retrieved for elements with nuclear charge smaller than 18 (Eickerling and Reiher 2008, Waber and Cramer 1965, Boyd 1976, Sen 1993).

Alternatives have been proposed: the Electron Localization Function (Becke and Edgecombe 1990), the Laplacian of the electron density (Bader and Essén 1984), and the average local electrostatic potential (Sen et al. 1995); these will not be considered here due to space limitations and their computational complexity.

These various examples of extracting information from $\rho(\underline{r})$ indicate that even before the advent of DFT people were convinced that $\rho(\underline{r})$ should contain basic information about atomic electronic structure—and, by extension, on periodicity (the radial distribution function had already been mentioned in the 1944 classical book by Eyring, Walter, and Kimball (Eyring et al. 1944).

When our group started thinking about this issue, in a time where the HK Theorem was already a few decades old and $\rho(\underline{r})$ was proven to contain the basic information, we adopted an approach which exploited the possibilities of new concepts in quantum chemistry: quantum similarity and information theory. This approach is treated in section 4.2.

4.2. Quantum Similarity and Information Theory as Applied to the Density and the Shape Function

4.2.1. Introduction

The basic idea of this subsection is no longer to consider the electron density (or a quantity derived from it) for a *single* gaseous atom on its own, but to investigate a function, which will then again be a functional, containing densities or related quantities of *two atoms*, in an attempt to disentangle the difference in information they contain, or to see which information the density of atom A contains about atom B. The first step is to investigate the quantum *similarity* (Carbo et al. 1980) between the densities of two different atoms; the second step is to use information theory (Shannon 1948), looking at the information deficiency as a tool to "read" the information difference in $\rho_A(\underline{r})$ and $\rho_B(\underline{r})$ (Kullback and Leibler 1951).

Sections 4.2.3 and 4.2.4 will be devoted to the construction of functionals

$$F_{AB} = F\left[\rho_A(\underline{r}), \rho_B(\underline{r})\right] \quad (9)$$

with the aim of disentangling differences and similarities between ρ_A and ρ_B when going through the PT, in other words, with the aim of retrieving periodicity. As we will see, an even simpler carrier of information than the density pops up in these approaches: the shape function, which we now introduce.

4.2.2. The Shape Function: An Even Simpler Carrier of Information

In 1983 Parr and Bartolotti introduced the shape function $\sigma(\underline{r})$, a renormalized version of the density function defined as

$$\sigma(\underline{r}) = \frac{1}{N}\rho(\underline{r}) \quad (10)$$

which characterizes the shape of the electron distribution (hence the name) and is, whatever the system being considered, normalized to 1:

$$\int \sigma(\underline{r})\, d\underline{r} = \frac{1}{N}\int \rho(\underline{r})\, d\underline{r} = \frac{1}{N} N = 1 \tag{11}$$

At first glance $\rho(\underline{r})$ may be thought to contain less information on the system (by dividing $\rho(\underline{r})$ by N) but it was shown by Ayers that $\sigma(\underline{r})$ is a full-fledged carrier of information: it contains all information about $v(\underline{r})$ (it shows cusps at the same positions in space as $\rho(\underline{r})$, determining the position of the nuclei, and from the behavior of $\sigma(\underline{r})$ near these nuclei their nuclear changes can be derived). Most remarkably however, $\sigma(\underline{r})$ also determines N on the basis of the convexity postulate and the long-range behavior of $\rho(\underline{r})$ (Ayers 2000). Consequently $\sigma(\underline{r})$ determines $v(\underline{r})$ and N, and so H_{op} and therefore, "everything." This unexpected result can be derived if eqn. (10) is rewritten as $\rho(\underline{r}) = N\sigma_N(\underline{r})$ where the N dependence of $\sigma(\underline{r})$ is denoted explicitly. In a proof-of-principle Ayers, De Proft, and Geerlings showed that the information content of $\sigma(\underline{r})$ is enough to predict atomic Ionization Energies and that in a moment expansion even a faster convergence in the case of $\sigma(\underline{r})$ was obtained than for $\rho(\underline{r})$ (Ayers et al. 2007).

Why this digression (Geerlings et al. 2007)? In the first functional we will consider in section 4.2.2 Z_{AB} and in several others, $\sigma(\underline{r})$ will be a natural result of Z_{AB}.

4.2.3. Quantum Similarity of Atomic Electron Densities

The first functional that we propose to compare the densities $\rho(\underline{r})$ of two atoms A and B is the quantum similarity index launched by Carbo (Carbo et al. 1980). Looking for similarity of molecules, he proposed to minimize the integrated square of the difference in densities $\int |\rho_A - \rho_B|^2 d\underline{r}$ leading to the demand that $\int \rho_A(\underline{r})\rho_B(\underline{r})d\underline{r}$ should be maximal. Normalizing this index leads us naturally to propose a similarity index Z_{AB} defined as

$$Z_{AB} = \frac{\int \rho_A(\underline{r})\rho_B(\underline{r})d\underline{r}}{\left[\left(\int \rho_A^2(\underline{r})d\underline{r}\right)\left(\int \rho_B^2(\underline{r})d\underline{r}\right)\right]^{1/2}}, \tag{12}$$

where Z_{AB} lies between 0 and 1.

Introducing the shape function Z_{AB} via eqn. (10), a particular choice of F_{AB} (eqn. (9)), turns out to be not only a functional of $\rho_A(\underline{r})$ and $\rho_B(\underline{r})$ but also a similar functional of $\sigma_A(\underline{r})$ and $\sigma_B(\underline{r})$:

$$Z_{AB} = \frac{\int \sigma_A(\underline{r})\sigma_B(\underline{r})d\underline{r}}{\left[\left(\int \sigma_A^2(\underline{r})d\underline{r}\right)\left(\int \sigma_B^2(\underline{r})d\underline{r}\right)\right]^{1/2}} \tag{13}$$

In other words, the shape functions of A and B do contain enough information to judge the (dis)similarity of the electronic distribution in A and B.

This approach to compare electron density functions of atoms in our group was first used in 2003, starting from Numerical Hartree-Fock wavefunctions (Borgoo et al. 2004). The result in Figure 8.3 shows the Similarity Index of the noble gas atoms He, Ne, Ar, Kr, Xe, Rn compared with all other atoms until

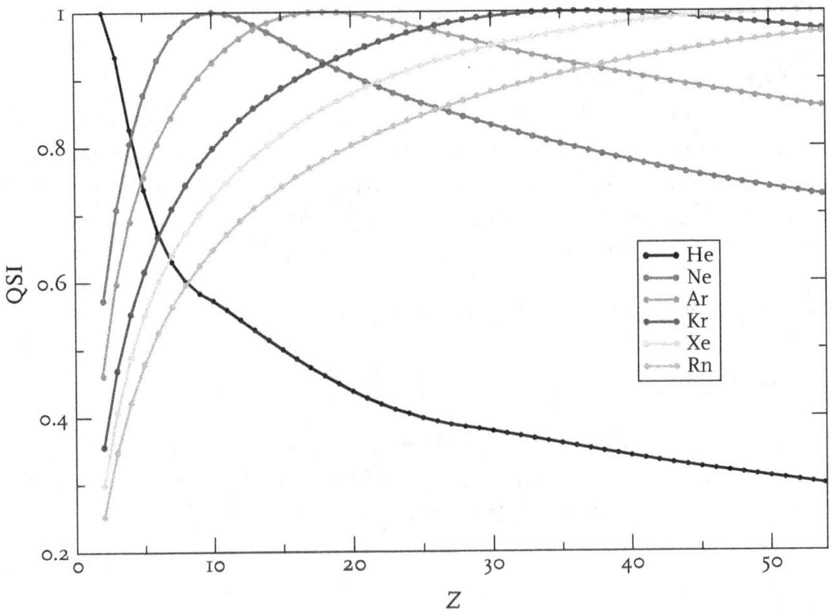

FIGURE 8.3 Quantum Similarity Index for noble gases. Each noble gas corresponds to a curve comprising similarity indices with all other atoms (Z< 54) (PCCP 13, 911(2011)).

Z=54. A nearest neighbor similarity pattern emerges: the most similar atoms to an atom with nuclear charge Z are its neighbors Z-1 and Z+1. Periodicity is completely lacking in this picture and apparently the first functional (cfr. eqn 9) $F_{AB} = F[\rho_A(\underline{r}), \rho_B(\underline{r})] = Z_{AB} = Z_{AB}[\sigma_A(\underline{r}), \sigma_B(\underline{r})]$ isn't capable of retrieving the fundamental aspect of periodicity of the Periodic Table.

We therefore looked at an alternative functional, we borrowed from Information Theory.

4.2.4. An Information Theory based Functional.

Information Theory was pioneered in the studies of the electronic structure of atoms and molecules by Sears, Parr, and Dinur (Sears et al. 1980) and witnessed a real breakthrough in quantum chemistry in the last 15 years through work by Gadre, Sen, Parr, Nalewajski, Esquivel, and Nagy (Gadre 2002, Sen 2005, Nalewajski 2006, Nagy and Parr 1996, Lopez-Rosa 2010, Geerlings and Borgoo 2011) (for a review see Geerlings (Geerlings and Borgoo 2011)). Our group chose to concentrate on the Kullback Leibler (Kullback and Leibler 1951) information deficiency ΔS, introduced in 1951 as a generalization of Shannon's information entropy (Shannon 1948). The Kullback Leibler information deficiency ΔS_{KL} measures the distance in information between two (equally) normalized probability distributions, $p_k(x)$, the one under consideration, and $p_o(x)$, a reference distribution also called "prior" distribution.

Δs_{KL} is then written as

$$\Delta s_{KL}(p_k / p_o) = \int p_k(x) \log \frac{p_k(x)}{p_o(x)} dx \tag{14}$$

and "measures" the information present in $p_k(x)$, distinguishing it from $p_o(x)$.

If we now replace p(x) by the electron density function ρ(r) (putting aside the question of normalization), we can introduce a quantity that characterizes $\rho_A(\underline{r})$, the distance in information between the density of a given atom A and a reference density $\rho_o(\underline{r})$. $\rho_o(\underline{r})$ was taken as the renormalized density of a hypothetical gas atom with equal number of electrons as atom A (see the demand for equal normalization mentioned above).

$$\rho_o(\underline{r})=\frac{N_A}{N_o}\rho_o'(\underline{r}) \quad (15)$$ with $\rho_o'(\underline{r})$ the noble gas atom density distribution.

$$\text{Evidently} \int\rho_o(\underline{r})d\underline{r}=\frac{N_A}{N_o}\int\rho_o(\underline{r}')d\underline{r}=\frac{N_A}{N_o}N_o=N_A \quad (16)$$

so that $\rho_o'(\underline{r})$ and $\rho_A(\underline{r})$ are both normalized to N_A. This choice for the prior is reminiscent of Sanderson's electronegativity scale (Sanderson 1951) based on the compactness of the electron cloud of a given atom

$$ED_A=\frac{Z_A}{\frac{4}{3}\pi r_A^3} \quad (17)$$

where r_A is the covalent radius. Sanderson's electronegativity scale is based on the stability ratio

$$S_A=\frac{ED_A}{ED_o} \quad (18)$$

with ED_o the compactness of a hypothetical noble gas atom with the same number of electrons (18) finally relates electronegativity to a ratio of electron densities. In Figure 8.4 we depict the KL information deficiency analogue

$$\Delta S_A^\rho=\int\rho_A(\underline{r})\log\frac{\rho_A(\underline{r})}{\rho_o(\underline{r})}d\underline{r} \quad (19)$$

(we will drop the KL index from now on). It is clear that as compared to the similarity index (eqn. 13) this functional already reflects periodicity to a much higher extent. In fact, one correction should be added, namely the normalization constraints in the KL expression where a 1-normalization is imposed. As the 1-normalized density is nothing other than the shape function (eqn 11) this step again introduces the shape function in a natural way leading to

$$\Delta S_A^\sigma=\int\sigma_A(\underline{r})\log\frac{\sigma_A(\underline{r})}{\sigma_o(\underline{r})}d\underline{r}. \quad (20)$$

Figure 8.5 shows, in an even more direct way, that not only is periodicity regained but also that the evolution of atomic properties throughout the rows of the periodic table now shows a clear trend, more obvious than in the density representation. The variation in the information discrimination is much larger for second row elements than for third row elements, which is much larger still than for fourth row (which, however, is in the same order as for fifth row).

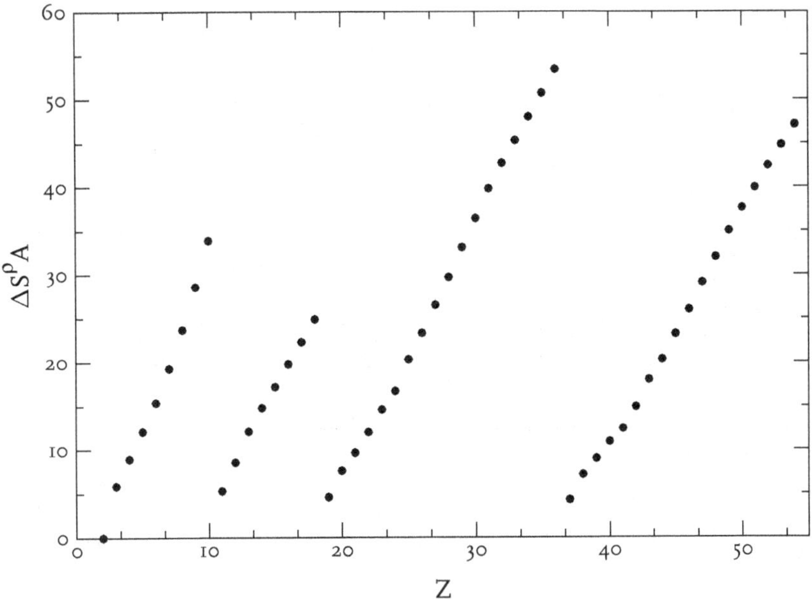

FIGURE 8.4 Information discrimination vs. Z for atomic densities with the noble gas of the previous row as reference (PCCP 13, 911(2011)).

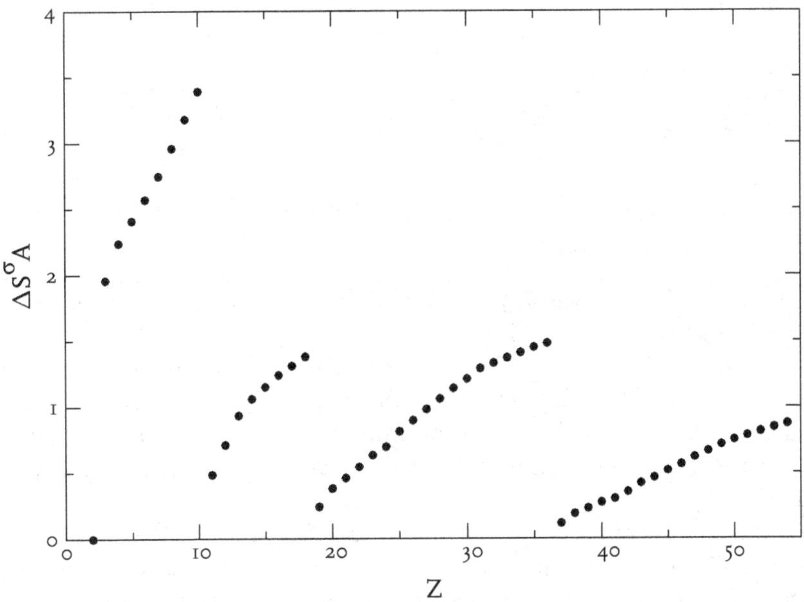

FIGURE 8.5 Information discrimination for atomic shape functions with the noble gas of the previous row as reference (PCCP 13, 911(2011)).

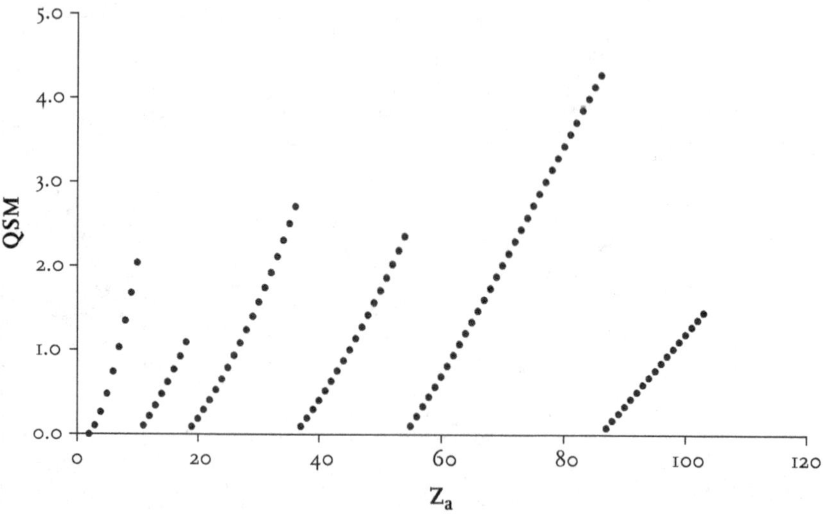

FIGURE 8.6 Cross section of the non-normalized QSM surface (JCP 126, 044102(2007)).

This evolution is in agreement with the "slower" evolution of many properties in a row when going down in the periodic table (e.g., f vs. d elements and d vs. s- and p-block elements).This aspect is moreover "prior" independent, whereas the periodicity retrieval might be subject to some criticism on the prejudice in the choice of the prior. Going one step further and combining the two above mentioned approaches we constructed a similarity measure based on a local version of Kullback's Information Discrimination (Borgoo et al. 2007). We therefore considered

$$\Delta S^{\rho}_{A}(\underline{r}) = \rho_A(\underline{r}) \log \frac{\rho_A(\underline{r})}{\rho_o(\underline{r}) \frac{N_A}{N_o}} \tag{21}$$

as a local analogue of eqn. (19) and plugged it in a normalized similarity index Z_{AB} of the type eqn. (12) with

$$Z_{AB} = \frac{Z'_{AB}}{\sqrt{Z'_{AA}}\sqrt{Z'_{BB}}} \text{ and } Z'_{AB} = \int \Delta s^{\rho}_{A}(\underline{r}) \Delta s^{\rho}_{B}(\underline{r})\, d\underline{r} \tag{22}$$

In this way an Information Theory–based Quantum Similarity Index (ranging from 0 to 1) is obtained. In Figure 8.6 the results of relativistic Dirac Fock calculations (in view of the increasing importance of relativistic effects when higher Z values are considered) are given as the cross-section of Z'_{AB} for $Z_B = 82$.

This cross-section of the Quantum Similarity Index (QSM) surface for Z = 82 (Pb) reflects the periodicity in Mendeleev's table at a relativistic level, and shows again that for each period the QSM gradually increases from the first column to the last one and that the slope of the corresponding curve decreases when going down in the periodic table.

To conclude this Information Theory–based approach we take an alternative view combining information transmitted through both rows and columns. Considering a given row, say Al, Si, P, S, Cl... we can ask ourselves about the information the elements of the row above contain on the elements of this next row. Specifically: what is the (lack of) information contained in the density or shape function of N about Al, Si, P, S, Cl,...? We constructed the following shape function–based functional:

$$F[\sigma_X,\sigma_N;\sigma_P]=\frac{\int \sigma_X(\underline{r})\ln\frac{\sigma_X(\underline{r})}{\sigma_N(\underline{r})}d\underline{r}}{\int \sigma_P(\underline{r})\ln\frac{\sigma_P(\underline{r})}{\sigma_N(\underline{r})}d\underline{r}} \qquad X= \text{Al, Si, P, S, Cl,...} \tag{23}$$

where we compare the information in the shape function of N on P, the element just above it, to that on elements to the left and the right of P. Note that by construction the value of the functional for X=P is equal to 1. Scheme 2 shows that the information content of N on the other elements on the next row is gradually decreasing both to the left and right sides of P. We hereby retrieve a different, "triangular" aspect of periodicity. (The same exercise has been done successfully with the (P, As), (As, Sb), and (Sb, Bi) couples).

				N				
Al	<	Si	<	P	>	S	>	Cl
0.9865		0.9969		1.0		0.9973		0.9903

SCHEME 2 Triangular Aspect of Periodicity regained through the functional $F[\sigma_X,\sigma_N;\sigma_P]$

In conclusion we can state that periodicity can be regained through the density (or shape function)—without invoking the orbital concept—by "reading" the density in different ways such that a combination of Quantum Similarity and Information Theory yields the most attractive results. The role of the "prior" in the first IT based approach, which might introduce a kind of bias, is avoided in the "triangular" construction in the final approach. Comparing two or more atoms via a single functional counteracts the considerations mentioned by Ostrovsky that in the traditional HF type approach only piecewise calculations for individual atoms are involved while an overview of the system's structure is not proved (Ostrovsky 2005).

5. *Perturbing the Electron Density Function: Sharpening Its Information through the Linear Response Function*

In this section we investigate the way the electron density $\rho(\underline{r})$ responds to a perturbation in external potential. This approach might already be closer to the real periodic table because until now, we have presented variations on the theme of "isolated" gas-phase atom densities, whereas the real periodic table is based largely on the "chemistry" of the elements (say their valency and the stoichiometry of their compounds), in which the atom under consideration is (strongly) perturbed by its environment, in this case its reaction partners. Therefore the study of $(\delta\rho(\underline{r})/\delta v(\underline{r}'))_N$, the linear response function, might be rewarding. Until recently this kernel, as it is called mathematically, received little attention except in the physics and chemical physics communities due to the complexity of evaluating it (and probably also because the interpretation of such a six-variable function is far from trivial). Complicating things even more, the frequency-dependent form of the kernel appears in a more prominent way—albeit only in a computational context—to obtain excitation energies (Martin 2004). The chemistry present in $\chi(\underline{r},\underline{r}')$ thus remained "hidden" with some exceptions, the calculations being done at a highly approximate level (Geerlings et al. 2014). Recently several computational techniques to obtain trustworthy but tractable expressions for $\chi(\underline{r},\underline{r}')$ were developed and moreover several representation/visualization schemes were presented by our group both for atoms and molecules (Yang et al. 2012, Sablon et al. 2007, Sablon et al. 2010a, Sablon et al. 2010b, Sablon et al. 2010c, Sablon et al. 2012, Fias et al. 2013, Boisdenghien et al. 2013, Boisdenghien et al. 2014, Fias et al. 2014). In this section we will limit ourselves to the simplest expression for $\chi(\underline{r},\underline{r}')$ which can be obtained via first-order perturbation theory in the context of a single Slater determinantal function of the Hartree-Fock or Kohn-Sham type. Taking the closed shell system as a case study, the advantage of such a function is that the density can easily be represented as

$$\rho(\underline{r}) = 2\sum_{i}^{N/2}\varphi_i^2(\underline{r}) \tag{24}$$

the sum of the squares of the orbitals φ_i constituting the determinant. (The extension to open-shell systems has recently been presented by Boisdenghien et al. 2014). Perturbing $\rho(\underline{r})$ by an external potential variation δv_{ext} (and neglecting the influence of this external potential variation on the Coulomb and exchange potentials; see Yang et al. 2012, Geerlings et al. 2014), one obtains an expression, known in the physics community as the Independent Particle Approximation (Martin 2004, Ayers 2007, Jaworski et al. 1996):

$$\chi(\underline{r},\underline{r}') = \left(\frac{\delta\rho(\underline{r})}{\delta v(\underline{r}')}\right)_N = -4\sum_i\sum_a \frac{\varphi_i(\underline{r})\varphi_a(\underline{r})\varphi_a(\underline{r}')\varphi_i(\underline{r}')}{\varepsilon_a-\varepsilon_i} \tag{25}$$

The summation over "i" runs over the occupied orbitals (i.e., those present in the Slater determinant), and the summation over "a" runs over the unoccupied or virtual orbitals resulting from a Kohn-Sham or Hartree-Fock calculation (Yang et al. 2012, Sablon et al. 2010, Ayers 2007). Expression (25), once calculated however, still asks for adequate representation or visualization as it contains six variables. Whereas for molecules an atom-condensed form is usually adopted (Geerlings et al. 2014, Sablon et al. 2007) we investigated for isolated atoms the non-integrated kernel in as much detail as possible (Boisdenghien et al. 2013, Boisdenghien et al. 2014). Starting from an averaged atomic electron density $\rho(\underline{r}) = \rho(r)$, placing a perturbations anywhere else than in the origin (where we locate the atomic nucleus), say in $\underline{r}'$, then breaks the spherical symmetry along the $\underline{r}'$ axis. However, the rotational symmetry around this line is not broken, leaving us with the cylindrical symmetrical quantity $\chi(r, r', \vartheta)$, which is cylindrically symmetrical around the $\underline{r}$ and $\underline{r}'$ axes so that we are finally left with three variables: r, r′, and the angle ϑ between $\underline{r}$ and $\underline{r}'$. We then integrate out this remaining angular dependence, leaving us with $\chi(r, r')$, which is easily visualized. In Figure 8.7 we plot the radial distribution analogue of $\rho(r)$ $\left(4\pi r^2\rho(r)\right)$, namely $r^2\chi(r,r')r'^2$ for helium and beryllium, the two atoms for which earlier results in the literature were available (Savin et al. 2001). Positive and negative regions appear where it is easily seen what the negative regions should be situated along the diagonal: a positive $\delta v(\underline{r})$ perturbation will indeed lead to an electron depletion in neighboring positions. As moreover it can easily be proven (Liu et al. 2009) that

$$\int \chi(\underline{r},\underline{r}')d\underline{r} = 0, \tag{26}$$

this negative region is compensated by positive regions to the left and right of the diagonal region.

Passing now to the evolution of the plots along a period, Figure 8.8 shows that when going through the period from Li to Ne a gradual evolution takes place, with the general structure being more complicated than for the first-row atoms (second red region close to the axis). When moving along this second period the region between the two red regions close to the axes diminishes abruptly when reaching boron (appearance of the p subshell). When passing from neon to sodium a third red region appears and the overall structure collapses near the axes with the black region between the second and third red region gradually diminishing. In Figure 8.9 we depict the cases He, Ne, Ar, and Kr indicating that the overall structure in the part that is not too close to the axes is highly similar, whereas the region closer to the axes shows increasing

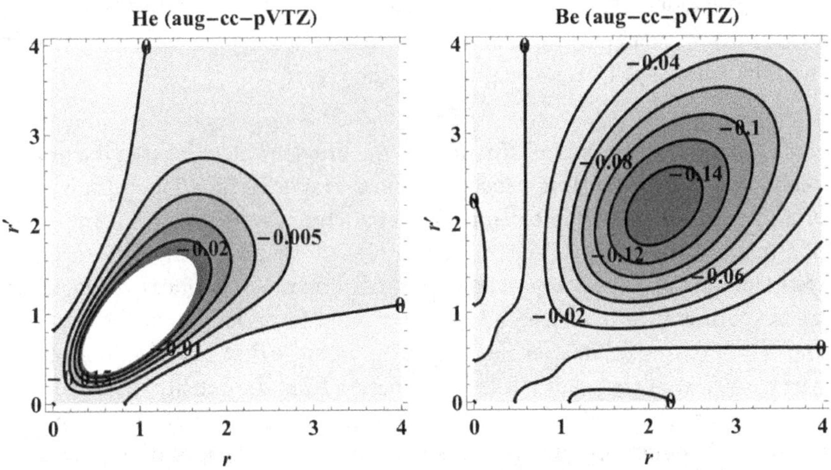

FIGURE 8.7 Contour plots for He and Be. Values, in a.u., obtained at the PBE-DFT level with a aug-cc-pVTZ basis set (J. Chem. Theor. Comp. 9, 1007 (2013)).

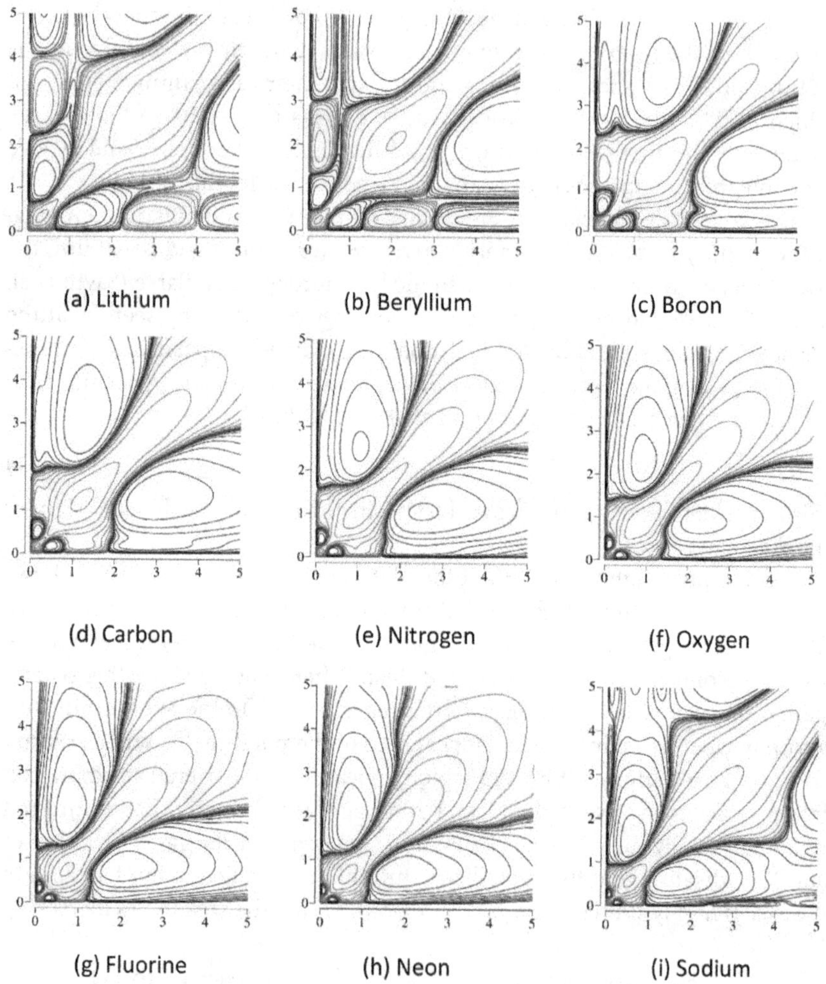

FIGURE 8.8 Contour plots for Li through Na. Values, in a.u., obtained at the PBE-DFT level with a aug-cc-pVTZ basis set (PCCP, 16, 1007 (2013)).

complexity retaining the information on the previous element (see the argon-neon comparison). In this sense the linear response function, without any prejudice, retrieves periodicity (and shell structure). As stated in the introduction this feature is of great chemical importance because it means that when responding to an external (potential) perturbation the differences/analogies in the responses turn out to be intimately related to periodicity without at any moment invoking orbitals, shells, and so on. In these first investigations of this type we observe that "reading" the information from the density yields sharper and more unbiased patterns when the reading is performed on a "perturbed" density, as is done with the linear response function. The idea of perturbation is an important issue when discriminating the "real" periodic table of the elements from the table of gaseous ground state atoms.

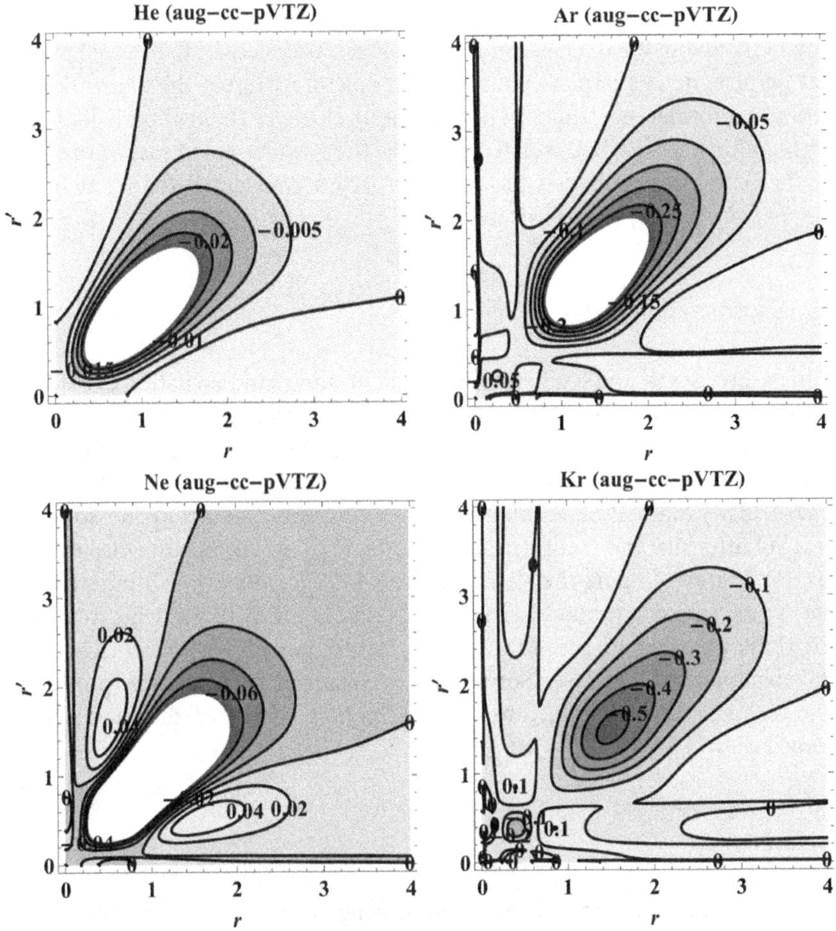

FIGURE 8.9 Contour plots for He, Ne, Ar, Kr. Values, in a.u., obtained at the PBE-DFT level with a aug-cc-pVTZ basis set (J. Chem. Theor. Comp. 9, 1007 (2013)).

6. *Conclusions*

The current "interpretation" of the periodic table is largely based on a wavefunction, orbital-based quantum mechanical approach known as the "orbital paradigm." Using the electron density, a much simpler information carrier which is moreover experimentally observable—or even simpler the 1-normalized density, the shape function—is an alternative approach. The reason either of these approaches may lead to possible successes lies in the Hohenberg-Kohn Theorem, proving that the electron density "determines everything." Reading the information content from the density with sufficient resolution is not obvious as implied by the passage from $\rho(\underline{r})$ to the Radial Distribution Function. The latter still shows deficiencies and is limited to studying only one type of atom at a time. Techniques based on Quantum Similarity and Information Theory permit pairwise comparisons between atoms of two (or even three) elements and determining periodicity with less and less bias. Finally, studying the response of the atomic electron density to external potential perturbations sharpens the periodicity picture without the distraction of orbitals and reference atoms. This moves us closer to the real periodic table, with no further handling isolated atoms in the gas phase, but just as in many of the (chemical, and even physical) properties considered for an element, atoms perturbed by their environment.

7. Acknowledgments

I thank Eric Scerri and Guillermo Restrepo for their kind invitation to contribute to this book. I am indebted to the many coworkers and visitors of the ALGC group in the past 25 years for their contribution to several aspects of conceptual DFT as mentioned in the references which are at the heart of this review. I particularly thank Prof. Frank De Proft, co-director of the group, for so many years of intensive and fruitful collaboration and many stimulating discussions on this chapter. I thank the VUB and the FWO-Flanders for continuous support to my research group and highly appreciates the VUB Strategic Research Program awarded to the ALGC group since January, 2013. Finally Zino Boisdenghien and Diane Sorgeloos are warmly thanked for providing Figure 8.3, and for taking care of the complete styling of the manuscript, respectively.

References

Allen, L.C., Knight, E.T. 2001. The Löwdin challenge: origin of the n+l, m (Madelung) rule for filling the orbital configurations of the periodic table. Int. J. Quant. Chem. *90*, 80–88.

Atkins, P.W., Jones, L., Laverman, L. 2013. *Chemical Principles. The Quest for Insight*, 6th ed. W.H. Freeman, New York.

Atkins, P.W., De Paula, J. 2009. *Atkins' Physical Chemistry*, 9th ed. Oxford University Press, Oxford.

Ayers, P.W. 2000. Density per particle as a description of Coulombic systems. Proc. Natl. Acad. Sci. USA *97*, 1959–1964.

Ayers, P.W., Parr, R.G. 2003. Sufficient condition for monotonic electron density decay in many electron systems. Int. J. Quant. Chem. *95*, 877–881.

Ayers, P.W., Anderson, J.S.M., Bartolotti, L.J. 2005. Perturbative perspectives on the chemical reaction predicting problem. Int. J. Quant. Chem. *101*, 520–534.

Ayers, P.W. 2007. The Physical basis of the hard/soft acid/base principle. Faraday Discuss. *135*, 161–190.

Ayers, P.W., De Proft, F., Geerlings, P. 2007. Comparison of the utility of the shape function and electron density for predicting periodic properties: atomic ionization potentials. Phys. Rev. A *75*, 012508.

Bader, R.F., Essén, H. 1984. The characterization of atomic interactions. J. Chem. Phys. *80*, 1943–1960.

Bartell, L.S., Brockway, L.O. 1953. The investigation of electron distribution in atoms by electron diffraction, Phys. Rev. *90*, 833–838.

Becke, A.D., Edgecombe, K.E. 1990. A simple measure of electron localization in atomic and molecular systems. J. Chem. Phys. *92*, 5397–5403.

Boisdenghien, Z., Van Alsenoy, C., De Proft, F., Geerlings, P. 2013. Evaluating and interpreting the chemical relevance of the linear response function for atoms. J. Chem. Theory Comp. *9*, 1007–1015.

Boisdenghien, Z., Fias, S., Van Alsenoy, C., De Proft, F., Geerlings, P. 2014. Evaluating and Interpreting the Chemical Relevance of the Linear Response Kernel for Atoms II Open Shell. Phys. Chem. Chem. Phys. *16*, 14614–14624.

Borgoo, A., Godefroid, M., Sen, K.D., De Proft, F., Geerlings, P. 2004. Quantum similarity of atoms: a numerical Hartree-Fock and information theory approach. Chem. Phys. Lett. *399*, 363–367.

Borgoo, A. Godefroid, M., Indelicato, P., De Proft, F., Geerlings, P. 2007. A quantum similarity study of atomic density functions: insights from information theory and the role of relativistic effects. J. Chem. Phys. *126*, 044102.

Boyd, R.J. 1976. Shell structure of atoms. J. Phys. B 9, L69–L72.

Bransden, B.H., Joachain, C.J. 2003. *Physics of Atoms and Molecules*, 2nd ed., Pearson Education, Harlow, UK.

Carbo, R., Leyda, L., Arnau, J. 1980. How similar is a molecule to another: an electron density measure of similarity between 2 molecular structures. Int. J. Quant. Chem. *17*, 1185–1189.

Chermette, H. 1999. Chemical reactivity indexes in density functional theory. J. Comp. Chem. *20*, 129–154.

Eickerling, G., Reiher, M. 2008. The shell structure of atoms. J. Chem. Theor. Comp. 4, 286–290.

Eyring, H., Walter, J., Kimball, G.E. 1944. *Quantum Chemistry*. John Wiley, New York.

Fias, S., Boisdenghien, Z., Stuyver, T. Audiffred, M., Merino, G., Geerlings, P., De Proft, F. 2013. Analysis of aromaticity in planar metal systems using the linear response kernel. J. Phys. Chem. A *117*, 3556–3560.

Fias, S., Boisdenghien, Z., De Proft, F., Geerlings, P. 2014. The spin polarized linear response from density functional theory: theory and application to atoms. J. Chem. Phys. *141*, 184107.

Fukui, K., Yonezawa, Y, Shingu, H. 1952. A molecular orbital theory of reactivity in aromatic hydrocarbons. J. Chem. Phys. *20*, 722–725.

Gadre, S. 2002. Information theoretical approach to quantum chemistry. In: *Reviews of Modern Quantum Chemistry*, vol. 1, K.D. Sen, ed. World Scientific Publishing Company, Singapore, 108–147.

Geerlings, P., De Proft, F., Langenaeker, W. 2003. Conceptual density functional theory. Chem. Rev. *103*, 1793–1873.

Geerlings, P., De Proft, F., Ayers, P.W. 2007. Chemical reactivity and the shape function. In: *Theoretical Aspects of Chemical Reactivity*, A. Toro Labbé, ed. Elsevier, Amsterdam.

Geerlings, P. De Proft, F. 2008. Conceptual DFT: the chemical relevance of higher response functions. Phys. Chem. Chem. Phys. *10*, 3028–3042.

Geerlings, P., Borgoo, A. 2011. Information carriers and (reading them through) information theory in quantum chemistry. Phys. Chem. Chem. Phys. *13*, 911–922.

Geerlings, P., Fias, S., Boisdenghien, Z., De Proft, F. 2014. Chemistry from the linear response function. Chem. Soc. Rev., *43*, 4989–5008.

Hartree, D.R. 1957. *Calculation of Atomic Structures*. Wiley, New York and references therein.

Hohenberg, P., Kohn, W. 1964. Inhomogeneous electron gas. Phys. Rev. B *136*, 864–871.

Jaworski, C., Casida, M., Salahub, D.R. 1996. Dynamic polarizabilities and excitation spectra from a molecular implementation of time-dependent response theory: N-2 as a case study. J. Chem. Phys. *13*, 5134–5147.

Koch, W., Holthausen, M.C. 2001. *A Chemist's Guide to Density Functional Theory*, 2nd ed. Wiley-VCH, Weinheim.

Kohn, W., Sham, L.J. 1965. Self consistent equations including exchange and correlation effects. Phys. Rev. A *140*, 1133–1138.

Kohn, W., 1999. Nobel Lecture: electronic structure of matter—wave functions and density functionals. Rev. Mod. Phys. *71*, 1253–1266.

Kullback, S., Leibler, R.A.1951. On information and sufficiency. Ann. Math. Stat. *22*, 79–86.

Levine, I.N. 2014. *Quantum Chemistry*, 7th ed., Prentice Hall, New York.

Liu, S., Li, T., Ayers, P.W. 2009. Potential philicity and potential phobicity: reactivity indicators for external potential changes from density functional reactivity theory. J. Chem. Phys. *131*, 114106.

Lopez-Rosa, S., Esquivel, R.O., Angulo, J.C., Antolín, J., Dehesa, S., Flores-Gallegos, N. 2010. Fisher information study in position and momentum spaces for elementary chemical reactions. J. Chem. Theory Comput. *6*, 145–154.

Löwdin, P.O. 1969. Some comments on the periodic system of the elements. Int. J. Quant. Chem. *111S*, 331–334.

Martin, R.M. 2004. *Electronic Structure: Basic Theory and Practical Methods.* Cambridge University Press, Cambridge Chapter 4.

McWeeny, R., Sutcliffe, B.T. 1969. *Methods of Molecular Quantum Mechanic.* Academic Press, London.

Mulder P. 2011. Are orbitals observable? HYLE, *17*, 24–35 and references therein.

Mulliken, R.S. 1934. A new electroaffinity scale; together with data on valence states and on valence ionization potentials and electron affinities. J. Chem. Phys. *2*, 782–793.

Nagy, A., Parr, R.G. 1996. Information entropy as a measure of the quality of an approximate electronic wave function. Int. J. Quant. Chem. *58*, 323–327.

Nalewajski, R.F. 2006. *Information Theory of Molecular System*, Elsevier Science, New York.

Ostrovsky, V.N. 2001. What and how physics contributes to understanding the periodic law, Found. Chem. *3*, 145–182.

Ostrovsky, V.N. 2005. On recent discussion concerning quantum justification of the periodic table of the elements. Found. Chem. *7*, 235–239.

Parr, R.G., Donnelly, R.A., Levy, M., Palke, W.E. 1978. Electronegativity—density functional viewpoint. J. Chem. Phys. *66*, 3801–3807.

Parr, R.G., Pearson, R.G. 1983. Absolute hardness: companion parameter to absolute electronegativity. J. Am. Chem. Soc. *105*, 7512–7516.

Parr, R.G., Bartolotti, L.J. 1983. Some remarks on the density functional theory of few-electron systems. J. Phys. Chem. *87*, 2810–2815.

Parr, R.G., Yang, W.T. 1984. Density functional-approach to frontier-electron theory of chemical reactivity. J. Am. Chem. Soc. *106*, 4049–4050.

Parr, R.G., Yang, W.T. 1989. *Density Functional Theory of Atoms and Molecules*. Oxford University Press, New York.

Parr, R.G., Yang, W.T. 1995. Density functional theory of the electronic structure of molecules. Annu. Rev. Phys. Chem. *46*, 701–728.

Parr, R.G. 2005. Some late-term thoughts of a density functional theorist. J. Chem. Sci. *117*, 613–615.

Pearson, R.G. 1997. *Chemical Hardness*. Wiley-VCH, Weinheim.

Pyykkö, P. 2010. A suggested periodic table up to Z=172 based on Dirac-Fock calculations on atoms and ions. Phys. Chem. Chem. Phys. *10*, 161–168.

Sablon, N., De Proft, F., Ayers, P.W., Geerlings, P. 2007. Computing Fukui functions without differentiating with respect to electron number. II. Calculation of condensed molecular Fukui functions. J. Chem. Phys. *126*, 224108.

Sablon, N., De Proft, F., Geerlings, P. 2010a. The linear response kernel: inductive and resonance effects quantified. J. Phys. Chem. Lett. *1*, 1228–1234.

Sablon, N., De Proft, F., Ayers, P.W., Geerlings, P. 2010b. Computing second-order functional derivatives with respect to the external potential. J. Chem. Theory Comput. *6*, 3671–3680.

Sablon, N., De Proft, F., Geerlings, P. 2010c. The linear response kernel of conceptual DFT as a measure of electron delocalization. Chem. Phys. Lett. *498*, 192–197.

Sablon, N., De Proft, F., Sola, M., Geerlings, P. 2012. The linear response kernel of conceptual DFT as a measure of aromaticity. Phys. Chem. Chem. Phys. *14*, 3960–3967.

Sanderson, R.T. 1951. An interpretation of bond lengths and a classification of bonds. Science *114*, 670–672.

Savin, A., Colonna, F., Allavena, M. 2001. Analysis of the linear response function along the adiabatic connection from the Kohn-Sham to the correlated system. J. Chem. Phys. *115*, 6827–6833.

Scerri, E.R. 2001. The recently claimed observation of atomic orbitals and some related issues. Phil. Sci., *68*, 576–588.

Scerri, E.R. 2006. Commentary on Allen and Knight's response to the Löwdin challenge. Found. Chem. *8*, 285–292.

Scerri, E.R., 2007. *The Periodic Table: Its Story and Its Significance*. Oxford University Press, New York, xiii.

Scerri, E.R. 2012. What is an element? What is the periodic table? And what does quantum mechanics contribute to the question? Found. Chem. *14*, 69–81.

Sears, S.B., Parr, R.G., Dinur, U. 1980. On the quantum-mechanical kinetic energy as a measure of the information in a distribution. Isr. J. Chem. *19*, 165–173.

Sen, K.D., Slamet, M., Sahni, V.1993. Atomic shell structure in Hartree-Fock theory. Chem. Phys. Lett. *205*, 313–316.

Sen K.D. 2005. Characteristic features of the Shannon information entropy of confined atoms. J. Chem. Phys. *123*, 074110.

Sen, K.D., Gayatri, T.V., Krisnaveni, R. Kakkar, M. Toufar, H., Janssens, G.O.A, Backclandt, B.G., Schoonheydt, R.A., Mortier, W.J. 1995. Average local electrostatic potential and the core-valency separation in atoms. Int. J. Quant. Chem. *56*, 399–408.

Shannon, C.E. 1948. A mathematical theory of communication. Bell Syst. Tech. *27*, 379–423.

Sperber, G. 1971. Analysis of reduced density matrices in the coordinate representation ii. The structure of closed shell atoms in the restricted Hartree-Fock approximation. Int. J. Quant. Chem. 5, 189–214.

Toussant, M. 2009. A scientific milestone. Chem. Eng. News, *3*, September 14.

Waber, J.T., Cramer, D.T. 1965. Orbital radii of atoms and ions. J. Chem. Phys. *42*, 4116–4123.

Wang, W.P., Parr, R.G. 1977. Statistical atomic models with piecewise exponentially decaying electron—density, Phys. Rev. A *16*, 891–902.

Wang, S.G., Schwarz, W.H.E. 2009. Icon of chemistry: the periodic system of chemical elements in the new century. Ang. Chem. Ent. Ed. *48*, 2–14.

Weinstein, J., Politzer P., Srebrenik, S. 1975. Misconception concerning electrons density distribution of an atom. Theor. Chim. Acta *38*, 159–163.

Yang, W.T., Cohen, A.J., De Proft, F., Geerlings, P. 2012. Analytical Evaluation of Fukui functions and real-space linear response function. J. Chem. Phys. *136*, 144110.

Zuo, J.M., Kim, M., O'Keeffe, M., Spence, J.C.H. 1999. Direct observation of d-orbital holes and Cu–Cu bonding in Cu_2O. Nature *401*, 49–52.

CHAPTER 9

Resemioticization of Periodicity

A Social Semiotic Perspective

YU LIU
College of International Education, Sichuan International Studies University, China

Chemical Periodicity and Its Diverse Visual Representations

Chemical periodicity is arguably one of the most important ideas in science, and it has profoundly influenced the development of both modern chemistry and physics (Scerri 1997, 229). While the definition of periodicity has remained largely stable in the past 150 years,[1] the periodic system has been visualized in a wide range of forms including (to name just a few) tables, spirals, and zigzags. Furthermore, information technology makes it much easier, and offers innovative ways, to produce new versions of periodic depictions (e.g., WebElements (Winter 1993)).

The multitude of periodic visualizations arouses growing interest among scholars with different academic backgrounds. For instance, educational researchers and practitioners (e.g., Waldrip et al. 2010) wrestle with the question of which visual representation will most effectively help students master the subject content of periodicity. Likewise, philosophers tend to identify the ultimate display of the periodic system, which they use as evidence to support a realistic view of periodicity (Scerri 2007, 21).[2]

Other researchers, however, take a different attitude toward the stunning diversity of periodic depictions. In a seminal paper, Marchese (2013) examines the visualization of periodicity at different stages of history from the perspectives of tabular, cartographic, and hypermedia design. His analysis illuminates the

---

[1] One exceptional change is that atomic numbers are identified as the ordering principle for the elements instead of atomic weights.

[2] However, Professor Scerri (Personal communication, 2013) recently abandoned the idea that there is one preexisting true table to be discovered.

periodic table's plasticity and endeavors to justify the constant transformation of the periodic displays as a necessary means to meet scientists' changing needs.

While all these studies generally emphasize the importance of periodic depictions in scientific research and education, they tend to give primacy to the notion of "periodic system."[3] By contrast, the periodic table seems to play a secondary role, which either passively reflects the chemical law or responds to the evolving knowledge of chemical elements. Such a view runs the risk of underestimating the significant function of the periodic table as a productive research tool, one which enabled Mendeleev to successfully predict the existence and the properties of undiscovered elements such as germanium in 1869 (Kibler 2007, 222).

It is important to note that science and technology are "both *material* and *semiotic* practices" (Halliday 1998, 228, italics in original). For instance, reading and writing texts are found to be equally as important as making experiments with sophisticated instruments at the laboratory as a means for scientists to undertake research and participate in a scientific community (Pearson et al. 2010, 460). It therefore follows that effective engagement with periodic depictions constitutes one essential aspect of the scientific research on the periodic system.

This chapter aims to further clarify the relationship between chemical periodicity and its diverse displays, and explores the dynamic nature of the periodic table from a social semiotic perspective. The next section introduces several key tenets of social semiotics before proposing a theoretical framework. The framework is then used to analyze two selected periodic tables designed by the great chemist John Newlands (1837–1898). Based on the social semiotic analysis, the present study attempts to demonstrate the interaction between periodicity and its ever-changing depictions and discusses Newlands' contributions to the modern periodic table.

Key Tenets of Social Semiotics

Social semiotics identifies language as the means to explore the central question "how do human beings construe experience?" (Halliday and Matthiessen 1999, 1). From a functional perspective, the semiotic system of language plays a constitutive role in interpreting material and social phenomena as experience when people exchange meanings by systematically choosing linguistic signs in specific cultural and situational contexts. As a comprehensive theory, social semiotics develops sets of principles. Here only the key tenets of stratification,

[3] According to Scerri (2007, 18), the periodic system is an abstract notion describing the fundamental relationship among chemical elements, which can be concretely visualized in the form of periodic tables. This chapter uses the notion "the periodic system" to stand for a material phenomenon, which can be interpreted as a scientific concept "chemical periodicity" referring to a set of semantic patterns from a social semiotic perspective. More discussions can be found in the section on methodology.

metafunction, and multimodality are briefly introduced, as they are most relevant to the current study.

In order to show how human experience is semiotically constructed, social semiotics treats social context and language as two complementary semiotic strata, which operate interdependently at different levels of abstraction. Namely, social context and language determine each other in a probabilistic rather than a categorical manner. On the one hand, social context at a more abstract level sets conditions for meaning making through linguistic choices. For example, when scientists at a lab meeting talk about "alcohol," they will most probably refer to it as a homologous series of hydrocarbons in which a hydrogen atom is replaced by a hydroxyl group (–OH) (Earl and Wilford 2014, 233) rather than beverages that can make people drunk. On the other hand, language patterns at a more concrete level play an important role in building up social context. For instance, the term "sodium chloride" seems to be more closely associated with the specialized field of chemistry than with daily life experiences such as cooking.

Furthermore, language is internally stratified as the strata of semantics, grammar,[4] and phonology (in speech) or graphology (in written language) at increasingly concrete levels. The multilevel stratification model reveals the complex interplay between language and experiential construal. In this bidirectional process, human experience is transformed into meaning, which is further transformed into words, and finally transformed into sounds or graphics, and vice versa (Halliday and Matthiessen 2004, 24–26). For example, a change (e.g., the gain of electrons) is typically constructed as the semantic category of a process in the grammatical unit of a verb (e.g., "reduce"), which can be read out or written down. It is important to note that the grammar in a stratified system has the power of transforming experience into new meaning patterns (Halliday 1998, 55). For instance, when the gain of electrons is reworded as the noun "reduction," its scope of meaning becomes broader as the semantic junction of both a process and a virtual entity (Liu and Owyong 2011, 825). The added "entity" meaning makes it possible for scientists to set up technical taxonomies in modern chemistry (e.g., "reduction" is semantically in contrast with "oxidization" because reduction is the reverse of the process of oxidization).

Second, language simultaneously performs three metafunctions (i.e., generalized functions; see Halliday and Matthiessen 2004, 29–31): to construct a presentation of "goings-on" or relationships in the world (ideational meaning); to take a stance toward the presentation and to the reader/viewer (interpersonal meaning); and to organize related elements into a coherent whole (textual meaning). For example, the utterance "In the periodic table, metals and non-metals can be divided by a staircase line" bears three types of meaning at

[4] According to Halliday and Matthiessen (1999, 5–6), the concept of grammar in social semiotics also includes the phenomenon of lexis in the traditional sense, which is regarded as the most delicate end of grammar.

the same time. First, it construes the ideational meaning of "doing." Interpersonally, the speaker's proposition appears to be highly arguable, as the modal verb "can" indicates a low degree of probability. Textually, the prepositional phrase "in the periodic table" gains a special prominence, for it serves as the point of departure of the message. Apart from language, other semiotic resources are also assumed to construe ideational, interpersonal, and textual meaning (Kress et al. 2001, 13). The metafunction principle therefore provides a common platform for conceptualizing all forms of communication.

Since the early 1990s, the social semiotic approach to language (also known as systemic functional linguistics) has been usefully extended to investigate meaning-making patterns in individual modes other than language, including displayed art (O'Toole 1994) and visual design (Kress and van Leeuwen 2006); and the semantic orchestration across different semiotic resources such as image-text relations in mass-media communication (e.g., Liu and O'Halloran 2009) and students' co-deployment of verbal, diagrammatic, and gestural representations in their explanations of scientific phenomena (e.g., Tang 2013). These studies have led to an emerging theory of multimodality.

Pioneering linguistic research has made progress toward the representational nature of scientific knowledge. For example, by analyzing the meaning-making patterns of the canonical essays written by Newton and Priestly, Halliday (1993, 67) observes that highly metaphorical language provided these great scientists with the necessary semiotic resource to build up a different form of consciousness and develop new theories in the physical sciences.

Multimodal research has also made preliminary efforts to account for chemical periodicity. Inspired by Lemke's (1998) functional interpretation of scientific discourse and O'Toole's (1994) multimodal theories of displayed art, Liu and Dwi-Nugroho (2012) propose a framework for analyzing the visual design of periodic tables. Their analysis indicates that the visual resource such as alignment between cells in the periodic table provides clues for readers to explore the designers' different views of the relationship among chemical elements—for instance, whether the elements have the potential for transmutation (Liu and Dwi-Nugroho 2012, 144).

Toward a Theoretical Framework of Resemioticization

It is important to note that both the design and the content of the periodic table are constantly changing (Marchese 2013, 31–32). While existing social semiotic frameworks are useful to account for the functionality of individual periodic displays, further research is needed to reveal the dynamic nature of periodicity. For this purpose, this section draws on Iedema's (2003) notion of resemioticization and develops a framework for analyzing the transformative dynamics of periodicity.

According to Iedema (2003), existing multimodal research primarily analyzes semiotic objects as finished products in terms of their modal complexity, but neglects the need to examine representational artifacts as an ongoing process of resemioticization to trace how and why certain communicational forms are selected to make meanings at different stages. Taking a resemioticization perspective, Iedema (2003) compares two Apple illustrations produced in 1992 and 1999, respectively. He observes that the seven-year transition involves a change of the semiotic landscape from a "linguistic technical" form to a "visual simple" form; these do not have the same meaning. Rather, the crossmodal transposition has led to increasing simplicity and accessibility, thereby effectively helping Apple build a new relationship with its customers and reconfigure its user-friendly ethos (Iedema 2003, 45–47).

Despite a useful lens, the notion of resemioticization remains programmatic, which makes it difficult to offer a systematic account of semiotic practices. To address this issue, it seems that metafunction, stratification, and multimodality—the three key tenets of social semiotics introduced earlier—are compatible with the concept of resemioticization and provide a solid departure point for developing a theoretical framework for analyzing the semiotic (re)making process of periodicity.

First of all, resemioticization entails transpositions from one mode to another, or from one set of modes to another, so a multimodal analysis is not only necessary but also important to illuminate the semiotic complexity constructed in the dynamic process. For instance, periodic displays can be considered multimodal ensembles in which different visual forms are nested with language (e.g., the names of chemical elements), chemical notations (e.g., elemental symbols), and mathematical symbolism (e.g., the atomic number), all of which contribute to the shaping of periodicity in a componential manner.

In addition, resemioticization is referred to as "socially situated meaning-making processes" (Iedema 2003, 30), and thus follows the metafunction principle. The metafunction hypothesis is especially effective in explaining why chemical periodicity takes a wide range of visual forms. As Kress and van Leeuwen (2006, 79) point out, different visuals have their own functional specialization. Therefore, no single depiction can make all the meanings chemical periodicity needs to transmit.

Third, given that social context and modal expressions play an equally important role as the semantic dimension in understanding resemioticization (Iedema 2003, 49), the stratification tenet is useful in capturing the varied dimensions of periodicity and exploring their relationships as chemical periodicity unfolded as a semiotic practice. Synthesizing the ideas of multimodality, metafunction, and stratification with Iedema's notion of resemioticization, this research proposes a social semiotic framework for analyzing modal transpositions, as presented in Figure 9.1.

At the bottom of this generalized resemioticization hierarchy is the stratum of expression where modalities are diversified at the most concrete level. Moving up the hierarchy is the stratum of grammar, which, however, is no longer treated as a set of formal linguistic rules of correctness, but now refers to "the structures of relations of elements in a specific mode and between

Field	F_1	Recontextualizing →	F_n
Semantics	S_1	Resemanticizing →	S_n
Grammar	G_1	Regrammaticizing →	G_n
Expression	E_1	Resemioticizing →	E_n

FIGURE 9.1 A social semiotic framework for analyzing resemioticization.

modes"(Kress et al. 2001, 12). For example, compositional symmetry is identified as a specific pattern of visual grammar to set up taxonomic relationships among participants (Kress and van Leeuwen 2006, 87).

Farther up the hierarchy is the more abstract stratum of semantics. Following the metafunction principle, there are three kinds of meanings at this level. However, in this study we restrict the discussion to the ideational meaning—not because it is more important than the other two meanings, but because ideational meaning has drawn most of the attention from both scientists and science educators.

The top level of the hierarchy is the most abstract stratum of field, a contextual dimension concerning the nature of social activities (Martin and Rose 2008, 11). For example, in contrast with commonsense knowledge, scientific knowledge is a specialized field (Martin 2007, 34). With respect to periodicity, this field is constituted by a variety of scientific theories and empirical research on chemical elements such as reactivity, valency, quantum mechanics, bonding patterns, atomic weights, and atomic numbers, to name just a few.

At every stratum of the hierarchy lies an arrow, visually representing the transformative dynamics of resemioticization. Keep in mind that there are four arrows in total, indicating that resemioticization does not merely occur at the expression stratum, although crossmodal transpositions at this level are the most noticeable. Rather, semiosis also operates simultaneously at the increasingly abstract levels of grammar, semantics, and field.

Drawing on Halliday's (1998) framework for analyzing language evolution in scientific discourse, the transformations at the four strata of the hierarchy in Figure 9.1 are referred to as the processes of resemioticizing, regrammaticizing, resemanticizing, and recontextualizing, respectively, to emphasize the process of resemioticization as a multilevel operation. Therefore, when semiotic artifacts such as the periodic table are shifted from one form of expression (E_1) to other forms (E_n),[5] analysis of resemioticization needs to take into account the corresponding grammatical changes (from G_1 to G_n), semantic changes (from S_1 to S_n), and contextual changes (from F_1 to F_n).

[5] The subscript "n" represents an integer greater than or equal to 2.

Methodology

The present study mainly analyzes two periodic tables, both of which were designed by John Newlands, one of the pioneers of the modern periodic system. Figure 9.2 is Newlands' first visual classification of elements (published in 1864) while Figure 9.3 is his famous display illuminating the law of octaves (presented two years later). The tables are selected for analysis to serve two interrelated purposes.

It seems that Newlands' contemporary scholars had an ambivalent attitude toward his research on the periodic system. On the one hand, Newlands' work was recognized by top chemists like Dmitri Mendeleev[6] and Newlands was awarded the prestigious Davy Medal in 1887 (Scerri 2007, 80–82). It therefore follows that analysis of Newlands' earlier subsidiary table and his more mature visual design may suffice to illustrate the transformative dynamics of periodicity. On the other hand, Newlands' law of octaves was dismissed and even ridiculed by other scientists (Scerri 2007, 78). A social semiotic account of

Member of a group		*One element immediately*			*Difference*
Having lowest equivalent		*Above the preceding one*			*H = 10 = 16*
Magnesium	24	Calcium	40	16	1
Oxygen	16	Sulphur	32	16	1
Lithium	7	Sodium	23	16	1
Carbon	12	Silicon	28	16	1
Fluorine	19	Chlorine	35.5	16.5	1.031
Nitrogen	14	Phosphorus	31	17	1.062

FIGURE 9.2 Newlands' first table of 1864. Remade from Scerri 2007, 74).

	No.		*No.*		*No.*		*No.*		*No.*		*No.*		*No.*		*No.*
H	1	F	8	Cl	15	Co & Ni	22	Br	29	Pd	36	I	43	Pt & Ir	50
Li	2	Na	9	K	16	Cu	23	Rb	30	Ag	37	Cs	44	Os	51
G	3	Mg	10	Ca	17	Zn	24	Sr	31	Cd	38	Ba & V	45	Hg	52
Bo	4	Al	11	Cr	18	Y	25	Ce & La	32	U	39	Ta	46	Tl	53
C	5	Si	12	Ti	19	In	26	Zr	33	Sn	40	W	47	Pb	54
N	6	P	13	Mn	20	As	27	Di & Mo	34	Sb	41	Nb	48	Bi	55
O	7	S	14	Fe	21	Se	28	Ro & Ru	35	Te	42	Au	49	Th	56

FIGURE 9.3 Newlands' table illustrating the law of octaves in 1866. Remade from Scerri (2007, 79).

6 According to Scerri (2007, 80), Mendeleev considered Newlands' research on periodicity more valuable than that of Lothar Meyer.

his periodic displays will thus provide further clues for addressing the question of whether Newlands deserved credit for about 20 years' efforts to "establish his priority in arriving at the first successful periodic system" (Scerri 2007, 80), although Mendeleev's name is synonymous with the periodic system (Scerri 2007, 101).

Following Lemke (1990), a scientific concept is a specialized semantic pattern recognized and shared by a scientific community. In the stratified model of Figure 9.1 for analyzing resemioticization, the concept of chemical periodicity can thus be assumed to operate at the stratum of semantics as a meaning construct, which interacts with the stratum of field as scientific knowledge of chemical elements at the more abstract level and with semiotic expressions at the more concrete levels.

So periodicity can be explored both with a top-down approach by investigating how it was historically shaped against a backdrop of social contexts and with a bottom-up approach by examining how the semiotic transition of visual displays reconstructed periodicity. Considering that in-depth philosophical research (e.g., Scerri 2007) has elaborated on the historical evolution of periodicity, complementary studies are needed to illuminate its representational dimensions.

This research starts the analysis from the expression stratum of periodicity and highlights a grammatical account of the two selected periodic displays given that grammar functions as the powerhouse to produce meaning and construe knowledge (Halliday and Martin 1993, 12). At the same time, our discussion is not confined to the strata of expression and grammar. Rather, drawing on Liu and Dwi-Nugroho's (2012) social semiotic interpretation of tabular modes, the present study considers resemioticization of periodicity as a four-level operation as presented in Figure 9.1. The detailed semiotic analysis aims to explore what new meaning patterns were created to update scientific knowledge on the classification of chemical elements when Newlands redesigned his periodic tables.

Analysis of Resemioticization

Expression Features of Figures 9.2 and 9.3

Although both displays take a tabular form, they show significant differences at the expression stratum. In terms of the multisemiotic landscape, Figure 9.2 is dominated by language in that Newlands not only used long nominal groups (e.g., "Member of a group having lowest equivalent") as column heads in the table proper, he also listed the chemical elements by their full names. By contrast, Figure 9.3 mainly consists of chemical and mathematical symbols with minimal use of language. Also noteworthy is that the two tables employ different numerical notations: cardinal numbers in Figure 9.2 but ordinal numbers in Figure 9.3.

From a social semiotic perspective, the two changes at the expression stratum have significant effects on the semantic dimension of periodicity.

Compared with linguistic terms, chemical symbolism tends to encode more specialized meaning (Liu and Owyong 2011, 829). For example, it will be more challenging for a novice learner to understand the symbol "Na" in Figure 9.3 as representing a chemical element than the word "sodium" in Figure 9.2, for the former has a Latin origin (i.e., "natrium").

In addition, by deleting all the elements' atomic weights, Newlands seemed to have presumed that the target readers of his octaves table must have learned the numerical values of atomic weights and the relevant physical theories. It therefore follows that periodicity developed a higher level of technicality when it was resemioticized from Figure 9.2 to Figure 9.3.

Furthermore, the modal shift from cardinal numbers to ordinal numbers indicates that in Figure 9.2 Newlands employed arithmetic notations with the aim of describing quantitative attributes of the chemical elements whereas in Figure 9.3 his main purpose in using numbers changed to place the elements in an ordered sequence.

Grammatical Patterns of Figure 9.2

Apart from the change of the semiotic landscape, Newlands' two tables present different grammatical patterns. Following Liu and Dwi-Nugroho (2012), a scientific table can be analyzed in terms of its constituents at four grammatical levels. The basic constituent is a *cell* in the form of a square or a rectangle, which contains linguistic and symbolic expressions at the lower rank of *part* such as numbers representing the elements' atomic weights. Additionally, cells can be stacked with each other to form a larger unit at the higher rank of *episode* while episodes make up a *table* at the highest rank.

Following a functional standpoint, cells need to be aligned meaningfully though they appear to be "basic building blocks" in any geometrical pattern (Kress and van Leeuwen 2006, 54). Parallels can be found in language. For example, when the three words "the," "periodic," and "table," are put together to form a word group as a larger unit, there will be six possible combinations, including "table periodic the" and "periodic the table." However, only the fixed sequence of "the periodic table" is grammatically acceptable in that "table" as the element of a thing is followed by "the" and "periodic" as the elements of deictic and classifier, respectively (Halliday and Matthiessen 2004, 312).

Likewise, inter-cell relation (i.e., semantic links between cells; see Liu and Dwi-Nugroho 2012, 144–145) plays a crucial role in the grammatical organization of visual rectangles representing chemical elements. For instance, in Figure 9.2 cells in the same horizontal row present two significant semantic patterns. The first is the inter-cell relation of hyponymy (i.e., class/subclass relation; see Martin 1992, 297–299) in that elements such as magnesium and calcium share a range of chemical properties and thus both of them are co-hyponyms of alkali earth metals from a modern standpoint. The second pattern is the inter-cell relation of cycles (Martin 1992, 303)—like the relationship between Monday and Tuesday in a week, because elements in the same row (e.g., lithium and sodium in Row 3) are ordered with a regular interval of about 16 atomic equivalents[7].

The two semantic relations provide the underlying principle for organizing the cells representing elements into six horizontal rows at the higher rank of episode. From a modern periodic table perspective, these episodes in Figure 9.2 correspond to six of the eight total main groups of elements.[8] Although each episode was far from a complete group (just a fraction), Newlands' first table of 1864 provided a preliminary yet powerful classification scheme to accommodate most of the already discovered main-group elements.

A similar table (Figure 9.4) was independently designed by another distinguished chemist, William Odling (1829-1921) in the same year. Obviously, both Newlands and Odling realized that quite a few pairs of analogous elements had a difference in atomic weights of about 16. According to Scerri (2007, 74), Figure 9.4 seems to have an advantage over Figure 9.2 because Odling recognized ten such cases in contrast to Newlands' six. However, further scrutiny is needed to compare the two tables.

First, Newlands actually discovered at least eight sets of similar elements with a gap of 16 in atomic weights. In addition to the six pairs listed in Figure 9.2, he clearly realized that such relationships also existed in two more sets of elements (i.e., sodium and potassium, lithium and magnesium). This finding was recorded in Newlands' other table (Scerri 2007, 75), which, together with

Cl	–	F	or	35.5	–	19	=	16.5
K	–	Na		39	–	23	=	16
Na	–	Li		23	–	7	=	16
Mo	–	Se		96	–	80	=	16
S	–	O		32	–	16	=	16
Ca	–	Mg		40	–	24	=	16
Mg	–	G		24	–	9	=	15
P	–	N		31	–	14	=	17
Al	–	B		27.5	–	11	=	16.5
Si	–	C		28	–	12	=	16

FIGURE 9.4 Odling's third table of 1864. Remade from Scerri (2007, 84).

[7] Scerri (2007, 44) points out that scientists' use of the terminology "atomic weights" and "equivalent weights" in the 19th century was not without controversy. It seems that "equivalents" in Newlands' 1864 table actually referred to atomic weights.

[8] From a modern standpoint, Figure 9.2 does not include the elements of group III (e.g., boron and aluminum) and group VIII (e.g., helium and neon), the latter of which (later known as the noble gases) were not discovered until the 1890s.

Figure 9.2, appeared in Newlands' 1864 article - to illustrate his classification of chemical elements.

Thus it is more likely that Newlands deliberately selected six pairs of elements in Figure 9.2 to best represent his elemental classification system. This kind of visual design suggests that Newlands had anticipated the existence of groups in the modern periodic table. For example, he seemed to realize that both the sodium and lithium pair and the lithium and magnesium pair could be classified in the same group, thereby only keeping the former pair in Figure 9.2 to avoid repetition. Moreover, despite the great similarities between lithium and magnesium,[9] Newlands might have known that the two elements did not in fact belong to the same group and thus should be excluded from Figure 9.2.

In contrast, Odling' compositional arrangement of analogous elements in Figure 9.4 shows a different grammatical pattern. For instance, in terms of inter-cell relations, only cycles dominate all the 10 horizontal lines of chemical elements in Figure 9.4, which follow a difference in atomic weights of 16. The inter-cell relation of hyponymy, however, does not exist in the fourth row in Figure 9.4 between molybdenum and selenium. The former is a transition metal whereas the latter is a main-group element, so they do not belong to the same group from a modern standpoint.

Considering that Newlands even excluded the lithium and magnesium—both of which are actually the main-group elements with highly similar properties and both of which have a numerical gap of 16 in atomic weights—from Figure 9.2, it seems that from a functional perspective Newlands insisted that both the inter-cell relations of cycles and hyponymy should play an equally important role in horizontally arranging chemical elements. On this point, one might argue that Newlands' classification is more consistent with the modern concept of groups than is Odling's.

Another significant difference between the two tables lies in the vertical grouping of cells representing chemical elements. In Figure 9.4, Odling put 10 pairs of elements in two columns: the 10 elements in each pair with the higher atomic weights were placed in the first column; the 10 with the lower atomic weights were placed in the second column. Such an organization made it very difficult to identify any specialized inter-cell relations between all the elements in each of the two columns. For instance, in the first column, it will be a serious challenge to recognize any specific semantic link between sodium and molybdenum either in terms of their chemical properties or in terms of their atomic weights.[10] Therefore, the two columns in

[9] Lithium and magnesium show greater similarity than that between the elements in the same group. This extraordinary kinship is termed "diagonal behavior" (Scerri 2007, 265–266).

[10] From a modern standpoint, sodium is an alkali metal whereas molybdenum is a transition metal, and they have different chemical properties. Further to this, sodium has a low atomic weight of 23 while molybdenum has a high atomic weight of 96—so the two elements have a large gap of 73 in atomic mass.

Figure 9.4 might not be considered well-organized episodes from a social semiotic perspective.

The vertical arrangement of elements in Figure 9.2, however, is much more meaningful. It is important to note that Newlands used the column heads "*Member of a group having lowest equivalent*" and "*One element immediately above the preceding one*" (italics in original) as a powerful grammatical device in his classification system. As a multimodal design, the heads not only attract readers' attention and give relative prominence to the elements in the two columns, they also make it explicit that the elements in the same column interrelate with each other due to their similar atomic weights.

To illustrate, there is a small difference in atomic weights between any two elements in each of the columns. In fact, even the interval between sodium, the element with the lowest atomic weight (7), and magnesium, the element with the highest atomic weight (24), in the first column is no greater than that between any two analogous elements in the horizontal row at about 16.

It would appear that Newlands realized that the difference of about 16 in atomic weights was a watershed in classifying the six pairs of elements in Figure 9.2. When the gap is no larger than about 16, the elements will share little similarity in their chemical properties. However, when the interval reaches the point of about 16, there will be an appropriate repetition in the elements' chemical properties.

When Newlands used the range of numerical values in atomic weights (e.g., from 7 to 24) in Figure 9.2 as a crucial criterion to group elements with different chemical properties, the device of meronymy (e.g., part/whole relations, see Martin 1992, 304–305) appeared to function cohesively between these elements from a social semiotic perspective. To illustrate, the first 31 days of one year fall under the temporal category of January, so any of these days is one part of the month of January. By analogy, when the six elements in the first column are similarly among the smallest ones in atomic weights, any of them is one part of the range "lowest equivalent" in Newlands' terms.

Also noteworthy is that except for magnesium, the elements in the first column belong to the same period and are the first member of the main groups from a modern standpoint.[11] Hence, the visual design of Figure 9.2 indicates that in 1864 Newlands began to realize the fundamental principle of periodicity: Atomic weights were not only the key to classify analogous elements, but were also the key to compare chemically dissimilar ones. At this point, Newlands even outdid Mendeleev, who was credited for making similar observations five years later in 1869 (Scerri 2007, 116).

[11] Scerri (2007, 84) correctly observes that seven out of the ten elements in the second column of Figure 9.4 are the first member of the main groups. However, as previously explained, the difference in atomic weight between elements in the same column can be as large as 73. It is thus difficult to use similar atomic weights as a meaningful way to vertically organize the cells as episodes.

Grammatical Patterns of Figure 9.3

The classification system in Figure 9.2 seems to be mature enough to lay a solid grammar foundation for Newlands to design Figure 9.3 to illustrate the law of octaves. Despite some changes at the expression stratum discussed earlier, the two tables share great similarities in their grammatical configurations—except for one major difference.

To illustrate: although Newlands included 44 more cells in Figure 9.3, the horizontal arrangement of cells largely follows the same inter-cell relations of cycles and hyponymy. By using the ordinal number, Newlands made each adjacent pair of elements in the same row have a regular interval of seven, thereby interrelating them into cycles from a functional perspective. Meanwhile, the elements, especially the main-group (from a modern standpoint) elements, in the same row are similar in terms of their chemical properties. For example, five out of the six alkali metals can be found in the second row of Figure 9.3; the exception is francium, which had not yet been discovered in 1866. Accordingly, the relation of hyponymy seems to play an important role in the horizontal grouping of chemical elements.

On the other hand, Newlands revised the principle for the vertical organization of the elements. As previously explained, Newlands paid attention to the relationship between chemically different elements at a very early stage and creatively lined them up based on their similar atomic weights. However, the inter-cell link of meronymy has limitations in illuminating more subtle relationships among different elements.

For example, meronymy is not sufficient to explain why Newlands placed the element of oxygen in the *second* cell of the first column in Figure 9.2. It appears that Newlands lined the elements up in ascending order according to the difference in atomic weights between each pair. The elements of oxygen and sulfur were put in the second cells of the two columns respectively in Figure 9.2 possibly because they have the smallest interval (16) in atomic weights. However, considering that the difference in atomic weights between oxygen and sulfur is exactly the same as that between magnesium and calcium, lithium and sodium, and carbon and silicon, the elemental location of oxygen and sulfur seems rather random and lacks sufficient semantic motivations.

In contrast, when Newlands designed Figure 9.3 in 1866, all the elements in the same column were assigned an ordinal number one by one following an increase in atomic weights (Scerri 2007, 76). From a functional perspective, this kind of rearrangement sets up the inter-cell relation of cycles in all the columns. Therefore, oxygen is repositioned in the sixth cell of the first column based on its ranking in atomic weights among all 62 elements. By introducing the semantic principle of cycles to organize the vertical cells (i.e., seven elements in one full cycle), Newlands made it explicit that the elements' chemical properties change according to increasing atomic weights and show an approximate repetition after a regular interval of seven. In this way, the special design of Figure 9.3 not only best visually represents the law of octaves, it also captures the core of periodicity from a modern standpoint.

Summary and Discussion of Findings

As might be clear from the preceding analysis, a social semiotic comparison of two highly similar visual displays (Figures 9.2 and 9.4) provides evidence that Newlands and Odling actually conceptualized periodicity in different ways. In a parallel way, a resemioticization analysis of Figures 9.2 and 9.3 reveals that even the same scientist might not hold to a fixed classification system of elements at all times. It therefore follows that chemical periodicity is far from a resolved set of meanings but is to some extent negotiable and "in process" like other scientific concepts (Hubber et al. 2010, 19).

Although four more pairs of analogous elements can be found in Figure 9.4, Figure 9.2 is arguably a more effective classification scheme from a functional perspective because Newlands employed more grammatical resources to organize the cells into two kinds of meaningful episodes, in contrast to Odling's one. Like Odling, Newlands paid attention to the semantic link between analogous elements and arranged them horizontally through the inter-cell relations of hyponymy and cycles. Thus the rows in Figures 9.2 and 9.4 served a classification function similar to modern groups, though in an early stage. On the other hand, Newlands began to identify the semantic relation between chemically dissimilar elements and interrelate them with each other through the resource of meronymy based on their similar atomic weights. Newlands' use of meronymy might count as evidence that he anticipated the existence of periods even earlier than Mendeleev; this finding also seems to go beyond the research scope of Odling's system in Figure 9.4.

When Newlands redesigned his classification of elements to illustrate the law of octaves, significant transmutations took place at the strata of expression and grammar, both of which had important impacts upon the semantic construction of periodicity. For example, the semiotic landscape of the display was changed from a "language intensive" expression in Figure 9.2 to a "symbolism intensive" form in Figure 9.3. This resemioticizing operation led to a higher level of technicality and thus only those with expert knowledge of chemistry were able to engage with the new classification system. Further to this, Newlands adopted the ordinal numbers in Figure 9.3 to replace the cardinal numbers used in Figure 9.2, thereby making it explicit that his new display aimed to serve the main function of ordering the elements rather than describe their quantitative attributes.

At the level of grammar, Figures 9.2 and 9.3 show great similarities in their horizontal arrangement of elements where both the inter-cell relations of hyponymy and cycles function to organize the cells into meaningful episodes. However, the two tables differ in their vertical grouping. Whereas the elements in Figure 9.2 were loosely lined up following the relation of meronymy based on their similar atomic weights, Newlands designed the octaves table in which each of the elements was assigned an ordinal number according to increasing atomic weights. Accordingly, the relation of cycles replaced the relation of meronymy and operated in Figure 9.3 to link the elements together in the same column.

This regrammaticizing operation had significant effects upon the semantic construction of periodicity. As previously discussed, while elements in the columns of Figure 9.2 were more randomly placed, Newlands introduced the relation of cycles by assigning a fixed yet highly semantically motivated position to all the elements in the columns, thereby formulating a more delicate classification system.

As different as Figure 9.3 might look from the modern periodic table (e.g., Scerri 2007, 12), the two actually share the same general semantic patterns, especially when groups and periods, the two most meaningful units in the modern periodic table, are conceptualized from a functional perspective. For example, in the modern periodic table elements of the same group (e.g., Group I) have notable parallels in chemical properties and thus they are co-hyponyms (of the alkali metals). At the same time, every pair of adjacent elements in the same group (like lithium and sodium) have a regular interval of one period, so they are in a cycle. In a similar vein, elements across the same period interrelate to each other through the link of cycles. For instance, lithium is to beryllium as boron is to carbon in that they have a regular difference of one in atomic number. Accordingly, I might argue that Newlands recognized periodicity as early as in 1866 and developed a relatively mature classification system of chemical elements.

The present study, however, does not claim that Newlands' octaves table is perfect to guide scientific research or equally as valid as the modern periodic table. On the contrary, a functional analysis of Figure 9.3 also reveals that it has quite a few defects. For instance, it can be argued the inter-cell relation of hyponymy does not function cohesively between all the elements in the same row. To illustrate, primary chemical kinships can be found to exist between oxygen, sulfur, selenium, and tellurium, but not between them and the other elements such as iron in the seventh horizontal line. It would appear that Newlands did not recognize the need to organize cells as different episodes to accommodate main-group elements and transition metals respectively.

Given that cells in the form of a rectangle constitute a geometric primitive (Marchese 2013, 44), tabular modes develop a powerful resource of grammar to allow cells to be organized in an unlimited number of patterns. Due to this great potential, the periodic table can be creatively manipulated by scientists as effective research tools to explore different trends of chemical elements. Yet it should be kept in mind that a random arrangement of cells, for example, ordering chemical elements alphabetically based on the first letter of each of their full names,[12] can hardly contribute to the maturity of periodicity.

---

12 George Carey Foster, a chemist who attended the London Chemical Society in 1866 used this example to ridicule Newlands' law of octaves (Scerri 2007, 78).

As Figure 9.1 indicates, contextual changes at the stratum of field, especially the development of scientific theories and empirical findings about elements have significant effects on the shaping of periodicity by providing semantic constraints on the grammatical organization patterns of cells. When Döbereiner insightfully discovered in 1817 that various groups of three elements with similar chemical properties had numerical relationships in atomic weights, other scientists soon took up the powerful notion of triads, to great effect (Scerri 2007, 43). However, it was not possible for those pioneers to design a modern periodic table in that specific sociohistorical context when equivalent weight was still confused with atomic mass and scientists were not able to estimate the atomic weights of liquids and solids (Scerri 2007, 44). It would appear that at the contextual level there were neither mature atomic theories nor sufficient empirical data to develop a specialized field in the early 1800s; both were necessary for chemists to effectively manipulate tabular forms to classify chemical elements.

It seems that only after the Karlsruhe conference was convened in 1860 to resolve the remaining theoretical issues of atomic weights and the technique of spectroscopy was developed to discover new elements did the mature periodic tables develop rapidly (Scerri 2007, 66–67). From a social semiotic perspective, the rationalization of atomic mass and novel empirical findings gave birth to a specialized field of chemistry concerning the classification of elements, which guided pioneering chemists such as Newlands, De Chancourtois, Odling, Heinrichs, Meyer, and Mendeleev to design the periodic table in a specific direction, thereby narrowing down the almost endless possibilities of elemental locations.

Apart from the impacts of context, the grammar of tables has its own limitations in classifying elements. For instance, horizontal and vertical arrangements of cells remain the two major patterns of composition in tabular forms (Liu and Dwi-Nugroho 2012, 142). So it will be difficult to use the periodic table to highlight the kinship between some elements such as zinc and tin, which are located in a special relationship termed "Knight's Move" (Scerri 2007, 273).

Closing Remarks

This chapter is meant to serve two main purposes, and to set up a dialogue between scholars from different communities. The first purpose targets functional linguists and semioticians. While Iedema's (2003) notion of resemioticization provides a needed dynamic view of multimodal artifacts, it remains programmatic and has limitations in its analytical capability. By incorporating the social semiotic tenets of stratification, metafunction and multimodality, the present study proposes an expanded framework for exploring semiotic practices as a multilevel operation when they unfold through time.

The second target audience includes historians and philosophers of science. The resemioticization framework is used to analyze several periodic displays designed in the 1860s to illuminate the transformative dynamics of periodicity and the periodic table's plasticity. The social semiotic analysis provides further evidence that Newlands deserves to rank among the very few pioneering discoverers of the periodic law (Scerri 2007, 82). It also indicates that there exists a dialectic relationship between the periodic table and periodicity, whose constant interaction enables scientists to interpret more and more yet not all about the complex nexus between the elements' chemical properties and physical attributes.

Acknowledgments

I am grateful to Prof. Eric Scerri, who in his publications and personal correspondence contributed much to my understanding of the development of periodic tables as discussed in this chapter. All the remaining errors are of course my own. This research is supported by "Research on Scientific Texts from a Multimodal Perspective" (SISU201207), granted by Sichuan International Studies University, China.

References

Earl, B., Wilford, D. 2014. *Cambridge IGCSE Chemistry*, 3rd ed. Hodder Education, London.

Halliday, M.A.K. 1993. On the language of physical science. In: Halliday, M.A.K., Martin, J.R. (eds.), *Writing Science: Literacy and Discursive Power*, 54–68. Falmer, London.

Halliday, M.A.K. 1998. Things and relations: Regrammaticising experience as technical knowledge. In: Martin, J.R., Veel, R. (eds.), *Reading Science: Critical and Functional Perspectives on Discourses of Science*, 185–235. Routledge, London.

Halliday, M.A.K., Martin, J.R. (eds.). 1993. *Writing Science: Literacy and Discursive Power*. Falmer, London.

Halliday, M.A.K., Matthiessen, C.M.I.M. 1999. *Construing Experience through Meaning: A Language-based Approach to Cognition*. Cassell, London.

Halliday, M.A.K., Matthiessen, C.M.I.M. 2004. *An Introduction to Functional Grammar*, 3rd ed. Arnold, London.

Hubber, P., Tytler, R., Haslam, F. 2010. Teaching and learning about force with a representational focus: pedagogy and teacher change. Res. Sci. Educ. 40, 5–28.

Iedema, R. 2003. Multimodality, resemioticization: Extending the analysis of discourse as multisemiotic practice. Vis. Commun. 2, 29–57.

Kibler, M.R. 2007. From the Mendeleev periodic table to particle physics and back to the periodic table. Found. Chem. 9, 221–234.

Kress, G., Jewitt, C., Ogborn, J., Tsatsarelis, C. 2001. *Multimodal Teaching and Learning: The Rhetorics of the Science Classroom*. Continuum, London.

Kress, G., van Leeuwen, T. 2006. *Reading Images: The Grammar of Visual Design*, 2nd ed. Routledge, London.

Lemke, J.L. 1990 *Talking Science: Language, Learning, and Values*, Ablex Publishing Corporation, Norwood, NJ.
Lemke, J.L. 1998. Multiplying meaning: Visual and verbal semiotics in scientific text. In: Martin, J.R., Veel, R. (eds.), *Reading Science: Critical and Functional Perspectives on Discourses of Science*, 87–113. Routledge, London.
Liu, Y., O'Halloran, K.L. 2009. Intersemiotic texture: Analyzing cohesive devices between language and images. Soc. Semiotics 19, 367–388.
Liu, Y., Owyong, Y.S.M. 2011. Metaphor, multiplicative meaning and the semiotic construction of scientific knowledge. Lang. Sci. 33, 822–834.
Liu, Y., Dwi-Nugroho, A. 2012. The social semiotic construction of chemical periodicity: A multimodal view. Semiotica 190, 133–151.
Marchese, F.T. 2013. Periodicity, visualization, and design. Found. Chem. 15, 31–55.
Martin, J.R. 1992. *English Text: System and Structure*. Benjamins, Amsterdam.
Martin, J.R. 2007. Construing knowledge: A functional linguistic perspective. In: Christie, F., Martin, J.R. (eds.), *Language, Knowledge and Pedagogy: Functional Linguistic and Sociological Perspectives*, 34–64. Continuum, London.
Martin, J.R., Rose, D. 2008. *Genre Relations: Mapping Culture*, Equinox, London.
O'Toole, M. 1994. *The Language of Display Art*, Leicester University Press, London.
Pearson, P.D., Moje, E., Greenleaf, C. 2010. Literacy and science: Each in the service of the other. Sci. 328, 459–463.
Scerri, E. 1997. Has the periodic table been successfully axiomatized? Erkenntnis 47, 229–243.
Scerri, E. 2007. *The Periodic Table: Its Story and Its Significance*, Oxford University Press, New York.
Tang, K.S. 2013. Instantiation of multimodal semiotic systems in science classroom discourse. Lang. Sci. 37, 22–35.
Waldrip, B., Prain, V., Carolan, J. 2010. Using multimodal representations to improve learning in junior secondary science. Res. Sci. Educ. 40, 65–80.
Winter, M.J. 1993. WebElements. http://www.webelements.com/ (1993). Accessed 20 November 2013.

CHAPTER 10 | Organizing the Transition Metals

GEOFF RAYNER-CANHAM
Chemistry Department, Memorial University Corner Brook, Canada

THOUGH THE PERIODIC TABLE has been a centerpiece of chemistry since the 1870s [1], it is in recent years that a more detailed study of patterns and trends has become of interest. In part, this study reflects the accumulated wealth of data on comparative formulas, structures, and properties of compounds across the periodic table. Traditionally, patterns in the periodic table have been examined through defined criteria of groups and periods. Such a mindset immediately obscures other linkages.

I have previously critically analyzed and extended nontraditional connections. Such connections have been

Isoelectronic [2]: similarities between species (molecules or ions) which have the same total number of electrons and of valence electrons together with the same structure (number and connectivity of atoms);

Isodiagonal [3]: similarities between an element of group (n) and period (m) with an element of group (n+1) and period (m+1).

Knight's move [4]: similarities between an element of group (n) and period (m) with the element in group (n+2) and period (m+1) in the same oxidation state (in the lower-right of the periodic table);

Group (n) and (n+10) [5]: similarities between elements of Group (n) with elements of Group (n+10) in the same highest oxidation state.

It is such other, less-obvious patterns that enrich the study of the elements and lead to additional possibilities for exploration. Here, we will deconstruct the monolithic block of transition metals and see how assigning an element to a specific cluster provides a more flexible approach to appreciating the complexity of the transition series than simple use of group or period assignment [6].

Which Are the Transition Metals?

The first task is to define the transition metals, which is not quite as obvious as it may seem. The main group elements are always identified as the members of groups 1–2, and 13–18. In addition, about half of sources include group 12 [7]. One might justifiably conclude, therefore that groups 3–11 or 3–12 would be the transition metals. However, identifying a "transition metal" is not simply a question of location in the periodic table, but also one of chemical behavior. A definition common among inorganic chemists is: a *transition metal* is an element that has at least one simple ion with an incomplete outer set of d electrons.

Using this criterion, the group 3 metals are excluded as their common chemistry is all based on the d° 3+ ion, Sc^{3+} and Y^{3+}. In fact, the chemistry of these two metals more closely resembles that of the lanthanoids. The linkage is particularly apparent for yttrium, which, like the lanthanoids, often adopts eight-coordination and is found in such ores as: bastnäsite, $(Ce,La,Y)CO_3F$.

The group 12 metals are also excluded. For them, the predominant ions, Zn^{2+}, Cd^{2+}, and Hg^{2+} have d^{10} configurations. The isolation of d^8 mercury(IV) fluoride, HgF_4, at very low temperatures [8], initially provoked the claim that mercury should be redesignated as a transition metal. However, if one adds to the definition, "which is stable under ambient conditions," then one does not need to be endlessly adding to the list of elements called transition metals as ever-more exotic and fragile species are identified [9]. In this chapter, we will discard this definition based strictly on electron configurations and instead develop a hybrid approach which is more chemically satisfying. Before doing so, it is appropriate to review the previous categorizations of transition metals.

Previous Classifications of Transition Metals

Of the two classic works on inorganic chemistry, Greenwood and Earnshaw [10] in *Chemistry of the Elements* treat each of the transition metal groups as individual entities, devoting a chapter to group 4, one to group 5, and so on (Fig. 10.1).

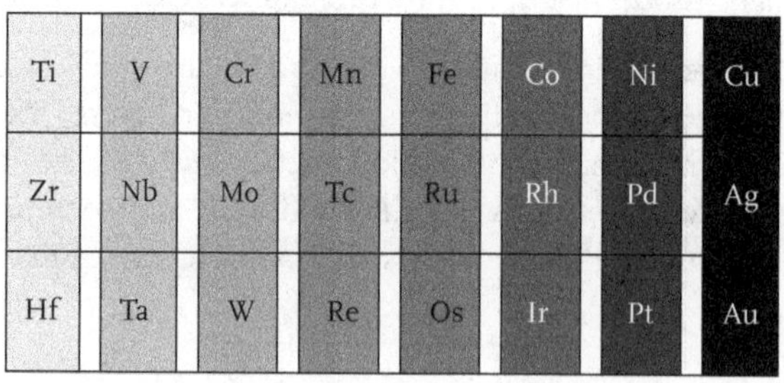

FIGURE 10.1 The traditional group-by-group study of the transition metals.

Another common approach is adopted by Cotton, Wilkinson, Murillo, and Bochmann [11], with each of the elements of the first transition series being treated individually, then the 4d-5d pairs of elements being covered in a subsequent section. This arrangement of material is also adopted by Housecroft and Sharpe [12] and by Rayner-Canham and Overton [13].

A third approach is to cover groups 4–7 and 11 individually, then the Fe–Co–Ni triad together, and the platinum metals as a separate entity (Fig. 10.2). This format was adopted by Nicholls [14], and by Gopalan [15]. Historically, the platinum group metals have, since their discovery, always been considered as a related "cluster" [16]. Lee [17] has commented that, for the Groups 8, 9, and 10: "... the horizontal similarities between these elements are greater than anywhere in the periodic table except among the lanthanides." He notes that the similarities "are sometimes emphasized by considering these nine elements as two horizontal groups: the three ferrous metals Fe, Co, and Ni, and the six platinum metals Ru, Rh, Pd, Os, Ir, and Pt."

Each of these classification systems has one flaw—they organize the transition metals largely according to one strategy and they define the trends according to that organization. Thus linkages, relationships, patterns, or similarities outside of that framework are ignored. Two exceptions have been the proposals by Habashi and by Schweitzer and Pesterfield.

Habashi [18] has identified three categories of transition metals and named them as follows:

The *vertical similarity transition metals*: [Zr–Hf]; [Nb–Ta]; [Mo–W]; and [Tc–Re].
The *horizontal similarity transition metals*: [Ti–V–Cr–Mn–Fe–Co–Ni].
The *horizontal-vertical transition metals*: [Ru–Os–Rh–Ir–Pd–Pt].

Thus Habashi considered the 3d metals as separate entities from the 4d and 5d metals (Fig. 10.3). Incidentally, he did not include the group 11 metals.

A recent book, the *Aqueous Chemistry of the Elements* (Schweitzer and Pesterfield), includes an updated series of Pourbaix diagrams [19]. In this book, the authors assign the 3d transition metals to two horizontal triads: [V–Cr–Mn] which form compounds in the maximum oxidation states; and [Fe–Co–Ni] for

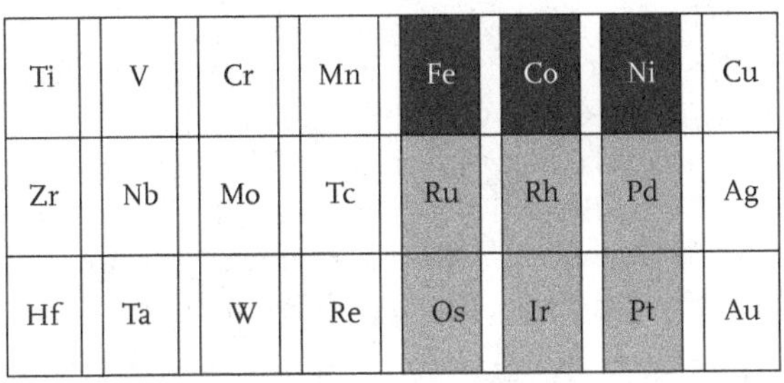

FIGURE 10.2 Combining discussion of the Fe–Co–Ni triad together, and considering the platinum group metals as a separate entity.

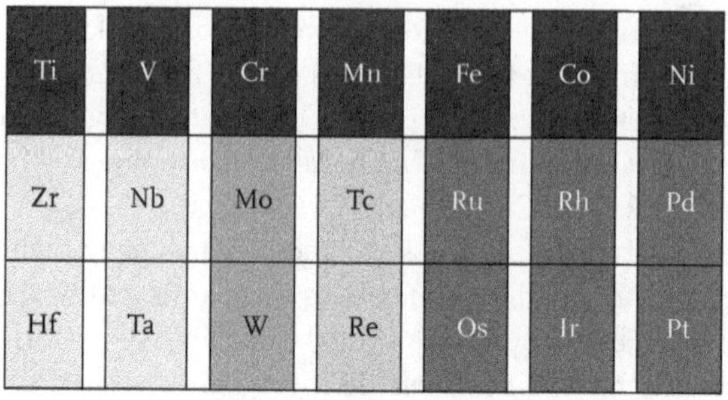

FIGURE 10.3 The transition metal classification according to Habashi [18].

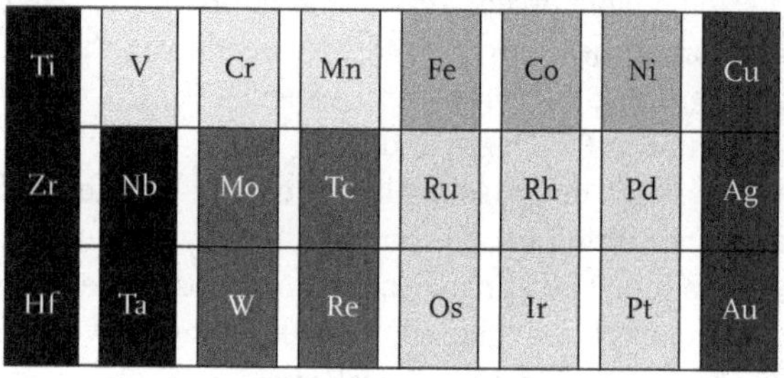

FIGURE 10.4 The transition metal classification according to Schweitzer and Pesterfield [19].

which +2 and +3 oxidation states predominate. They treat copper with the other two group 11 metals as the [Cu–Ag–Au] vertical triad, while titanium is placed in a chapter with all of the 4d and 5d transition metals. Schweitzer and Pesterfield divide the heavy transition metals (plus titanium) into three sub-categories:

The elements for which insoluble oxides dominate [Ti–Zr–Hf–Nb–Ta].
The elements with high oxidation state oxoanions [Mo–W–Tc–Re].
The platinum metals [Ru–Os–Rh–Ir–Pd–Pt].

In recent years, using chemotopological methods, there have been new attempts at classifications of the elements. Sneath's study [20] divided the heavy transition metals into one cluster and the later ones (plus gold) into a separate cluster. The analysis by Leal et al. [21] suggested that, for the 3d metals, there was a [Cr–Fe–Co–Ni] linkage, while titanium belonged to group 4 as [Ti–Zr–Hf] and manganese and vanadium were unique in their chemistry. For the 4d-5d transition metals, they proposed the following linkages: [Zr–Hf (with Ti)]; [Mo–W (with Ge)]; [Nb–Ta–Tc–Re]; [Ru–Os]; [Rh–Ir–Pd–Pt]; and [Ag–Au].

Here, then, a schema will be deduced from first principles using chemical criteria that have similarities to, but also differences from, those of Habashi and of Schweitzer and Pesterfield (and, like Sneath, that consider gold to be a member of the platinum metal group). Though we will conclude that copper and gold fit the chemical transition-metal criteria, silver does not. In developing our criteria, we will point out the richness of the relationships and the necessity to have flexibility in any classification and, indeed, see that at least two elements might be considered to have "secondary allegiances."

The 3d Period Patterns

It has always been recognized that the metals of the 3d transition series differ significantly from those of the 4d and 5d series. The 3d metals are more commonly found in lower oxidation states and they can form high-spin compounds as a result of the lower crystal field stabilization energy. But the 3d metals do not form a homogeneous series. This differentiation can be seen by looking at the formulas of the common oxides (Table 10.1).

Among the many oxides, for the group 4 to group 7 metals, there is an oxide series for which the metal has its maximum oxidation state (TiO_2, V_2O_5, CrO_3, and Mn_2O_7). For the same set of metals, there is also an oxide series MO_2 (with M = Ti, V, Cr, and Mn). On the basis of the oxides, it could be considered that there is a [Ti–V–Cr–Mn] tetrad sub-group of the 3d metals. Likewise, the next three of the 3d transition metals, the [Fe–Co–Ni] triad, are characterized by having +2 and +3 oxidation states in their common oxides. Copper is unique among the 3d metals in exhibiting an oxidation state of +1 in an oxide (and in its chemistry in general). Of course, the divisions are not clear-cut. For example, there is the oxide series of M_3O_4 which encompasses manganese through to nickel.

TABLE 10.1 The common oxides of the 3d transition metals

OX. NO.	TITANIUM	VANADIUM	CHROMIUM	MANGANESE	IRON	COBALT	NICKEL	COPPER
+7				Mn_2O_7				
+6			CrO_3					
+5		V_2O_5						
+4	TiO_2	VO_2	CrO_2	MnO_2			NiO_2	
+3	Ti_2O_3	V_2O_3	Cr_2O_3	Mn_2O_3	Fe_2O_3			
+2 & +3				Mn_3O_4	Fe_3O_4	Co_3O_4	Ni_3O_4	
+2	TiO	VO		MnO	FeO	CoO	NiO	CuO
+1								Cu_2O

TABLE 10.2 Comparative species for the Ti–V–Cr–Mn tetrad over the pH range under strongly oxidizing conditions

	VERY ACIDIC	ACIDIC	BASIC	VERY BASIC
Titanium	$TiO^{2+}(aq)$		$TiO_2(s)$	
Vanadium	$VO_2^+(aq)$	$H_2VO_4^-(aq)$	$HVO_4^{2-}(aq)$	$VO_4^{3-}(aq)$
Chromium	$H_2CrO_4(aq)$	$HCrO_4^-(aq)$	$CrO_4^{2-}(aq)$	
Manganese		$MnO_4^-(aq)$		

The decision of which elements belong in what groupings will always be subjective. This author places the greater weight on the species under strongly oxidizing conditions. To illustrate this point, Table 10.2 shows that, under strongly oxidizing conditions, manganese completes the set of isoelectronic highly oxidizing anions. Whereas vanadium, chromium, and manganese all form soluble tetroxoanions, titanium forms an insoluble oxide. For this reason, on balance, titanium is the "weakest link" in this set. In addition, the tetroxoanions of the V–Cr–Mn form a series of increasing acid strength.

The Trouble with Titanium

In the previous section we saw that, over the pH range, titanium does not have an aqueous chemistry like its group 5, 6, and 7 neighbors. In fact, titanium is a troublesome element in the context of placement. Should it be considered the beginning of the 3d Period series or the top member of group 4? Let us next consider the linkage with the other members of group 4 (Table 10.3).

But before congratulating ourselves, we should look at the comparative species at very low pH while altering the potential. Slicing the Pourbaix diagram vertically, titanium chemistry matches more with that of vanadium and less with zirconium. In fact, as we see in Table 10.4, there is a remarkable similarity in oxidation states and species, taking into consideration that the maximum oxidation state of vanadium is +5 while that for titanium is +4:

Thus titanium lays claim both to be the top member of the group 4 triad (Ti–Zr–Hf) *and* the first member of the early 3d transition series tetrad (Ti–V–Cr–Mn). On balance, because of the dominance of the insoluble +4 oxidation state oxide, this author leans toward including titanium with zirconium and hafnium.

TABLE 10.3 Comparative species for the Ti–Zr–Hf triad over the pH range under strongly oxidizing conditions

	VERY ACIDIC	ACIDIC	BASIC	VERY BASIC
Titanium	$TiO^{2+}(aq)$		$TiO_2(s)$	
Zirconium	$ZrO^{2+}(aq)$		$ZrO_2(s)$	
Hafnium	$HfO^{2+}(aq)$		$HfO_2(s)$	

TABLE 10.4 A Comparison of species of titanium, vanadium, zirconium, and hafnium species at very low pH over the redox range

	STRONGLY REDUCING	REDUCING	NEAR-ZERO POTENTIAL	OXIDIZING	STRONGLY OXIDIZING
Titanium	$Ti(s)$	$Ti^{2+}(aq)$	$Ti^{3+}(aq)$	$TiO^{2+}(aq)$	
Vanadium	$V(s)$	$V^{2+}(aq)$	$V^{3+}(aq)$	$VO^{2+}(aq)$	$VO_2^{+}(aq)$
Zirconium	$Zr(s)$		$ZrO^{2+}(aq)$		
Hafnium	$Hf(s)$		$HfO^{2+}(aq)$		

Manganese Muddies Things

Just as titanium has two allegiances at the beginning of the 3d row, so manganese in the middle also presents a dilemma. Table 10.2 shows that, under highly oxidizing conditions, manganese completes the set of isoelectronic highly oxidizing anions. However, under normal conditions of aqueous chemistry, manganese favors the +2 oxidation state and its species match well with the following member of the 3d row, iron (see Table 10.5). And, as alluded to earlier, manganese forms Mn_3O_4—part of the series running from manganese to nickel. So manganese, like titanium, has a "dual identity."

TABLE 10.5 Comparative species for the Mn and Fe over the pH range under neutral to mild oxidizing conditions

	VERY ACIDIC	ACIDIC	BASIC	VERY BASIC
Manganese	$Mn^{2+}(aq)$		$Mn(OH)_2(s)$	
Iron	$Fe^{2+}(aq)$		$Fe(OH)_2(s)$	

Dividing the Early 4d-5d Elements

A unifying feature of the early 4d-5d elements is the larger size of the metal species, enabling them to have coordination numbers up to eight. An excellent example is the series of isostructural (though not all valence-isoelectronic) octafluoro-complexes as shown in Table 10.6.

On this basis, should we therefore consider a Zr–Hf–Nb–Ta–Mo–W–Tc–Re octad of elements? One pattern does not make a cluster, but on the other hand, if we took every single piece of evidence, we would consider each element completely unique.

In this author's view, a defining difference between the Zr–Hf–Nb–T tetrad and Mo–W–Tc–Re tetrad is the fundamental difference in aqueous chemistries. The simple chemistry of the first tetrad across most of the pH range is defined by the insoluble oxides: ZrO_2, HfO_2, Nb_2O_5, and Ta_2O_5. For the second tetrad, it is the soluble tetroxoanions which dominate: MoO_4^{2-}, WO_4^{2-}, TcO_4^-, and ReO_4^-.

TABLE 10.6 The isostructural octofluoro-complexes of the early 4d-5d transition elements

	GROUP 4	GROUP 5	GROUP 6	GROUP 7
4d-Period	ZrF_8^{4-}	NbF_8^{3-}	MoF_8^{2-}	TcF_8^{2-}
5d -Period	HfF_8^{4-}	TaF_8^{3-}	WF_8^{2-}	ReF_8^{2-}

The Platinum Metals

The platinum metal group, sometimes called the platinoids [22], consists of the Ru–Os–Rh–Ir–Pd–Pt hexad of elements. Their simple chemistry is characterized by insoluble oxides under strongly oxidizing conditions. For Ru–Os, the +8 oxidation state is favored: RuO_4 and OsO_4, while for Rh–Ir–Pd–Pt it is the +4 oxidation state: RhO_2, IrO_2, PdO_2, and PtO_2. There is a general acceptance that these six elements form a cluster.

Is There in Fact a Group 11?

Using the word "group" implies similarity between members. For example, group 4 was simple. The oxidation state of +4 dominated all of the three elements. The three elements, copper, silver, and gold of so-called group 11, can scarcely be said to form a "group." As shown in Table 10.7, for copper, the +2 oxidation state dominates; for silver, the +1 state; and for gold, the +3 state.

In fact, in terms of its coordination chemistry, copper fits better with the later 3d elements. For example, in the species across the oxidizing pH range, we see a strong similarity (Table 10.8).

The dominance of the +1 d^{10} state for the normal aqueous chemistry of silver [23] make it more appropriately considered as a main group metal. As discussed elsewhere [4], silver is the "classic" example of the "knight's move" relationship, showing a startling similarity to thallium. A unique parallel is that they are the only two metal ions to form brick-red insoluble chromates, Ag_2CrO_4 and Tl_2CrO_4. Under oxidizing conditions, we have the parallel shown in Table 10.9.

TABLE 10.7 Comparative species for the Cu–Ag–Au elements over the pH range under oxidizing conditions

	VERY ACIDIC	ACIDIC	BASIC	VERY BASIC
Copper	$Cu^{2+}(aq)$		$Cu(OH)_2(s)$	$Cu(OH)_4^{2-}(aq)$
Silver	$Ag^{+}(aq)$		$Ag_2O(s)$	$AgO^{-}(aq)$
Gold		$Au_2O_3(s)$		$Au(OH)_4^{-}(aq)$

TABLE 10.8 Comparative species for the Fe–Co–Ni–Cu tetrad over the pH range under neutral to mild oxidizing conditions

	VERY ACIDIC	ACIDIC	BASIC	VERY BASIC
Iron	$Fe^{2+}(aq)$		$Fe(OH)_2(s)$	
Cobalt	$Co^{2+}(aq)$		$Co(OH)_2(s)$	$Co(OH)_4^{2-}(aq)$
Nickel	$Ni^{2+}(aq)$		$Ni(OH)_2(s)$	
Copper	$Cu^{2+}(aq)$		$Cu(OH)_2(s)$	$Cu(OH)_4^{2-}(aq)$

TABLE 10.9 Comparative species for the Ag–Tl diad over the pH range under oxidizing conditions

	VERY ACIDIC	ACIDIC	BASIC	VERY BASIC
Silver	$Ag^{+}(aq)$		$Ag_2O(s)$	$AgO^{-}(aq)$
Thallium		$Tl^{+}(aq)$		$TlO^{-}(aq)$

TABLE 10.10 Comparative species for Au–Pt over the pH range under oxidizing conditions

	VERY ACIDIC	ACIDIC	BASIC	VERY BASIC
Gold		$Au_2O_3(s)$		$Au(OH)_4^{-}(aq)$
Platinum		$PtO_2(s)$		$Pt(OH)_6^{2-}(aq)$

That gold does not fit with copper and silver has been discussed previously and it has been referred to as the "gold anomaly" [24]. Gold is actually the easiest one of this column to assign to a cluster—to the platinum metals. In fact, gold is sometimes found in nature as an alloy with platinum and palladium [25]. Though the term "noble metals" has been used to describe many combinations and permutations of the heavier unreactive metals [26], this author would contend that the platinum metals plus gold best fit this category.

Just as there are unusual species linking silver(I) and thallium(I), so gold(V) resembles platinum(V). A good example is the pair of compounds: $[O_2]^+[PtF_6]^-$ and $[O_2]^+[AuF_6]^-$. If we compare the species for platinum and gold under oxidizing conditions, and allowing for the fact that +4 is the predominant oxidation state of platinum and six its common coordination number, we again see a much closer parallel of behavior of gold with platinum than with silver or copper (Table 10.10).

A Concluding Synthesis

So is there a classification of the transition elements which better reflects the linkages? Obviously, we cannot satisfy all the many similarities, but the argument is made here that the best fit is not accomplished by either the Group or the Period approach. Instead, a hybrid combination generates the clusters of elements which have similarities most worthy of highlighting (Fig. 10.5).

To review, this author considers the most logical clustering as listed below, showing the "secondary allegiances" of titanium and of manganese:

The Ti–Zr–Hf–Nb–Ta pentad whose simple chemistry is dominated by insoluble oxides.

The V–Cr–Mn triad which exhibits isoelectronic soluble oxidizing isoelectronic tetroxoanions plus a stable +3 oxidation state, to which Ti can be appended for other aspects of common chemistry.

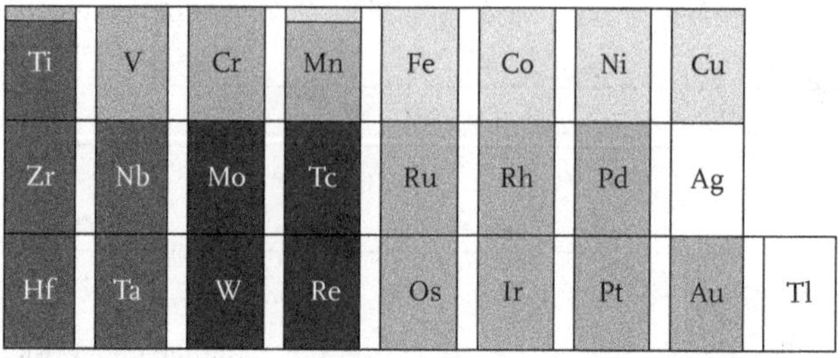

FIGURE 10.5 A hybrid approach to transition metal classification.

The Fe–Co–Ni–Cu tetrad for which the +2 aqueous ion is a major component of simple chemistry, to which Mn can be appended for some commonalities.

The Mo–W–Tc–Re tetrad for which non-oxidizing soluble valence-isoelectronic tetroxoanions exist.

The Ru–Os–Rh–Ir–Pd–Pt–Au heptad which can be defined as the "noble metals."

The Ag–Tl diad which reminds us to "think outside the (transition metal) box."

References

1. E. Scerri, *The Periodic Table: Its Story and Its Significance*, Oxford University Press, New York, 2007.
2. G. Rayner-Canham, "Isoelectronic series: a fundamental periodic property," *Found. Chem.*, 11, 123–129 (2009).
3. G. Rayner-Canham, "Isodiagonality in the periodic table," *Found. Chem.*, 13, 121–129 (2011).
4. G. Rayner-Canham and M. Oldford, "The chemical 'Knight's Move' relationship: what is its significance?" *Found. Chem.*, 9, 119–125 (2007).
5. G. Rayner-Canham, "Periodic patterns: the group (*n*) and group (*n*+10) linkage," *Found. Chem.*, 15, 229–237 (2013).
6. A preliminary study was published as G. Rayner-Canham, "Relationships among the transition elements," *Found. Chem.*, 13, 223–232 (2011).
7. W.B. Jensen, "The place of zinc, cadmium, and mercury in the periodic table," J. Chem. Educ., 80, 952–961 (2003).
8. X. Wang, L. Andrews, S. Riedel, and M. Kaupp et al, "Mercury is a transition metal: the first experimental evidence for HgF_4," *Angew. Chem. Int. Ed.* 46, 8371–8375 (2007).
9. W.B. Jensen, "Is mercury now a transition element?" *J. Chem. Educ.*, 85, 1182–1183 (2008).
10. N.N. Greenwood and A. Earnshaw, *Chemistry of the Elements*, 2nd ed., Butterworth-Heinemann, Oxford, 1997.
11. F.A. Cotton, G. Wilkinson, C. A. Murillo, and M. Bachmann et al., *Advanced Inorganic Chemistry*, 6th ed., Wiley-Interscience, New York, 1999.

12. C.E. Housecraft and A.G. Sharpe, *Inorganic Chemistry*, 4th ed., Pearson, Harlow, 2012.
13. G. Rayner-Canham and T. Overton, *Descriptive Inorganic Chemistry*, 6th ed., W. H. Freeman, New York, 2014.
14. D. Nicholls, *The Chemistry of Iron, Cobalt and Nickel* (Pergamon Texts in Inorganic Chemistry, volume 24), Pergamon Press, Oxford, 1975; S.E. Livingstone, *The Chemistry of Ruthenium, Rhodium, Palladium, Osmium, Iridium and Platinum* (Pergamon Texts in Inorganic Chemistry, volume 25), Pergamon Press, Oxford, 1975.
15. R. Gopalan, *Textbook of Inorganic Chemistry*, CRC Press, Boca Raton, 2012.
16. W.P. Griffith, "The periodic table and the platinum metals," *Platinum Metals Rev.* 52, 114–119 (2008).
17. J.D. Lee, *Concise Inorganic Chemistry*, 5th ed., Chapman & Hall, London, 1996.
18. F. Habashi, "Metals: typical and less typical, transition and inner transition," *Found. Chem.*, 12, 31–39 (2010)
19. G.K. Schweitzer and L.L. Pesterfield,: *Aqueous Chemistry of the Elements*, Oxford, Oxford University Press, 2010.
20. P.H.A. Sneath, "Numerical classification of the chemical elements and its relation to the periodic system," *Found. Chem.*, **2**, 237–263 (2000).
21. W. Leal, G. Restrepo, and A. Bernal, "A network study of chemical elements: from binary compounds to chemical trends," *MATCH Commun. Math. Comput. Chem.*, 68, 417–442 (2012).
22. See, for example, "Platinum Group," http://en.wikipedia.org/wiki/Platinum_group, accessed 22 December 2014.
23. J.A. McMillan, "Higher oxidation states of silver," *Chem. Revs.*, 62, 65–80 (1962).
24. P. Schwerdtfeger, "Relativistic effects in properties of gold," *Heteroatom Chem.*, 13, 578–584 (2002).
25. A. Knopf, "A gold-platinum-palladium lode in Southern Nevada," *Contr. Econ. Geol.*, 1–19 (1915).
26. See, for example, "Noble Metals," http://en.wikipedia.org/wiki/Noble_metal, accessed 22 December 2013. This author would contend the silver is so much more chemically-reactive and has such a different chemistry, that it should not be considered as a "noble metal." Simon Cotton has used the term "precious metals" to cover the six platinum metals plus silver and gold, see: S. Cotton, *Chemistry of Precious Metals*, Springer, New York, 1997.

CHAPTER 11

The Earth Scientist's Periodic Table of the Elements and Their Ions

A New Periodic Table Founded on Non-Traditional Concepts

L. BRUCE RAILSBACK
Department of Geology, University of Georgia, USA

A New Periodic Table in the Twenty-First Century

The Earth Scientist's Periodic Table of the Elements and Their Ions is a fundamentally new table that was first published in 2003 in the Geological Society of America's (GSA) prominent journal *Geology* (Railsback 2003). The new table was reported in *Nature*, it was featured in a cover article by *Science News*, it was included among *Discover* magazine's 100 Top Science Stories of 2003, and its publication was noted in many other magazines and online outlets. GSA sold a large number of reprints of the 2003 paper and then, in 2004, published a revised version of the table in GSA's Map and Chart Series (Railsback 2004). When GSA's printed stock ran low, the Society published a further revised version of the table in its Map and Chart Series in 2011 (Railsback 2011). The table has been translated into Chinese (Jin 2006), Spanish (Bernal and Railsback 2008), Portuguese (Franco de Souza Lima and Railsback 2012), and German.[1] The original 2003 paper has been cited in journals ranging from *Journal of Mathematical Chemistry* to *Carbohydrate Research* to *Geomicrobiology Journal* to *Journal of Arid Environments* to *Resource Geology* to *Reviews in Geophysics*, and it has proven useful in understanding the topology of the periodic table (Restrepo et al. 2006).

The success of the new Earth Scientist's Periodic Table of the Elements and Their Ions across the past decade suggests that the periodic table, as a general

[1] The new table is also available electronically in all five languages on Railsback's website titled "An Earth Scientist's Periodic Table of the Elements and Their Ions." http://www.gly.uga.edu/railsback/PT.html

concept, is not a static document but instead is still subject to evolution, especially as scientific fields beyond traditional chemistry increasingly use chemical perspectives. It further suggests that volumes like this one are not simply retrospective ruminations on a nineteenth-century invention, but instead they can be part of an ongoing process to find new meaning in the periodic concept and to make it more applicable in broader contexts in the twenty-first century.

Fundamental Considerations of Purpose and Concept

Despite the diversity of periodic tables produced over the last 140 years (e.g., Mazurs 1974), the Earth Scientist's Periodic Table of the Elements and Their Ions differs both in conceptual origin and in form from almost all previous versions.[2] Its differences arise from the table's purpose of serving natural scientists (those who study matter in its many natural contexts and environments) rather than laboratory scientists or theorists (who can work with or envision chemical entities that are unstable in nature). The goals of the natural scientists are simultaneously more restrictive and more ambitious. Their goals are more restrictive because the natural scientists need only concern themselves with forms found in nature, rather than the diversity of forms sustainable in the controlled environment of a laboratory. Their goals are nonetheless more ambitious, because they would like a table that contextualizes all the occurrences of matter in its natural environments, rather than settling for a periodic table that, while theoretically elegant, does little to explain the variety and complexity of matter in nature and has nothing to say about context or environment.

These differences in purpose lead to two major differences in conceptual origin and thus in form for the new table. First, most if not all previous periodic tables have sought to represent each element in its zero-oxidation state or elemental form (for example, Na with its eleven electrons and Si with its 14 electrons). The arrangement of periodic tables has consequently focused largely on presenting a coherent explanation of the arrangements of electrons in outer shells. All this has little meaning to natural scientists, who never encounter elemental Na (there is no elemental Na in nature on Earth) and for whom elemental Si is virtually unknown (elemental Si is not found in materials at or near Earth's surface; it has been reported only in deep-crustal to mantle materials (Novgorodova et al. 1989)). For scientists of natural settings, a more useful table is arranged according to non-elemental states of matter where oxidation state is not zero.[3] In such a table, Al and Si are not on the opposite side of the table from Na and Mg but instead are found in a contiguous row of Na^{I+}, Mg^{II+},

[2] The new table bears some similarity to those of Grimm (1922), Cartledge (1928), Jaeger (1957), Szádeczky-Kardoss (1959), Rayner-Canham (2002), and Laing (2004), all of which showed some elements more than once; and to those of Mason (1952) and Heald (1954), which showed all elements only once but had at least Al on the left side of the table.

[3] In this respect, the new table follows the practical approach to categorization of chemical entities that Restrepo and Villaveces (2008) attributed to Dmitri Mendeleev.

Al^{III+}, Si^{IV+}, P^{V+}, and S^{VI+}. Similarly, Th^{IV+} and U^{VI+} (representatives of two elements that do not exist in elemental form in nature) are not relegated to the bottom of the table with the rare earths to which they are not related, but instead they are located in a contiguous row of Fr^{I+}, Ra^{II+}, Ac^{III+}, Th^{IV+}, Pa^{V+}, and U^{VI+}, and thus just below Hf^{IV+} and W^{VI+}, respectively.[4] These new positions, and those of many other entities, not only make more sense arithmetically but also provide a much better framework for contextualizing the behavior of matter in nature, as discussed below.

This approach leads to the second major conceptual difference between the new table and previous periodic tables. Almost all previous tables have shown each element exactly once, a logical result of treating each element in its elemental and thus invariant form. That approach is compatible with a Platonist notion of ideal forms, where the observed and varied forms of matter in tangible nature (for example, the sulfate of seawater and gypsum, the sulfite of some solutions and of the mineral scotlandite, and the sulfide of anoxic solutions and of sulfide minerals like sphalerite) are thought to be better represented by the ideal and complete elemental forms (in this case, the elemental sulfur found less commonly in nature). The new table, and any table arranged according to non-elemental states of matter, must show many elements more than once, because many elements have more than one oxidation state. This is logical to natural scientists, who encounter different forms of matter in different natural contexts. Just as one person may be scientist, parent, and activist with different roles in different settings, any given element may have oxidized forms and/or an elemental form and/or reduced forms that play such different roles in different natural contexts that they are best understood by considering those different roles, rather than one idealized form. Thus the new table presents 44 elements more than once, and sulfur is a prominent but not unique example in its four forms (sulfate, sulfite, elemental, and sulfide).

These two conceptual differences lead to a table inevitably different in form from the conventional periodic table of the early twenty-first century (Figs. 11.1 and 11.2). A third conceptual difference superposed on that form, and arising from the purposes of natural scientists, is the notion that the table might be better viewed not as columns and rows but as a matrix across which oblique gradients can better contextualize the behavior of matter. For example, the natural behaviors of Be^{II+}, Al^{III+}, Ti^{IV+}, Zr^{IV+}, and Ta^{V+} are all similar, but different from the behaviors of their neighbors to the left and right in rows or above and below in columns. One parameter useful in understanding these oblique patterns is ionic potential (Cartledge 1928; Goldschmidt 1937). Ionic potential, sometimes incorrectly termed "density of charge," is most simply quantified as ionic charge or oxidation number divided by ionic radius[5]. Thus for Si^{IV+}, the

---

[4] In this respect, the Earth Scientist's Periodic Table of the Elements and Their Ions is a "pre-Seaborg table" (Scerri 2007, 23).

[5] Geochemists also speak of "field strength" of an ion in parallel with ionic potential, so that cations with ionic potential great than 2.0 are considered "high field strength cations" (e.g., Rollinson 1993, 103–104).

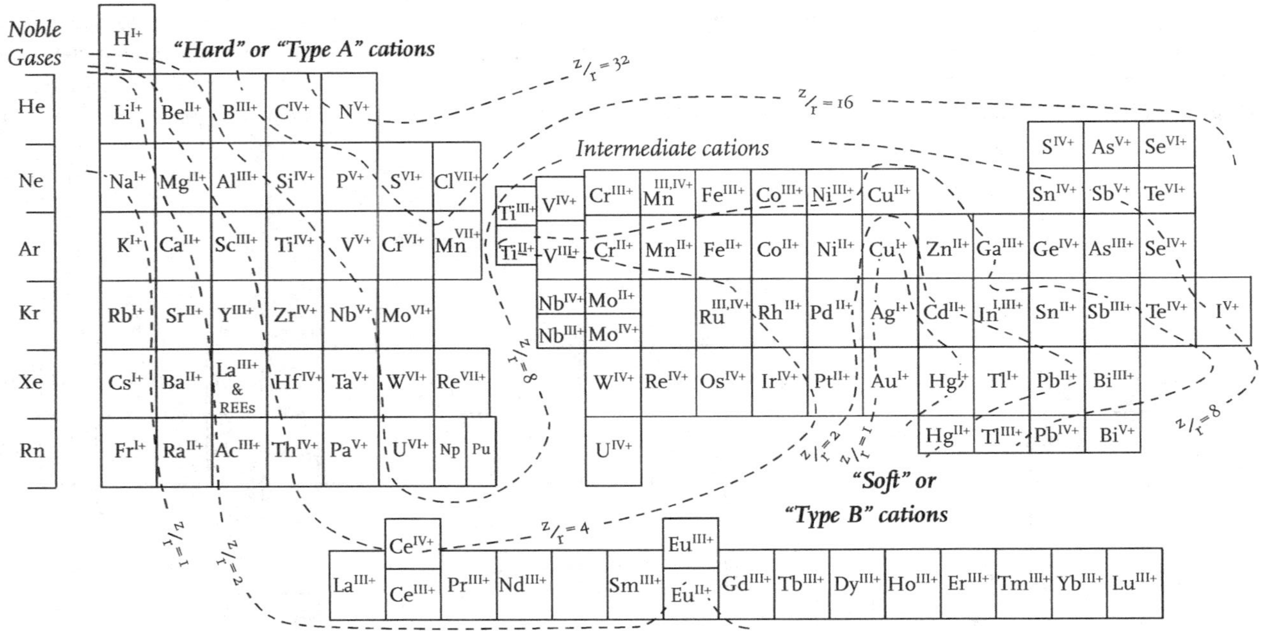

FIGURE 11.1 Skeletal framework underlying the left ("cation") side of the Earth Scientist's Periodic Table of the Elements and Their Ions. Actual table (Railsback 2003, 2011) has many colored symbols indicating occurrences in minerals, natural waters, soils, etc., which fall in discrete swaths across this table. Contours show values of ionic potential (oxidation number divided by radius, with the latter in angstroms). Fig. 11.2 shows the right side of the Table, and Fig. 11.3 shows a more complete representation of the hard cations.

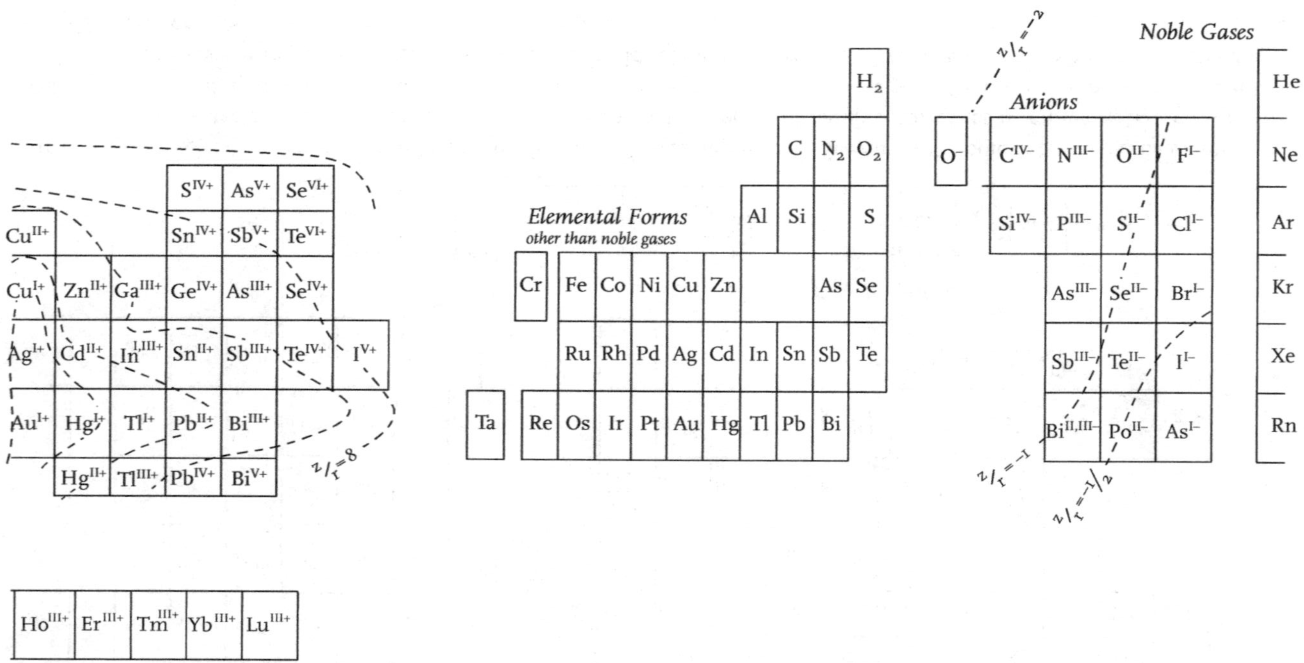

FIGURE II.2 Skeletal framework underlying the right ("anion") side of the Earth Scientist's Periodic Table of the Elements and Their Ions. Actual table (Railsback 2003, 2011) has many colored symbols indicating occurrences in minerals, natural waters, soils, etc., which fall in discrete swaths across this table. Contours show values of ionic potential (oxidation number divided by radius, with the latter in angstroms). An inset on the original table shows the range of oxidation number possible for C and N. Fig. II.1 shows the left side of the table.

oxidation number of IV+ and the ionic radius of 0.4 Å (0.04 nm) yields a value of ionic potential of +10. Values of ionic potential can be calculated for each non-elemental form in the new table, and those values can be contoured, yielding curves that sweep across the table independent of columns and rows. The significance of these curves is that they isolate ions with similar natural behavior (for example, members of the oblique group of Be^{II+}, Al^{III+}, Ti^{IV+}, Zr^{IV+}, and Ta^{V+} discussed above all have ionic potential between 4 and 8), providing a contextual framework in which to understand the known behavior of chemical matter and to predict the behavior of analogs about which less is presently known (e.g., Sc^{III+}).

Patterns in the New Table

The Earth Scientist's Periodic Table of the Elements and Their Ions consists of four blocks of information. The left presents the naturally occurring hard cations,[6] the positively charged entities for which a naïve understanding of oxidation state would suggest loss of all outer-shell electrons[7], ranging from H^{+} to U^{6+} (Pearson 1963, 1968) (Fig. 11.3). The central block presents the naturally occurring cations ranging from intermediate (those with at least a few outer-shell electrons, as in Fe^{II+}) to soft (those with many outer-shell electrons, as in Au^{I+}). A block farther to the right presents the elements that occur in nature in elemental form, which range across the conventional table from Ta to O_2 (if we for the moment ignore the noble gas elements). The rightmost block presents the naturally occurring anions, which range from I^{I-} to C^{IV-}. At the margins of the table, to both left and right, are the noble gas elements.

Decorating these blocks are symbols indicating natural occurrences of the various chemical entities, as reported from previous geochemical (and other) studies (Fig. 11.3). Red symbols indicate entities early to form, or stable in, solids at high temperature. For example, three different red symbols label cations common in phases that formed early in the cooling of the solar system, cations that form minerals early in the cooling of magmas, and cations that are not depleted from Earth's mantle but instead remain in Earth's hot interior.

---

6 Pearson (1963, 1968) proposed a scheme to categorize ions by the ligand atoms with which they make their most stable complexes, which is largely a function of the polarizability of the base. Table 1 of Pearson (1963) and Tables 3 and 4 of Pearson (1968) categorized acids and bases (largely cations and anions) as "hard," "borderline" or "soft." Figure 8 of Railsback (2007) characterized ions in a gradient from hard to soft on the basis of their associations in minerals.

7 The more common viewpoint, following Fajans' Rules (Fajans, 1923), is that bonding of species with high formal oxidation numbers is essentially covalent. In this respect, the Earth Scientist's Periodic Table of the Elements and their Ions would more correctly be named the "Earth Scientist's Periodic Table of the Elements and Their Non-elemental Forms." However, the fact that high-oxidation-state entities (e.g., N^{V+}) combine only with low-oxidation-state cations of low ionic potential (e.g., K^{I+} and Na^{I+}) (Railsback 2005, Fig. 5) to form simple minerals (e.g., nitrates) suggests that the positive charge of those high-oxidation-state entities is sufficiently concentrated to reject the positive charge of higher-oxidation-state cations of greater ionic potential like Mg^{II+}.

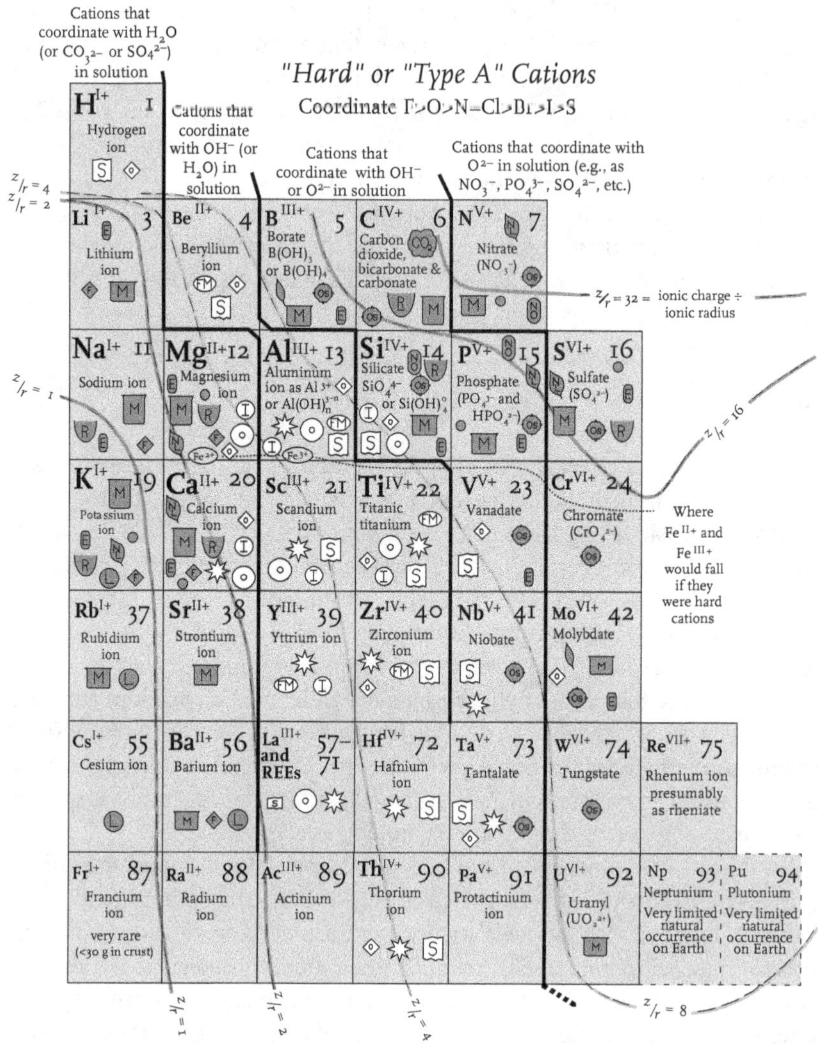

FIGURE 11.3 Gray-scale presentation of the leftmost part of the Earth Scientist's Periodic Table of the Elements and Their Ions. White-filled symbols generally indicate ions that readily enter into, or stay in, natural solids, whereas dark-filled symbols indicate ions that are less likely to enter or stay in solids. Specifically, white-filled symbols indicate ions that are enriched in soils, that form deep-sea ferromanganese nodules, that are not depleted from the mantle, that are part of calcium-aluminum inclusions (early-forming phases in the solar nebula), and that occur in minerals that form early in the crystallization of igneous rocks. Dark-filled symbols indicate natural occurrences as atmospheric gases, in solution in natural waters, as limiting nutrients, and in minerals that form late in the crystallization of igneous rocks. Actual table (Railsback, 2003, 2004, 2011) uses color to distinguish further between symbols, and much information on the actual table (atomic masses, ionic radii, isotopes, decay paths) has been deleted here for the sake of clarity. Font sizes for chemical symbols are keyed to abundance of elements in Earth's crust. Other forms of elements (e.g., C^{IV-}, Mo^{IV+}, and U^{IV+}) are shown farther right in the table, as shown in Figs. 11.1 and 11.2.

Brown symbols mark cations that form or stay in solids at low temperatures—both the cations enriched in residual soils during chemical weathering and the cations that leave seawater to enter ferromananganese nodules on the seafloor. Reds and browns thus combine to label cations that enter into or stay in natural solids, which are largely solids where negative charge is supplied by O^{II-}. With that said, the utility of the new table to earth scientists emerges, because these red and brown symbols fall in a distinct swath across the leftmost and central blocks of the table, and in a swath bounded by the contours of ionic potential with values of 4 and 8 (Fig. 11.3). In short, a pattern of predictable behavior emerges, and a pattern that is not coherent in the conventional periodic table.

In the original publications (see above), complementing the red and brown symbols are blue and green symbols. Blue symbols mark the cations and anions that, on land, have been leached from soils and are abundant in groundwater, and that at sea are abundant in seawater rather than precipitating readily from it. Other blue symbols indicate cations that are present in magmas but only enter solids late in the cooling of magmas. Blues thus collectively indicate ions that enter late into, or leave readily from, solids in which negative charge is supplied by O^{II-} (which is to say most of the solids of Earth). Green symbols mark ions that are nutrients for organisms, a biological phenomenon but one relevant to inorganic chemistry because a nutrient must be sufficiently soluble to pass from soil waters to roots and through cell membranes. Blues and greens thus combine to mark ions that enter solids late or leave them readily. Again, the utility of the new table to earth scientists emerges, because these blue and green symbols fall in distinct swaths across the table, most notably on each side of the red-and-brown swath in the block of hard cations (Fig. 11.3).

Other symbols continue this theme. Small yellow diamonds, which mark cations that form simple sulfide minerals, all fall in the central block of intermediate and soft cations, presumably because the outer-shell electrons of those cations allow somewhat covalent bonding with S^{II-} rather than O^{II-}. Larger yellow diamonds, which mark cations that form simple bromide and iodide minerals, fall in an even more restricted part of the central block delimited by the contour marking ionic potential of 1.0. Thus patterns of natural occurrence develop across the entire table and provide widespread predictability for earth scientists. Further patterns in the compositions, properties, and structures of minerals that become apparent with application of the new table were demonstrated by Railsback (2005, 2007).

New Neighborhoods and New Distinctions in the Periodic Table

Restrepo and Villaveces (2008) pointed out the usefulness of considering the periodic table in terms of neighborhoods. Historically, those neighborhoods have either been columns of the table (e.g., the alkali metals or alkaline earth metals), or broad areas of the table (e.g., metals, semi-metals, and non-metals).

The Earth Scientist's Periodic Table of the Elements and Their Ions is very much a document about neighborhoods of chemical entities, but different neighborhoods than those of the conventional periodic table. For example, Al^{III+}, Ti^{IV+}, and Th^{IV+} become part of one neighborhood (the oxide-loving neighborhood) on the new table (Fig. 11.3), whereas they are relegated to very different parts of the conventional table. Further, the use of contours for ionic potential makes Fe^{III+} (also isolated on the conventional table) part of that neighborhood too (Figs. 11.1 and 11.2). Similarly, S^{VI+}, Mo^{VI+}, and U^{VI+} become part of one neighborhood (Fig. 11.3), again despite isolation from each other on the conventional table.

New neighborhoods also arise because the new table distinguishes between multiple forms of many elements. As noted above, sulfur becomes part of as many as four neighborhoods, depending on its oxidation number and thus its location. Perhaps more critically, the two oxidized forms of iron become part of two different neighborhoods, with Fe^{II+} linked to Ca^{II+} and Mg^{II+} in one neighborhood, and Fe^{III+} linked to Al^{III+} and Sc^{III+} in another. These distinctions help clarify issues that were at best obscure for at least eighty years. For example, the famous geochemist Victor Goldschmidt[8] recognized the importance of ionic potential in determining the behavior of many elements in nature, and he presented this concept in a famous plot of ionic charge and ionic radius (Goldschmidt 1937, 665). On this plot, he divided elements into "kations" (the left dark swath of Figure 11.3), "elements of hydrolysates" (the central light swath of the figure), and "soluble complex anions" (the dark swath of the right side of the figure). However, the entities on Goldschmidt's plot were elements rather than ions, each presented once. Iron, as Fe^{II+} and Fe^{III+}, could not be shown in one place, and thus Goldschmidt's focus on elements rather than ions remarkably forced him to omit iron, a geochemically abundant and critical entity, from his plot. More modern versions of his diagram corrected this shortcoming (e.g., figure 19 of Mason 1952), but even then without explicit acknowledgement of the difference in oxidation state. In contrast, the Earth Scientist's Periodic Table of the Elements and Their Ions explicitly displays both Fe^{II+} and Fe^{III+} in the context of their respective neighborhoods of soluble and insoluble cations.

If the conventional table can be seen in terms of two or three large neighborhoods (e.g., metals, semi-metals, and non-metals), the Earth Scientist's Periodic Table of the Elements and Their Ions can be seen in terms of larger neighborhoods or directions that have chemical underpinnings but geological meanings. In general, more oxidized forms occur to the left in the table, and more reduced are found to the right. That is true at large scale (S^{VI+} in sulfate, N^{V+} in nitrate, and C^{IV+} in carbonate are the forms of sulfur, nitrogen, and

---

[8] Victor Goldschmidt (1888–1947) is commonly recognized as the father (Mason 1992) or founder (Kauffman 1997) of modern geochemistry, and his work is so fundamental to geochemistry that the Geochemical Society, the world's foremost organization for the advancement of geochemistry, hosts an annual Goldschmidt Meeting as its main conference and presents a V.M. Goldschmidt Award as its foremost award.

carbon at the left side of the table, and S^{II-}, N^{III-}, and C^{IV-} are at the far right). It is also true at smaller scale (within the cations in the left half of the table, more oxidized forms like Ti^{IV+}, Cr^{VI+}, Mo^{VI+}, and U^{VI+} are in the leftmost block of hard cations, whereas Ti^{III+}, Cr^{III+}, Mo^{IV+}, and U^{IV+} are near the center in the intermediate cations). In the context of Earth's history, this makes the trend rightward across the table the "pre-solar to Archean to early Proterozoic" direction in the table, because O_2 was absent or very scarce in Earth's atmosphere until about two billion years ago, which was the time of Earth's "Great Oxidation Event" (Holland 2002). In contrast, the trend leftward is the "late Proterozoic to modern" direction of the Table, because O_2 has been sufficiently abundant in the atmosphere over the last two billion years to oxidize many elements, among them C, S, Fe, Mo, Cr, and U (e.g., Kump et al. 2011; Wille et al. 2013). In the context of Earth's structure, the same pattern in the table makes the trend leftward the direction relating to modern near-surface processes in which O_2 is more abundant, and rightward the direction more relevant to processes within the Earth (or on Earth's moon), where O_2 is more scarce and most elements are in their more reduced states.

The new table also elucidates understanding of old neighborhoods recognized by Viktor Goldschmidt (1937). Goldschmidt divided the elements into five groups: liithophile, chalcophile, siderophile, atmophile, and biophile. Those categories have often been shown on the conventional periodic table (e.g., fig. 8.3 of Faure 1998; fig. 7.2 of White 2013), although such tables commonly neglect Goldschmidt's biophile group. On the new table, the lithophile elements are almost all represented among the hard cations, and the chalcophile elements are almost all represented among the intermediate to soft cations. The new table's capacity to show elements in more than one place allows illustration of the dual role of elements like Cr, which is lithophile or chalcophile depending on its oxidation state. The new table's emphasis on gradients also allows illustration of why one oxidation state of iron, Fe^{II+}, can be either lithophile (i.e., bonding with the O^{II-} of silicates) or chalcophile (i.e., bonding with the S^{II-} of sulfides) depending not on redox conditions but simply on the abundance of its potential bonding partners.

Boundaries, and the Ambivalent Behavior of Si^{IV+}

In the conventional periodic table, the columns or "groups," the rows, and the neighborhoods all imply the utility of dividing chemical entities into categories with defined boundaries. With its contours emphasizing the gradient of ionic potential, the Earth Scientist's Periodic Table of the Elements and Their Ions also allows recognition of entities whose distinctive behavior arises from their position on transitions in the table. Ca^{II+} and Pb^{II+} are examples, but perhaps none is more striking than Si^{IV+}, the most abundant positively charged entity in Earth's crust (Si is second only to O in crustal abundance of the elements). With an ionic potential of 10, Si^{IV+} sits at the boundary between solids-loving cations like Al^{III+} and more geochemically mobile entities like P^{V+} and S^{VI+} (Fig. 11.3). Thus Si^{IV+} enters into solids like olivine and anorthite early in the cooling of magmas, and it stays in the solids of residuals soils and sediments, in the quartz and clay minerals of those materials. On the other hand, Si^{IV+} is also a comparatively abundant solute in groundwater and river water, and an important nutrient solute in seawater. In plant biology, Si^{IV+} is sufficiently soluble to move from soil water to roots and into plants (especially grasses), but sufficiently insoluble to then be precipitated as a solid in those plants, as phytoliths (Colin & Meunier 2001). Given the abundance of Si^{IV+}, one might attribute this diversity of its roles to its ubiquity, but its position at a gradational boundary in the periodic table illustrates that this diversity has a chemical, rather than simply probabilistic, origin. Even the ambivalent chemical expression of the speciation of Si^{IV+} in solution, sometimes as $Si(OH)_4^{\circ}$ like $Al(OH)_n^{3-n}$ and sometimes as $H_4SiO_4^{\circ}$ like $H_2CO_3^{\circ}$ (e.g., Lasaga 1998 p. 116), reflects a diversity of roles resulting from a transitional position in a periodic table where gradients, as well as sharp boundaries, can be seen.

Reduction, or Expansion?

Scerri (2007 pp. xvii–xviii) discussed the question of whether the periodic table, and thus chemistry, "reduces to physics" (e.g., Pyykkö 2012). The development and propagation of the Earth Scientist's Periodic Table of the Elements and Their Ions instead suggests that the period table can be "expanded" to other sciences beyond the rivalry between chemistry and physics. In this respect, our consideration of the periodic concept need not be purely historical, because it can also look forward to a future in which other fields similarly develop new tables that reveal new potential in the periodic concept.[9]

References

Bernal, J.P., Railsback, L.B. 2008. Introducción a la Tabla Periódica de los Elementos y sus Iones para Ciencias de la Tierra. Revista Mexicana de Ciencias Geológicas 25, 236–246.

Cartledge, G.H. 1928. Studies on the periodic system. I. The ionic potential as a periodic function. Journal of the American Chemical Society 50, 2855–2863.

Colin, F., Meunier, J.D. 2001. *Phytoliths: Applications in Earth Sciences and Human History*. Taylor & Francis, Washington, D.C.

[9] To extend Nikos Psarros's (2001) metaphor of the unhappy partnership of a wheelchair-bound physics providing the sight to direct a blind but mobile chemistry as the two move through the world together, the natural (non-laboratory) sciences might be additional players who know the broader landscape sufficiently to suggest new places to go—and new applications of the periodic concept.

Faure, G. 1998. *Principles and Applications of Geochemistry*. Prentice Hall, Upper Saddle River, NJ.

Fajans, K. 1923. Struktur und Deformation der Elektronenhüllen in ihrer Bedeutung für die chemischen und optischen Eigenschaften anorganischer Verbindungen. Naturwissenschaften 11, 165–172.

Franco de Souza Lima, R., Railsback, L.B. 2012. Uma tabela periódica dos elementos e seus íons para cientistas da Terra. Terrae Didatica 8, 73–82.

Goldschmidt, V.M. 1937. The principles of distribution of chemical elements in minerals and rocks. Journal of the Chemical Society 140, 655–673.

Grimm, H.G. 1922. Periodisches system der atomionen. Zeitschrift für Physykalische Chemie 101, 410–413.

Heald, M.T. 1954. A periodic table of elements for geologists. Journal of Geological Education 2, 19–23.

Holland, H.D. 2002. Volcanic gases, black smokers, and the Great Oxidation Event. Geochimica et Cosmochimica Acta 66, 3811–3826.

Jaeger, J.L. 1957. *La Géochimie*. Presses Universitaires de France, Paris.

Jin, C. 2006. Understanding an earth scientist's periodic table of the elements and their ions. Journal of Jinhua College of Profession and Technology 6, 71–75.

Kauffman, G.B. 1997. Victor Moritz Goldschmidt (1888–1947): A tribute to the founder of modern geochemistry on the fiftieth anniversary of his death. The Chemical Educator 2, 1–26.

Kump, L.R., Junium, C., Arthur, M.A., Brasier, A., Fallick, A., Melezhik, V., Lepland, A., Crne, A.E., Luo, G. 2011. Isotopic evidence for massive oxidation of organic matter following the Great Oxidation Event. Science 334, 1694–1696.

Laing, M. 2004. Patterns in the periodic table—old and new. In: King, R.B., Rouvray, D.H. (eds.) *The Periodic Table: Into the 21st Century*, pp. 123–141. Research Studies Press Ltd., Baldock, Hertfordshire, England.

Lasaga, A.C. 1998. *Kinetic Theory in the Earth Sciences*. Princeton University Press, Princeton, NJ.

Mason, B. 1952. *Principles of Geochemistry*. John Wiley & Sons, New York.

Mason, B.H. 1992. *Victor Moritz Goldschmidt: Father of Modern Geochemistry*. Geochemical Society, San Antonio, TX.

Mazurs, E.G. 1974. *Graphic Representations of the Periodic System during One Hundred Years*. University of Alabama Press, Tuscaloosa, Alabama.

Novgorodova, M.I., Boronikhin, V.A., Generalov, M.E., Kramer, H. 1989. On native silicon in association with native gold and other metals. Doklady Akademii Nauk SSSR 309, 1182–1185.

Pearson, R.G. 1968. Hard and soft acids and bases. Journal of the American Chemical Society 45, 581–587.

Pearson, R.G. 1963. Hard and soft acids and bases, HSAB, Part I. Journal of Chemical Education 85, 3533–3539.

Psarros, N. 2001. The lame and the blind, or how much physics does chemistry need? Foundations of Chemistry 3, 241–249.

Pyykkö, P. 2012. The physics behind chemistry and the periodic table. Chemical Reviews 112, 371–384.

Railsback, L.B. 2003. An earth scientist's periodic table of the elements and their ions. Geology 31, 737–740 + insert.

Railsback, L.B. 2004. An Earth Scientist's Periodic Table of the Elements and Their Ions. Geological Society of America Map and Chart Series item MCH 092.

Railsback, L.B. 2005. A synthesis of systematic mineralogy. American Mineralogist 90, 1033–1041.

Railsback, L.B. 2007. Patterns in the compositions of oxysalt and sulfosalt minerals, and the paradoxical nature of quartz. American Mineralogist 92, 356–369.

Railsback, L.B. 2011. An Earth Scientist's Periodic Table of the Elements and Their Ions. Geological Society of America Map and Chart Series item MCH 092Rev (2011)

Rayner-Canham, G. 2002. An inorganic chemist's periodic table. In: Rayner-Canham, G., Overton. T., *Descriptive Inorganic Chemistry* (3rd ed.). W.H. Freeman and Company, New York.

Restrepo, G., Llanos, E.J., Mesa, H. 2006. Topological space of the chemical elements and its properties. Journal of Mathematical Chemistry 39, 401–416.

Restrepo, G., Villaveces, J.L. 2008. Elements' neighbourhoods and the periodic table. Journal of Science Education 10, 166.

Rollinson, H.R. 1993. *Using Geochemical Data: Evaluation, Presentation, Interpretation.* John Wiley & Sons, New York.

Scerri, E. 2007. *The Periodic Table: Its Story and Its Significance.* Oxford University Press, New York.

Szádeczky-Kardoss, E. 1959. Seltene elemente und geochemie. Freiberger Forschungshefte C58, 5–19.

White, W.M. 2013. *Geochemistry.* John Wiley and Sons, Chichester.

Wille, M., Nebel, O., Van Kranendonk, M.J., Schoenberg, Kleinhanns, I.C., Ellwood, M.J. 2013. Mo-Cr isotope evidence for a reducing Archean atmosphere in 3.46-2.76 Ga black shales from the Pilbara, Western Australia. Chemical Geology 340, 68–76.

CHAPTER 12

The Origin of Mendeleev's Discovery of the Periodic System

MASANORI KAJI
Department of Industrial Engineering & Management, Tokyo Institute of Technology, Japan

1.

D.I. Mendeleev (1834–1907) was a 19th-century Russian chemist whose discovery of the periodic law is a milestone of 19th century chemistry and continues to occupy a central place in chemistry.

Despite an enormous amount of study on the origin of Mendeleev's discovery of the periodic law, relatively little research has been done on either the conceptual or social origin of the law. This chapter aims to fill that gap.[1]

B. M. Kedrov (1903–1985) was a pioneering historian on the origin of Mendeleev's discovery. Kedrov discovered Mendeleev's archived materials related to his first periodic table (dated February 17, 1869) at the Mendeleev Museum in the St. Petersburg State Museum at the end of the 1940s. He then began full-fledged research on Mendeleev's discovery of the periodic law, which culminated in the publication of *The Day of a Great Discovery* (Kedrov 1958) and other books which are generally reliable resources (Mendeleev and Kedrov 1958, 1960).

In his meticulous reconstruction of Mendeleev's compilation of the first periodic table, Kedrov emphasized the "one-day discovery" (February 17, 1869[2]), and paid only scant attention to both Mendeleev's earlier works and the social context of the discovery. At about the same time, R. B. Dobrotin analyzed Mendeleev's earliest works, especially those of the 1850s, and argued that essential elements of the future periodic system already existed (Dobrotin 1953;

[1] This is a revised and expanded essay based on earlier versions in Kaji 1997 and Kaji 2003b.
[2] The dates for events in Russia are given in the Julian calendar, which lags behind the Gregorian calendar by twelve days in the 19th century and thirteen days in 20th century.

Shchukarev and Dobrotin 1954). However, Kedrov's version of the discovery had a dominant influence on Soviet history of science.

In the 1970s, A. A. Makarenya cautiously expressed some doubts about Kedrov's description of Mendeleev's scientific works (Makarenya 1972). After Kedrov's death, D. N. Trifonov proposed in 1990 "Version-2" of the discovery and criticized Kedrov's "one-day version" (Trifonov 1990).

Starting from a critical reevaluation of Kedrov's research, very different reconstructions have recently been proposed. These reconstructions pay more attention to Mendeleev's earlier works and incorporate a broader historical context. For example, I. S. Dmitriev proposed a different origin of the periodic law in apaper based on a very detailed analysis of the first edition of the *Principles of Chemistry* (Mendeleev's famous chemistry textbook, hereafter referred to as *Principles*), as well as its manuscripts and related archival materials (Dmitriev 2001). He extended his analysis to Mendeleev's early period, beginning with his graduation papers in 1854–1855 (Dmitriev 2004b). M. D. Gordin also studied Mendeleev's early works, and he asserted that the periodic system has its roots in type-theory organic chemistry in the mid 19th century (Gordin 2002). N. Brooks explored the pedagogical origin of the discovery as the writing of *Principles* and carried out a detailed analysis of Mendeleev's work during 1869–1871 to refine and substantiate the periodic law (Brooks 2000, 2002). I also have been following this line of work (Kaji 1987, 1988, 1997, 2002, 2003).

Kedrov has also analyzed the reception in Russia of Mendeleev's discovery, (Kedrov 1959). Its reception in France (Urbain 1934; Nekoval-Chikhaoui 1994), the United States, and Britain (Kauffman 1970; Solov'ev 1984; Brush 1996) has already been studied. These studies concentrated mostly on Mendeleev's success in persuading scientists to accept the law[3], and only minimal attention was paid to the social context of reception, especially in the scientific community. Brush suggested the need to extend his reception study on the United States and Britain to other countries, especially Germany and France. Nekoval-Chikhaoui, who had studied the French case, also called on scientists to conduct a comparative analysis of the diffusion of Mendeleev's discovery in different European countries. Only recently have comprehensive comparative studies on the reception, response, and appropriation of the periodic system in various European countries and Japan been published (Kaji et al. 2015) to meet their call and suggestion.

In 2007 Scerri published a work on the periodic table on the centenary of Mendeleev's death (Scerri 2007), following J. W. van Spronsen in the study of the history of the periodic system (Spronsen 1969). Scerri's philosophical analysis can be appreciated, including his emphasis on Mendeleev's distinction between abstract elements and simple substances, but contained some mistakes

[3] There is a counter argument against exaggeration of the role of prediction (Scerri and Worrall 2001) and criticism to such an assertion (Akeroyd 2003, 2004).

in his description of historical facts, resulting from his lack of knowledge of the Russian language.[4]

This chapter focuses on the cognitive and social origins and reception of Mendeleev's discovery, taking into account three factors based on original Russian sources: Mendeleev's earlier works and their relevance to the discovery; his concepts of chemistry, especially regarding the elements, which were developed through his work in the 1850s and 1860s; and the social context of the path to the discovery as well as its reception at home. Mendeleev's path to the periodic system was not a sudden one-day inspiration as Kedrov claimed, nor even a slightly longer development over a month as Trifonov proposed, but; rather it was the product of more than 15 years of preparation in chemistry and physics. Dobrotin first hinted at this approach, but I significantly extended and showed the more complicated story. Dmitriev also showed a very detailed analysis of the development from 1854 to 1871; however his analysis was too detailed to make the path to the discovery clear.

2. *Mendeleev's Early Research in the 1850s*

Dmitrii Ivanovich Mendeleev was born in 1834 in Tobol'sk, a Russian town in western Siberia founded at the end of the 17th century. His father was the head of a local gymnasium (a secondary school for the wealthy), and his mother came from a family of once-prosperous local merchants. After graduating from the gymnasium, Mendeleev entered the Main Pedagogical Institute of St. Petersburg, where his father had +once studied, in 1850. In his student days. Mendeleev published his first paper on a chemical analysis of minerals from Finland (Mendeleev 1854a[5]), followed by two more papers on the same subject[6] (Mendeleev 1854b, 1856a). His graduation thesis on isomorphism can be considered as a development in mineral analysis (Mendeleev 1855; *Works* I, 7–137). Even though it is essentially a literature review, this thesis shows signs of Mendeleev's future line of research: first, it shows his talent for compiling and systematizing large amounts of data; second, it mentions Auguste Laurent (1808–1853) and Charles Gerhardt (1816–1856), the reformers of chemistry in the 1840s and 1850s;[7] and third, its theme, the relationship between crystal form and composition, led Mendeleev to seriously consider the problem of similarity of substances. This was the beginning of his involvement in the issue of classifying substances.

4 See my review of his book (Kaji, 2007).

5 This is in German. For a Russian translation, see *Works*, XV, pp. 16–19.

6 The latter was in *Works* XV, pp. 20–23.

7 On the role of Laurent and Gerhardt in the reform of the atomic weight-molecular formula problem, see Rocke 1984 (pp. 191–214). On their influence in Russia, including on Mendeleev, see Faershtein 1961 (pp. 290–321). For a more recent study on Mendeleev and Gerhardt, see Gordin 2002 and 2004.

Mendeleev taught for a while at gymnasiums in southern Russia before returning to the capital to defend his master's thesis and to become a lecturer at the St. Petersburg University. His thesis on specific volumes illustrates his later line of thought even more clearly (Mendeleev 1856b; rpt. in *Works* I, 139–323; *Works* XXV, 112–228). Mendeleev adopted the atomic-weight system of Gerhardt and Laurent as well as Avogadro's hypothesis, which Mendeleev called Gerhardt's law. The thesis also shows Mendeleev's interest in the natural classification of substances based on their specific volume.

In the 1850s, Mendeleev was already interested in departures from the law of definite proportions in some groups of substances. This can be considered the beginning of his interest in indefinite compounds, which played an important role. In his 1856 *pro venia legendi* dissertation, he discussed the structure of silicates (Mendeleev 1856c; rpt. in *Works* XXV, 108–228), arguing that such compounds must be a kind of "alloy" of oxides, because, like alloys "to some extent they can vary their composition (and formula) without changing their forms and main properties" (*Works* XXV, 220).

3. Karlsruhe Congress and Its Effect on Mendeleev

In April 1859, Mendeleev went to Western Europe to further his studies in chemistry. During his two-year stay, he studied the cohesion (the forces that hold molecules together) of various substances, especially organic liquids, by measuring their capillary action. Mendeleev tried to find a universal formula to explain the relationships between cohesion, expressed in terms of surface tension, composition, density, and molecular weight (Mendeleev 1860a,b,c,d). The instruments that he purchased in Heidelberg, Bonn, and Paris enabled him to make precise measurements. In September 1860, Mendeleev attended the first International Chemical Congress in Karlsruhe, where significant contemporary issues in chemistry, especially atomic weights, were discussed. Along with everyone else in attendance, Mendeleev received a copy of the famous paper on the new atomic-weight system, written and distributed by Cannizzaro (Cannizzaro 1858[8]). After reading the paper, Mendeleev immediately wrote to his teacher A.A. Voskresenskii (1808–1883), in St. Petersburg, with an informative report on both the Congress and the content of Cannizzaro's paper. The letter was published in a St. Petersburg newspaper and in a Moscow journal in the same year (Mendeleev 1860e; Mendeleev and Kedrov 1958, pp. 660–669). In pointing out the inconsistency of Gerhardt's atomic weights of metals and arguing that Cannizzaro corrected them with "multiatomicity of metals," Mendeleev clearly recognized Cannizzaro's systematic use of the laws of atomic heats and isomorphism.

[8] For an English translation, see Cannizzaro 1969. This translation is also reprinted in Mary Jo Nye 1984, pp. 31–87.

Mendeleev returned to Russia in early 1861. That same year, while teaching at various schools, he completed his first textbook, *Organic Chemistry* (Mendeleev 1861). As a guiding principle for the textbook, he used what he called "the theory of limits" (Mendeleev 1862; rpt. in *Works* VIII, 22–27), to classify organic compounds based on their degree of saturation and their substitution reactions. Although this theory was soon forgotten with the advent of the structural theory of organic compounds, Mendeleev's textbook was well received in Russia. In 1862, the St. Petersburg Academy of Sciences awarded him the Demidov Prize for the most outstanding scholarly book written in Russian in the previous year. In this textbook, Mendeleev defined atomic weight as "the minimum quantity of an element in the compound molecules of the element" (Mendeleev 1861, v), following Cannizzaro. He also explicitly distinguished between "bodies" and "radicals," terming the former "something divisible (molecule)" and the latter "the theoretical notion" and "indivisible whole (atom)" (Mendeleev 1861,36n).

After completing the textbook, Mendeleev intended to write other textbooks on inorganic and theoretical chemistry. He tried to extend the idea of saturation ("the theory of limits") to inorganic compounds, but with little success (Makarenya 1982, 92–100). He also wrote an 1864 lecture notebook on theoretical chemistry (Mendeleev 1968, 11–68).

During the 1860s, Mendeleev further developed the concepts from his *pro venia legendi* dissertation of 1856. He gave the term "indefinite compounds" to substances that had constant physical properties, but varied composition, such as solutions, alloys, isomorphous mixtures, and silicates. Such compounds had been little studied, and Mendeleev himself could not explain their formation properly. However, he emphasized the following points: they are not simple physical mixtures; some chemical energy must be involved in their formation; and they show some properties similar to those of definite compounds.[9] His doctoral thesis "On Compounds of Alcohol with Water," submitted in 1865, can be regarded as a study of solutions that arose from his interest in indefinite compounds (Mendeleev 1865; rpt. in *Works* VI, 1–152).

Underlying this interest was Mendeleev's concern that the formation or composition of indefinite compounds was difficult to explain in terms of atomic theory, which was based on the concept of definite proportions. Mendeleev was always cautious about atomic theory, and he made this clear in a lecture on theoretical chemistry published in 1864:

> In fact, while the atomic Theory was strongly supported by the law of definite chemical compounds, it was also challenged by the so-called indefinite compounds. (Mendeleev 1968, 24)[10]

[9] For his arguments during the 1860s, see Mendeleev 1968, pp. 9–69, especially pp. 11–14 and 26–59.

[10] Almost the same passage appears in the first part of the first edition of the *Principles of Chemistry* (Mendeleev 1868, pt. I, chap. 10; rpt. in *Works*, XIII. p. 337): "(Compounds with indefinite compositions... speak against the atomic doctrine as much as definite chemical compounds speak in its support."

4. *Mendeleev's Publications in the 1860s*

Mendeleev became a professor of chemistry at the St. Petersburg Technological Institute in 1864. He became an extraordinary professor of technical chemistry at the St. Petersburg University the following year, and was promoted to full professor at the end of the same year. In the fall of 1867 Mendeleev was transferred to the professorship of general chemistry to succeed Voskresenskii, his mentor, who left the university to become curator of the Kiev educational district the same year (Figurovskii and Elagina 1958, 228).

Mendeleev's research career in chemistry, which began in 1854, reached its first peak with the discovery of the periodic law in 1869. This discovery can also be considered the height of his social activity during this period. Those years, beginning in the middle of the 1850s after the Crimean War and running their course by the 1860s with the emancipation of the serfs in 1861, were a period of great change and reform in Russia. That was yet another attempt at social and economic change after the social and political reforms of Peter the Great in the early 18th century, and it has been called by some historians "the Great Reforms Era." It was also a time of change in chemistry: the dispute over the merits of different atomic-weight systems had finally been settled after the Karlsruhe Congress, and the classical organic structural theory had emerged.

The emergence of a new generation of chemists in Russia, who were eager to engage in original laboratory work in chemistry, was an important background to Mendeleev's activities during this period. The educational system, especially at the higher levels, was also reorganized during this period. Thanks to the large number of Russian chemists moving into posts in academic institutions, the Russian Chemical Society was founded in 1868 (Kozlov 1961, 11–30; Brooks 1989, 1998; Brooks et al. 2008), Mendeleev was one of the founding members.

Let us consider the objectives that the Russian chemists, including Mendeleev, expected to achieve during this period. These objectives can be classed as practical or theoretical. The practical objective was the education of qualified professionals for the new industrial production that Russia required. The theoretical objective was to deal with current theoretical problems, as well as experimental ones, in chemistry to meet the needs of a period when the classical foundations of chemistry were about to be laid down.

To illustrate this point, see the list of books Mendeleev published during the 1860s after returning from Europe (Table 12.1). The content of these books indicates that they all met the practical demands of Russian society. *Wagner's Technology*, for example, was initially the translation of German encyclopedic manuals on technology. As the editor, Mendeleev proposed translating the parts needed in Russia, that is, the parts on agricultural products and processing. Later, he expanded it by adding the translation of other related books and asked appropriate Russian specialists to write original texts. All were published

TABLE 12.1 Books published under Mendeleev's authorship or editorship

1861	*ORGANIC CHEMISTRY*, FIRST EDITION
1862	*Cahours' Textbook for Elementary General Chemistry*, the second pt. (translation)
	Wagner's Technology (1862–1869), 8 vol., (translation and compilation)
1863	*Organic Chemistry*, second edition
1864	*Gerhardt and Chancel's Analytical Chemistry, Qualitative Analysis* (translation)
1866	*Analytical Chemistry*, the second pt., vol. 1–3 (1866–1869)
1867	*Today's Development of Some Chemical Productions—From the Point of View of Application to Russia* (Report of the International Exposition at Paris in 1867)
1868	*The Principles of Chemistry*, 1st pt., 1st vol.

by the same company, "Obshchestvennaya pol'za" ["Social Benefit"], which published books and pamphlets on science and technology for the "social benefit and enlightenment of the people" (Tridtsatipyatiletie, 5).

Mendeleev started writing his textbook of inorganic chemistry, *Principles*, in 1868 for the freshman chemistry course in the university. Published between 1868 and 1871, *Principles* grew out of Mendeleev's need for a suitable textbook on chemistry in Russian, which was didn't exist when he began teaching at St. Petersburg University in the fall of 1867 as professor of general chemistry. As he wrote later:

> I began to write [*Principles*] when I started to lecture inorganic chemistry at the university after [the departure of] Voskresenskii and when, having looked through all the books, I did not find anything to recommend to the students. (Mendeleev 1951, 52–53)[11]

Therefore, Mendeleev's *Principles* was the culmination of his work to help satisfy his country's demands during that period, both theoretical and practical.

5. *The First Part of* The Principles of Chemistry *and the Classification of Elements: "Valence" as a Basic Principle*

Mendeleev's first table of elements, entitled "An Attempt at a System of the Elements Based on Their Atomic Weight and Chemical Analogies" (Mendeleev 1869a, 1958, 9) (hereafter referred to as "Attempt") was dated February 17, 1869. His first paper on the discovery, "The Relationship between Atomic Weight of Elements and Properties" (hereafter referred to as "Relationship") (Mendeleev 1869b, 1958, 10–31) was read at the meeting of the newly established Russian Chemical Society[12] on March 6 by N. A. Menshutkin (1842–1907),

[11] This is from an autobiographical essay in 1905, two years before his death.

[12] The inaugural meeting of the Russian Chemical Society was held on November 6, 1868. The first paper on Mendeleev's discovery was presented at the fifth meeting only 4 months after its founding.

the secretary of the society. The paper was published in the combined second/third issue of the first volume of the society's journal the following May (Mendeleev 1869b).

Mendeleev stated the close relationship between the discovery of the law and *Principles* in "Relationship":

> In undertaking to prepare a textbook called "Osnovy khimii" [*Principles*], I wished to establish some sort of system of simple bodies in which their distribution is not guided by chance, as might be thought instinctively, but by some sort of definite and exact principle (Mendeleev 1869b, 65).

Beginning with Kedrov, many historians emphasized the importance and coupled Mendeleev's first construction of the periodic table to the writing of *Principles*.

Let us examine here as well "Relationship" and *Principles*, especially the early chapters of its second part, which must have been written around the same time. As shown earlier, when Mendeleev started to write *Principles*, he was keenly aware of the need to seek "some sort of definite and exact principle" like "the theory of limits" in his textbook.

It seems natural, from retrospective reconstruction, that Mendeleev should pay attention to atomic weights as "definite and exact principles." However, Mendeleev paid very little attention to atomic weights in the first part of *Principles*. He mentioned the atomic weights of only 22 of the most familiar elements of the time (Mendeleev 1868, rpt. in *Works*, XIII. p. 342). In the second chapter of the first part, he listed 63 known elements in alphabetical order without mentioning their atomic weights[13] (77–82). It seems likely that the existence of indefinite compounds made Mendeleev accept the limitations of atomic theory and the narrow scope of atomic weights (340–341).

Even though Mendeleev was wary of atomic theory because of exceptions to the law of definite proportions, he insisted on the existence of distinct chemical elements that were clearly distinguished from simple bodies. He argued this point in his first series of lectures at St. Petersburg University in the fall of 1867[14]: "it is necessary to distinguish the concept of a simple body from that of all element. A simple body, as we already know, is a substance, which, taken individually, cannot be altered chemically by any means produced up until now

[13] The third edition has the same table, but it includes atomic weights.

[14] *Lektsii po obshchei khimii 1867/68 g.*, Lecture V, St. Petersburg, rpt. in *Works*, XV, pp. 381–382. A lithographic edition of these lecture notes was found in the library of the former Bestuzhev women's courses, one of the most important higher educational institutions for women in pre-revolutionary Russia. Consisting of 16 lectures, these notes are similar to the first half of part 1 of *Principles*, written in 1868. In the fifth lecture, there is a table of 63, ordered alphabetically by their Latin names. Twelve of these elements had incorrect atomic weights, which would have been unlikely after the discovery of the periodic law. All this evidence shows that these lecture notes are most probably the record of Mendeleev's lectures of general chemistry given at St. Petersburg University in the fall of 1867.

or be formed through the transformation of any other kinds of bodies. An element, on the other hand, is an abstract concept; it is the material that is contained in a simple body and that can, without any change in weight, he converted into all the bodies that can be obtained from this simple body."

A similar definition of an element and the same argument for the need to distinguish clearly between elements and simple bodies were also presented in the first part of *Principles* (Mendeleev 1868, pt. 1, chap. 2; rpt. in *Works*, XIII, 73–74; Mendeleev, 1869c, chap. 15; rpt. in *Works*, XIII, 488–490).

Thus, this distinction between "simple bodies" and "elements" is essentially the same as that between "bodies" and "radicals" in Mendeleev's 1861 organic chemistry textbook, but without any mention of atoms or molecules. Paradoxically, then, it appears that Mendeleev eventually came to believe that weight is an invariable characteristic of elements thereby leading to his periodic system—but not by adhering to the concept of atoms, because of the failures of the law of definite proportions. It is reasonable to suppose that he refined the concept of the elements to incorporate individual chemical entity without employing the notion of atoms because of the supposed limitations of atomic theory.

One should notice that Mendeleev organized the first part of *Principles* on the basis of the principle of valence in this stage; he clearly stated this at the end of part 1 of *Principles*. During the 1860s, the theory of valence enjoyed great success, contributing to the development of a new theory of organic chemistry, that is, a structural theory of organic compounds. At the time Mendeleev wrote his textbook, which was based on a pre-structural theory ("the theory of limits"), he did not hold a high opinion of the concept of valence and admitted its importance only as a means to simplify a system (Mendeleev 1861, xv).

However, maybe because of the great success of valence theory, Mendeleev for a brief time used a kind of structural formula in his lectures on organic chemistry at the Technological Institute at St. Petersburg (Mendeleev 1868b; rpt. in *Works*, XXV, 360–362). Thus, Mendeleev tentatively took the valences of the elements as a basic principle in writing the *Principles*:

> Though we do not give absolute significance to division of bodies by their atomicity [valence], we will use it as a principle for arranging other elements in the row, since there are no other better principles. (Mendeleev 1869c, part 1, chap. 19; rpt. in *Works*. XIII, 651–652)

First, he discussed univalent hydrogen, divalent oxygen, trivalent nitrogen, and tetravalent carbon (Mendeleev 1869c, chap. 19; rpt. in *Works*, XIII, 650–652). Mendeleev argued that their compounds can be "types" [Gerhardt's types] for all the other compounds. Obviously, Gerhardt's "type theory" was influential here since Mendeleev had been familiar with it since his student days. However, he did not mention Gerhardt and went directly to the concept of valence, for which he used the word "*atomnost'*" (atomicity).

6. *The Second Part of* Principles of Chemistry *and the Classification of Elements: Change of a Guiding Principle*

After his treatment of the univalent halogens, which concludes the first part of the textbook, Mendeleev began the second part with a description of the univalent alkaline metals. At the end of the chapter on heat capacity, which follows the alkaline metals, he explained that he would next describe the alkaline earth metals, which are divalent and not analogs of copper, which awkwardly exhibit both uni- and divalence (Mendeleev 1870a, chap. 3; rpt. in *Works*, XIV, 120–121). Although he had followed the principle of valence to this point in the textbook, he immediately began the next chapter on a different principle, comparing alkaline-earth metals with alkaline metals on the basis of their atomic weights.

In this connection, it should be noted that toward the end of "Relationship" Mendeleev stressed that the "purpose of my paper would be entirely attained if I succeed in turning the attention of investigators to the relationships in the size of the atomic weights of *non-similar* elements, which have, as far as I know, been almost entirely neglected until now" (Mendeleev 1869b, 77). He emphasized the word "non-similar." and alkaline metals and alkaline-earth metals were obviously such non-similar elements.

If Trifonov's modification of Kedrov's analysis of Mendeleev's process (Trifonov 1990) is accepted, then Mendeleev noticed this comparison of non-similar groups of elements in the middle of February 1869, and he first compiled the central part of the table based on this principle. According to Trifonov and Kedrov, with the help of cards for the chemical elements[15] which he made for this occasion, Mendeleev finally succeeded in organizing a table of all the known elements on the basis of their atomic weights. He completed this on February 17, 1869. Clearly, at that moment, Mendeleev had conceived that atomic weight might be the fundamental numerical property of the elements.

In "Relationship" Mendeleev wrote (Mendeleev 1869b, 66)

> No matter how properties of simple bodies may change in the free state, *something* remains constant, and when the element forms compounds, this *something* is material existence and establishes the characteristics of the compounds, which include the given element. In this respect we know only one constant peculiar to an element, namely, the atomic weight. The size [magnitude] of the atomic weight, by the very essence of matter, is common to the simple body and all its compounds. Atomic weight belongs not to coal or diamond but to carbon.

This "something" emphasized in the quotation above exactly corresponds to Mendeleev's definition of the elements. In other words, atomic weight belongs to elements!

As a result of this reconceptualization, or discovery, Mendeleev realized that he should use atomic weights, not valence, as the guiding principle for the

[15] There is a dispute on the existence of such cards among historians. See, for example, Gordin, 2001, 75n.95.

			Ti = 50	Zr = 90	? = 180.
			V = 51	Nb = 94	Ta = 182.
			Cr = 52	Mo = 96	W = 186.
			Mn = 55	Rh = 104,4	Pl = 197,4.
			Fe = 56	Ru = 104,4	Ir = 198.
			Ni = Co = 59	Pj = 106,6	Os = 199.
H = 1			Cu = 63,4	Ag = 108	Hg = 200.
	Be = 9,4	Mg = 24	Zn = 65,2	Cd = 112	
	B = 11	Al = 27,4	? = 68	Ur = 116	Au = 197?
	C = 12	Si = 28	? = 70	Sn = 118	
	N = 14	P = 31	As = 75	Sb = 122	Bi = 210?
	O = 16	S = 32	Se = 79,4	Te = 128?	
	F = 19	Cl = 35,5	Br = 80	J = 127	
Li = 7	Na = 23	K = 39	Rb = 85,4	Cs = 133	Tl = 204.
		Ca = 40	Sr = 87,6	Ba = 137	Pb = 207.
		? = 45	Ce = 92		
		? Er = 56	La = 94		
		? Yt = 60	Di = 95		
		? In = 75,6	Th = 118?		

Д. Менделеев.

{*Отдельное издание 1 марта 1869 г.*}

FIGURE 12.1 The first published version of Mendeleev's periodic system in "Attempt" and "Relationship" paper (Mendeleev 1958, 9).

remainder of his textbook. This was when he began the chapter on alkaline earth metals. He defined the concept of elements without a notion of atoms; he considered atomic weights to be the fundamental property of the elements, but they were not necessarily based on atomic theory, which was still speculative in some respects. Thus, the scope of atomic weights would have to be broader than that of definite proportions on which the atomic theory was thought to be based. Mendeleev even once suggested the use of the word "elementary weight" instead of "atomic weight" (Mendeleev 1871b, 136n).

Therefore, starting from the arrangement of non-similar elements in the size of their atomic weights, he finally reached a system of elements ("Attempt") on February 17 (Fig. 12.1).

This success assured Mendeleev that atomic weight was the basis for the arrangement of the "abstract elements." The comparison of the groups of elements according to the atomic weights led Mendeleev to conclude that the distribution of the elements according to their atomic weights "does not disturb the natural similarities which exist between the elements" (Mendeleev 1869b, 67; Mendeleev 1958, 18–20; Leicester and Klickstein 1952, 440). In the conclusion of "Relationship,", Mendeleev stated that the size of the atomic weight determines the nature of the element.

The recognition of the importance of the concept of chemical elements for the discovery of the periodic law is the key factor for a historical analysis of

Mendeleev's discovery. Mendeleev's 1871 paper, published in the German journal *Annalen der Chemie und Pharmacie*, was the best known and most comprehensive paper on the periodic law.[16] It is curious that the paper starts by referring to the need to make a distinction between the terms "simple body" and "element," which were often confused (Mendeleev. 1871b, 133–134). Some, such as Paneth (1965), Bensaude-Vincent (1986), and Scerri (2000) have discussed the relevance of this distinction. Paneth proposed that basic substance and simple substance were separate concepts. Scerri further discussed Paneth's distinction in identifying abstract elements with basic substances. Bensaude-Vincent argued that with Mendeleev the concept of element obtained a positive criterion different from Lavoisier's purely negative definition and showed Mendeleev's deep understanding of what became a systematic classification (Bensaude-Vincent 1986, 12, 14). I agree with those authors on the importance of the distinction between elements and simple bodies for Mendeleev's discovery. However, they all failed to elucidate the origin of Mendeleev's distinction. This chapter shows that the distinction originated from Mendeleev's earlier research in the 1850s and 1860s, especially from his work on indefinite compounds.

I admit that the distinction between elements and simple bodies is not the only conceptual origin of Mendeleev's discovery, but it is arguably the central and most important one.

7. Mendeleev's Work on the Periodic Law after His First Periodic Table and the Early Reception of the Law in Russia

After showing his "Attempt," in the second half of "Relationship" Mendeleev presented the following five problems or research proposals from his new classification:

(1) determining positions in the table for certain elements
(2) other possible forms of the periodic law of the elements
(3) the relationship between the chemical properties of elements and their groups
(4) the correction of some elements' atomic weights
(5) undiscovered elements

Mendeleev's research after the "Attempt" can be regarded as the development of problems proposed in "Relationship." He concentrated his efforts on solving these problems until the end of 1871.

"Relationship" was published in the new journal of the Russian Chemical Society in May. Here, one finds the first reaction to Mendeleev's table. F. N. Savchenkov (1832–1903), a chemist in the government's mining agency said

[16] Mendeleev had written the paper in Russian and his friend, Felix Wreden (1841–78), translated it. The English translation was published in the *Chemical News* in 1879–1880 (Mendeleev, 1880).

that there was a similar table of elements in a chemistry textbook written by Englishman William Odling, which was published in 1865 and translated into Russian by Savchenkov himself in 1867. Mendeleev maintained that since there was no explanation of the table, Odling himself attached no theoretical significance to his table.

The actual studies done by Mendeleev over the following three years can be divided into four categories:

(1) relationship between the periodic law and physical properties of chemical compounds and the correction of atomic weights using this relation.
(2) relationship between groups of elements and formula of chemical compounds
(3) possible forms of the periodic law of the elements
(4) undiscovered elements

Table 12.2 shows Mendeleev's papers, presentations, and manuscripts related to the periodic system. From this table, one can see that Mendeleev presented papers mainly to the Russian scientific community. Out of 21 pieces of work during this period, 19 were related to the periodic law. Besides writing other books, pamphlets, and manuscripts, he gave eight oral presentations in the Russian Chemical Society (six of them were published later as a single paper), published seven papers in the *Zhurnal russkogo khimicheskogo obshchestva* (*Journal of the Russian Chemical Society*), gave three oral presentations to the Congress of Russian Natural Scientists, and published one paper in the journal of the Academy of Sciences and two in German journals.

TABLE 12.2 Mendeleev's study on the periodic law of elements after 1869

Ann. =*Annalen der Chemie und Pharmacie*
AN =*Akademiya nauk* (St. Petersburg Academy of Sciences)
Ber. =*Berichte der Deutschen chemischen Gesellschaft*
RCS = *Russkoe khimicheskoe obshchestvo* (Russian Chemical Society)
SRE =*Sezd russkikh estelltvoispitatelei* (Congress of Russian Scientists)
TSRE =*Trudy sezda russkikh estestsvoispitatelei* (Proceedings of the Congress of Russian Scientists)
JRCS =*Zhurnal russkogo khimicheskogo obshchstva* (Journal of Russian Chemical Society). The numbers indicate the volume.
OR = Oral Report, P=Pamphlet, []: Only in manuscript
1869 Feb. 17 "Attempt of System of Elements on Atomic Weights and Chemical Similarity" (P) ["Attempt" in this paper]
Mar. 1 Preface for the 1st pt. of The Principles of Chemistry
Mar. 6 "The Relationship between Atomic Weight of Elements and Properties" (OR at RCS; JRCS vol. 1) ["Relationship" paper in this paper]
Mar. (second half) *The Principles of Chemistry*. 1st pt., 2nd vol.
Aug. 23 "On the Atomic Volume of Simple Bodies" (OR at SRE; TSRE)

(*Continued*)

TABLE 12.2 Continued

Oct. 2 "On the Quantity of Oxygen in Salified Oxycides and Valency of Elements" (OR at RCS; JRCS vol. 2)
Nov. 6 "On the Law of Heat Capacity and Complexity of the Carbon Molecule" (OR at RCS; JRCS vol. 2)
1870 Mar. 5 "On Metal Ammonia Compounds" (OR at RCS)
Feb. (end) or Mar. (early). *The Principles of Chemistry*, 2nd pt., 3rd vol.
Oct. 8 "On Thionic Acid" (OR at RCS; JRCS vol. 2)
Nov. 5 "On Compounds with NO_2 group" (OR at RCS: JRCS vol. 3)
Nov. 24 "Über die Stellung des Ceriums im System der Elemente" (OR at AN; Bulletin de l'Académie impériale des sciences de St.-Pétersbourg 1871)
Dec. 3 "The Natural System of Elements and Its Application to Show Properties of Undiscovered Elements" (OR at RSC; JRCS vol. 3)
1871 Feb. (end) *The Principles of Chemistry*. 2nd pl., 4/5th vol.
Mar. "Zur Frage über das System der Element" (Ber.)
Jul. Finished writing "Die periodische Gesetzmässigkeit der chemischen Elemente" (Ann.) and published in Germany on Nov. 6 (Gregorian calendar)
Aug. 21 "On Specific Volume of Chlorine Compounds" (OR at SRE)
Aug. 24 "On Crystal Water" (OR at SRE)
Oct. [On Some So-Called Molecular Compounds]
Nov. [On Polymerization in Mineral Compounds]
Nov. "Note on Peroxides" (JRCS vol. 3)
Dec. 2 "On the Atomic Weight of Yttrium" (OR at RCS)

While Mendeleev continued to pursue work on the periodic system, this work was either ignored or discounted by some Russian chemists. For example, in the fall of 1869, Nikolai Nikolaevich Zinin (1812–1880), a leading organic chemist and the first president of the Russian Chemical Society, advised Mendeleev to do "[real] work," meaning do something experimental, preferably on organic chemistry, which was the mainstream research discipline in Russia at that time. Mendeleev drafted a letter to Zinin in response, although apparently he did not send it.[17]

> December 24, 1869
> Dear Nikolai Nikolaevich [Zinin],
> When I was in your place last time, you advised me "to do work." Since I value your opinion, I would like to say a word. I did not say anything then, because I know that the matter on Butlerov[18] was not finished yet. But now it is possible, as I heard.

[17] Mendeleev's (undelivered) letter to Zinin was dated December 24th, 1869. Mendeleev Museum-Archive attached to St. Petersburg State University I Archive, 44, –1–A–6. See also Kedrov 1959 (pp. 243–244). There is a complete Japanese translation in Kaji, 1997 (223–226) and a partial English translation in Gordin 2004 (44).

[18] This likely refers to the proposed election of A. M. Butlerov to adjunct member of the St. Petersburg Academy of Sciences by Zinin with other members in December 1869 (Kedrov 1959, 244).

I write you directly: what do you want, that I leave my area [of study], that I busy myself with the discovery of new bodies, that I worry about how often people are citing me? I do not think so. I suppose that your words came from the fact that you do not know what I have done and you are not following what I am doing in my area. You cannot deny that I discovered the law of volumes and many people talked about it after me. You also cannot deny that I showed the law of limits for carbon compounds and introduced the first and what's more now existing concept of limits, which everybody is talking about. The first attempt and experiments on the relationship between composition and cohesion belong to me, which many started to work after me. My work on alcohol contains new ways, which provide the criteria of exactness to the research on the problem of unlimited compounds. The law of symmetry of simple body belongs to me, which promises great future. One can do without doing work?

I suppose that you do not know them, because you follow the different aspects of science. It is not vanity makes me speak this way. Believe me, it is not my advantage, but the right to defend in front of respected person. Even if Germans do not know my works, it is understandable and does not annoy me. I will not take any measure to make Germans know my works. When what I had done was stolen by others (for example, in 1856 I gave explanation of the anomaly of the vapor density and density formula, and Kopp did the same two year later. I established the concept of limits and Kekulé and Wurtz appropriated them), I would not say a word, because I do not have self-deception which is flagrant and harmful for science and because I despise priority disputes. I am not writing you to raise them, but to say directly and clearly the following:

I consider the elaboration of the facts of organic chemistry in our time as not leading to a goal as quickly as it did 15 years ago, and I'm not going to busy myself with the petty facts of this sprig of chemistry nor to worry about citations. These are none of my business. I ask either don't censure or judge me, or say already what the errors are in my works, and not that I am not working. I am asking you [to speak] about errors.

Believe me, Nikolai Nikolaevich, that even though this letter is rude, I would like to show you my admiration toward you clearly and defend myself. I think that you will understand only if you consider my creation. I would look at who would have done as much as I have in my position, and I attribute your words to a lack of attention to my works, which suffer precisely from the fact that they do not comprise only the one-sided interest found in studies today (not before nor after, I believe). I do not deny their benefit and as you know, I know them and only evaluate them by their merit.

Yours truly, D. Mendeleev.

In this letter, Mendeleev defended his current research activities, obviously stung by Zinin's criticisms. It is interesting to note that Mendeleev did not explicitly mention the periodic system by name, nor did he discuss his work on *Principles*. Instead, Mendeleev tried to downplay the importance of organic chemistry, arguing that at present, organic chemists were only concerned with "petty facts," not larger issues, as was the case fifteen years prior. Mendeleev then related an incident from 1856 where he had done some research but had not immediately published the results. Soon after, a German chemist had published the same results. What Mendeleev obviously was trying to argue in this letter was that his instincts about research topics were valid, since influential foreign chemists had also worked on them. It is clear from the tone of this letter that Mendeleev felt defensive about his current work on the periodic system but did not intend to give it up. It also appears that Zinin wanted Mendeleev to work on organic chemistry rather than the periodic system, which involved inorganic and physical chemistry. This implies that Zinin felt that organic chemistry, not inorganic or general chemistry was the most important field in chemistry at that time and that Mendeleev was wasting his time working on any other field.

Mendeleev's work on the periodic law was recognized by the majority of Russian chemists after two papers appeared on the "Natural System of Elements." They were the German paper on November 24 (Mendeleev 1870b)[19] and the Russian paper "The Natural System of Elements and Its Application to Show Properties of Undiscovered Elements" on December 3, 1870 (Mendeleev 1870c; Mendeleev 1958, 69–101). Mendeleev predicted properties of some undiscovered elements in detail. Here, one can also find a thoroughly refined, short form of the periodic table (Fig. 12.2) that divided all of the elements into eight groups. Even Zinin, who was once critical of Mendeleev's work, wrote a laudatory letter, dated February 18, 1871, to Mendeleev (Kedrov 1959, 246).

The first article on the periodic law by a Russian chemist other than Mendeleev was published in May 1871 in the Mining Journal (Savchenkov 1871,). The author was Savchenkov, who gave one of the earliest comments on Mendeleev's "Relationship." Savchenko's article was a long review of Mendeleev's work on the periodic system, based on the "Relationship" paper on the periodic law and on his paper "The Natural System of Elements" of December 3, 1870. Note that Mendeleev not only predicted properties of undiscovered elements, but for a while, he also tried to find them. He collected various minerals both in Russia and in Europe and often asked his friends and colleagues at St. Petersburg University, and some governmental institutions, to help him collect samples.[20]

By the end of 1871, Mendeleev had finished writing *Principles* and had summarized his work on the periodic law in a long article written for the German

---

[19] For a Russian version of this German paper, see Mendeleev 1953. pp. 151–157 or Mendeleev, 1958, pp. 59–68.

[20] Some private and official letters by Mendeleev requesting mineral samples remain (Mendeleev 1953, pp. 187, 649).

[31]	Группа I	Группа II	Группа III	Группа IV	Группа V	Группа VI	Группа VII	Группа VIII. переход к группе I
Типические . элементы	H = 1 Li = 7	Be = 9,4	B = 11	C = 12	N = 14	O = 16	F = 19	
Первый период: Ряд 1-й	Na = 23	Mg = 24	Al = 27,3	Si = 28	P = 31	S = 32	Cl = 35,5	
Первый период: – 2-й	K = 39	Ca = 40	– = 44	Ti = 50?	V = 51	Cr = 52	Mn = 55	Fe = 56, Co = 59, Ni = 59, Cu = 63
Второй период: – 3-й	(Cu = 63)	Zn = 65	– = 68	– = 72	As = 75	Se = 78	Br = 80	
Второй период: – 4-й	Rb = 85	Sr = 87	(?Yt = 88?)	Zr = 90	Nb = 94	Mo = 96	– = 100	Ru = 104, Rh = 104, Pd = 104, Ag = 108
Третий период: – 5-й	(Ag = 108)	Cd = 112	In = 113	Sn = 118	Sb = 122	Te = 128?	J = 127	
Третий период: – 6-й	CS = 133	Ba = 137	– = 137	Ce = 138?	–	–	–	– –
Четвертый период: – 7-й	–	–	–	–	–	–	–	
Четвертый период: – 8-й	–	–	–	–	Ta = 182	W = 184	–	Os = 199?, Ir = 198? Pt = 197?, Au = 197
Пятый период: – 9-й	(Au = 197)	Hg = 200	Tl = 204	Pb = 207	Bi = 208	–	–	
Пятый период: – 10-й	–	–	–	Th = 232	–	Ur = 240	–	
Высшая соляная окись	R^2O	R^2O^2 или RO	R^2O^3	R^2O^4 или RO^2	R^2O^5	R^2O^6 или RO^3	R^2O^7	R^2O^8 или RO^4
Высшее водородное соединениө			(RH^5?)	RH^4	RH^3	RH^2	RH	–

FIGURE 12.2 Short-form periodic system "the Natural System of Elements" in the Dec. 3, 1870 paper.

journal *Liebig's Annalen*. In this article, Mendeleev elaborated on his conception of the periodic law and showed how it could be used to correct the atomic weights of several little-studied elements, as well as predict the properties of some undiscovered elements. Later, Mendeleev looked back at this work and concluded: "This is the best collection of my views and reflections about the periodicity of the elements.... This is the main reason for my scientific renown because much was proved correct much later." This papers can be regarded as a kind of theoretical conclusion of his three years of research after drawing up his first table of elements ("Attempt").

Mendeleev's work on the periodic law was conveyed through German-speaking subjects of Imperial Russia, including translation of papers or summaries. For example, Viktor von Richter, a Baltic German who was a native speaker of German and also fluent in Russian, became the first correspondent-chemist of the German Chemical Society and sent many articles about the activities of the Russian Chemical Society to Germany, including Mendeleev's research. In 1874, von Richter published the first inorganic chemistry textbook in Russian, *A Textbook of Inorganic Chemistry Based on the Newest Point of View*, based on the periodic law (Richter 1874). Von Richter also refined the structure of his textbook on and added new information in every following edition up to the sixth.[21]

Mendeleev's main works on the periodic law were abruptly terminated by the end of 1871 (as it seems from outside). At the end of December 1871, he abruptly started a new research project on gas expansion. Even though there remained some problems, such as finding predicted undiscovered elements and the places of rare earths, both of which he engaged in for a while. He did not have sufficient patience to continue these chemical studies of the periodic law after publishing long theoretical papers on the law both in Russian and German. Mendeleev changed his attention to gas expansion. explaining that this new research as a search for a true physical foundation of the periodic law (Mendeleev 1953, 226; Mendeleev and Kedrov 1960, 671; Gordin 2004, chap. 3). Here again, he turned to a more fundamental problem, thinking now that inorganic chemistry is only concerned with "petty facts," using Mendeleev's own words in his letter to Zinin.

8. *Mendeleev and Lothar Meyer*

In November 1882, D. I. Mendeleev and Lothar Meyer (1830–1895)[22] both received the Davy Gold Medal from the Royal Society in London for "research on the classification of the elements" (Seubert 1896, 1119). This means that they were both recognized, at least in Europe, as the discoverers, independently, of the periodic law of the elements (Ihde 1964, 243–251). Although their

[21] For the details of von Richter's textbook, see Kaji et al. 2015, pp. 27–28.
[22] For a biography of Meyer, see, Bedson 1896; Seubert 1896; Benfy 1981.

research paths often crossed, their social backgrounds were leagues apart. Let us look at Meyer's research on classification of chemical elements briefly to illustrate the historical context of Mendeleev's discovery. For the more detailed, balanced analysis on Lothar Meyer's work on the periodic system one should consult M. Gordin (Gordin 2012, 2013).

Both Mendeleev and Meyer attended the First International Congress of Chemists at Karlsruhe in 1860. At that time, Mendeleev was on a two-year study stint in Heidelberg, and Meyer was a private docent at Breslau University. At the Congress, both of them received S. Cannizzaro's famous paper on the atomic weight system. They were both much impressed by the paper and each wrote a chemistry textbook based on Cannizzaro's system after returning home.

Meyer wrote a textbook on theoretical chemistry in 1864 (Meyer 1864),[23] while Mendeleev wrote an organic chemistry textbook. At the end of his book, Meyer arranged 50 elements in three separate tables. The arrangement was based on the valences and the differences in atomic weights of analogous elements. In 1866, Meyer left Breslau and became a professor of chemistry in the School of Forestry at Neustadt-Eberswalde. In 1868 when he was preparing a new edition of his textbook, he managed to combine the three tables. He arranged 52 elements in 15 columns. However, because publication of the new edition was delayed, this table was not published during his lifetime.[24]

As already shown in detail, in early February 1869, Mendeleev compiled his first table of the elements based on atomic weights (see "Attempt"). His first paper on the periodic law of the elements ("Relationship"), was published in May 1869.

It is interesting to note that Mendeleev asked Friedrich Konrad Beilstein (1838–1906), another Russian imperial subject of German ancestry (then professor of chemistry at St. Petersburg Technological Institute), to translate the summary of his first paper on the periodic law for publication in a German journal. One of Beilstein's students translated the paper but the translator mistakenly interpreted the word "periodical (periodichnyi) as "stepwise (stufenweise)" (Mendelejeff 1869d). The translated summary was sent in March or April, 1869, to *Zeitschrift für Chemie*, where Beilstein himself was one of the editors. According to the student who translated it, the summary was sent to the journal through none other than Lothar Meyer (Krolikov 1969, 129).

At the end of the same year, Lothar Meyer submitted his famous paper, "The Nature of the Chemical Elements as a Function of their Atomic Weights," which was published early in 1870 (Meyer 1870). Although Meyer admitted in the paper that his table was essentially the same as Mendeleev's,[25] his table of elements was more refined than Mendeleev's first table, especially in showing

23 This small book developed into Meyer's major work by being constantly revised throughout his life: the 2nd edition was published in 1872; the 3rd in 1876; the 4th in 1883; the 5th in 1884; and the 6th in 1896 (only the first volume was published).

24 This table was first published in 1895 (Seubert 1895).

25 Meyer quoted the abstract in *Zeitschrift für Chemie* (Mendeleev, 1869d).

the so-called transition metals clearly. Meyer also had the correct atomic weight of indium (it was incorrect in Mendeleev's first table). He succeeded in vividly conveying a periodic dependency of properties of the elements on their atomic weights by plotting the solid-state atomic volumes of those elements that could be isolated as elementary substances against their atomic weights. However, he started his paper quite differently from Mendeleev: Meyer started his paper with a speculation related to Prout's hypothesis (Meyer 1870. pp. 354–355), in which the existence of primordial matter was assumed. Meyer's conclusion was very tentative and cautious:

> It would be hasty to alter on such uncertain bases the previously accepted atomic weights. On the whole, one may not, for the time being, give much importance to arguments of the sort given, or expect to obtain reliable measurement of atomic weights determined from their specific heat or vapor density. They may, however, serve even now to turn our attention upon doubtful and uncertain assumptions and to challenge us to a renewed testing of them. And again, conversely, this testing will help to clarify and extend the meager beginnings of our knowledge of atoms. (Meyer 1870. p. 364)

Why did he lack confidence? One interpretation would be that this only shows Lothar Meyer's gentlemanly attitude as a researcher. On the other hand why was Mendeleev so bold? I will get back to this point soon.

Mendeleev began to be recognized in Western Europe after the appearance of his long paper on the periodic system in German in *Annalen der Chemie und Pharmacie* in 1871 (Mendeleev, 1871b). Thanks to this paper, Lothar Meyer became assured of the correctness of the periodic law and tried to apply it fully to systemize inorganic chemistry. Meyer's paper "For the Systemization of Inorganic Chemistry" in 1873 was one result (Meyer 1873).

Another factor in the change in attitudes toward Mendeleev in Russia in the late 1870s through the early 1880s was the failure of Mendeleev to be elected to full membership in the Academy of Sciences in 1880. The event became a huge scandal because it was seen as rejecting a worthy Russian, instead electing a foreigner (Meyer, a German) to continue a trend in which the Academy of Sciences was dominated by foreigners. Mendeleev became very well known to the entire country, not just among chemists or scientists, developing into a national icon (Dmitriev 2002; Gordin 2004, 113–141).

The priority dispute between Mendeleev and Meyer was one of several that Russian chemists had with foreign chemists in the years after the 1860s. These priority disputes stoked Russians' feelings of nationalism, which were growing very strong at this time in many other aspects of Russian life. Even some Russians who did not particularly like Mendeleev appeared to defend him in this priority dispute.

In 1880, Lothar Meyer (Meyer 1880a. 1880b) and Mendeleev (Mendeleev, 1880) had a disagreement over the priority and contributions toward the discovery and the development of the periodic law in the journal of the German Chemical Society. Here, Meyer suggested that there was an unfavorable atmosphere toward theoretical work in the German chemical community for "a paper without any new data" (Meyer 1880a, 263). However, Mendeleev was bold enough to send a very long paper "without any new data" to the same journal. The younger members of the editorial board of the journal[26] strongly supported the publication of Mendeleev's paper. Erlenmeyer wrote to Mendeleev on his long paper in *Annalen* on November 5, 1871 (using the Gregorian calendar), saying, "Thank you for your paper, which interests me very much. The paper contains many wonderful and great thoughts" (Mendeleev 1953,703–704). With powerful arguments in the paper itself—such as a detailed prediction of undiscovered elements which had persuaded Russian chemists to take Mendeleev's side—these favorable conditions in the German editorial board helped to promulgate Mendeleev's discovery.

On Mendeleev's side his concept of the chemical elements—abstract, stable, material ingredients in the form of atomic weights ("elementary weight"), not necessarily based on the speculative concept of atom—corresponded to the state of chemistry in the mid 19th century and gave Mendeleev a conceptual base to do research with confidence. Simultaneously, the support and encouragement of the Russian chemical community, as well as "German mediation" through the help of Baltic Germans and some Russophile editors in Germany[27], helped Mendeleev to concentrate his studies on the periodic law during this important period. These factors supported Mendeleev's boldness.

9. *Conclusion*

There are conceptual as well as social origins of Mendeleev's discovery of the periodic law. One must agree with Bensaude-Vincent's argument that the hypothesis of a primordial matter was the logical consequence of Lavoisier's negative definition of elements (Bensaude-Vincent 1986, 12). In Lavoisier's definition, one cannot distinguish simple bodies from elements. Lothar Meyer's speculative primordial idea of atoms was one of the natural consequences of Lavoisier's definition. On the other hand, Mendeleev's concept of elements, which gave a positive criterion of the elements, as an invisible abstract and yet material entity distinct from the empirical concept of simple bodies, is therefore, a clear departure from Lavoisier's concept of elements. However, Bensaude-Vincent's argument is based more on a retrospective reconstruction, saying nothing about the origin of Mendeleev's new concept of elements. By following Mendeleev's early works in the 1850s and 1860s, I encountered the problems of indefinite compounds as one of the sources of his new understanding, or at least as a stimulus for further understanding.

[26] After the retirement of Kopp from the editorship in March 1871, Emil Erlenmeyer (1825–1909), Jacob Volhard (1834–1910) and Liebig were the editors of *Annalen* (Rocke, 1993, 288). Erlenmeyer was very much a Russophile.

[27] For the "German mediation," see Kaji et al. 2015, pp. 24–26.

The social situation in 1860s Russia helped Russian chemists to establish a new specialized society, the Russian Chemical Society, and posed educational tasks to them that stimulated Mendeleev to write new textbooks, including *The Principles of Chemistry*. This newly formed society provided support to Mendeleev to carry out his research on the law, both as a first audience and as a means to promulgate his ideas outside of Russia, including "German mediation" in the form of Germans and German subjects of the Russian Empire.

References

The following abbreviations are used: *Works*, followed by volume number in roman numerals, D. I. Mendeleev, *Sochineniya* [Works], 25 vols., Leningrad: Izd. AN SSSR, 1934–1954; ZhRKhO, *Zhumal russkogo khimicheskogo obsbchestva* [Journal of Russian Chemical Society]; VIET, *Voprosy istorii estestvoznaniya i tekhniki* [Problems of History of Science and Technology].

The dates for events in Russia are given in the Julian calendar, used in Russia until January 1918. The Julian calendar lags twelve days behind the Gregorian calendar in the 19th century. The Gregorian dates are used for events outside Russia.

Transliterations follow a modification of the standard British Library formal.

Akeroyd, M. 2003. Predictions, Retrodictions and the Periodic Table. *Foundations of Chemistry*. 5, 85–88.

Akeroyd, M. 2004. Predictions and the Periodic Table: A Response to Scerri and Worrall. *Journal for General Philosophy of Science*. 34, 337–355.

Bedson, P. 1896. Lothar Meyer Memorial Lecture. *Journal of the Chemical Society*. 69, 1402–1439.

Beusaude-Vincent, B. 1986. Mendeleev's Periodic System of Chemical Elements. *British Journal for the History of Science*. 19, 3–17.

Brooks, N.M. 1989. The Formation of a Community of Chemists in Russia, 1700–1870. Unpub. Ph.D. dissertation, Columbia University.

Brooks, N.M. 1998. The Evolution of Chemistry in Russia during the Eighteenth and Nineteenth Centuries. In: D. Knight and H. Kragh (eds.), *The Making of the Chemist: The Social History of Chemistry in Europe*, 1789–1914, 163–176. Cambridge University Press, Cambridge.

Brooks, N.M. 2000. Dmitrii Mendeleev's Principles of Chemistry and the Periodic Law of the Elements. In: A. Lundgren and B. Bensaude-Vincent (eds.), *Communicating Chemistry: Textbooks and Their Audiences*, 1789–1939, 295–309. Science History Publication, Canton, MA.

Brooks, N.M. 2002. Developing the Periodic Law: Mendeleev's Work During 1869–1871. Foundation of Chemistry, 4, 127–147.

Brooks, N.M., Kaji, M., and Zaitseva E. 2008. Russia: The Formation of the Russian Chemical Society and Its History until 1914. In: Anita Kildebæk Nielsen and Sona Strobanova (eds.), *Creating Networks in Chemistry: The Founding and Early History of Chemical Societies in Europe*. RSC Publishing, Cambridge, 281–327.

Brush, S.G. 1996. The Reception of Mendeleev's Periodic Law in America and Britain. *Isis*, 87, 595–628.

Cannizzaro, S. 1858. Sunto di un corso di filosofia chimica. Il nuovo cimento, 7, 321366.

Cannizzaro, S. 1969. *Sketch of a Course in Chemical Philosophy. Alembic Club Reprints*, no. 18. Alembic Club, Edinburgh, re-issue.

Dobrotin, R.B. 1953. Rannii period nauchnoi deialel' nosti D. I. Mendeleeva kak etap na puti k otkrytiyu periodicheskogo zakona. Candidate dissertation, Leningrad State University.

Dmitriev, I.S. 2001. Nauchnoe otkrytie in Statu Nascendi: periodicheskii zakon D.I. Mendeleeva. *VIET* (l), 31–82.

Dmitriev, I.S.: Skuchnaya istoriya (o neizbrannii D. I. Mendeleeva v Imperatorskuyu akademiyu nauk v 1880 g). *VIET* (2)., 231–280 (2002).

Dmitriev, I.S. 2004b. Nauchnoe otkrytie in Statu Nascendi. in *Chelovek epokhi peremen: Ocherki o D. I. Mendeleeve i ego vremeni* [A Man in a Changing Epoch: A Treatise on D. I. Mendeleev and his Times] Khimizdat, St. Petersburg, 90–207.

Faershtein, M.G. 1961. *Istoriya ucheniya o molekule v khimii (do 1860g.)*. Izd. Akademii Nauk SSSR, Moscow.

Figurovskii, N.A., Elagina, K.Ts. 1958. Aleksandr Abramovich Voskcesenskii (1809–1880). *Trudy institltta istorii estestvoznaniya i tekhniki* 18, 213–235.

Gordin, M.D. 2001. The Ordered Society and Its Enemies: D.I. Mendeleev and the Russian Empire, 1861–1905. Ph.D. dissertation, Harvard University.

Gordin, M.D. 2002. The Organic Roots of Mendeleev's Periodic Law. *Historical Studies in the Physical and Biological Sciences*, 32(1), 263–290.

Gordin, M.D. 2004. *A Well-Ordered Thing: Drimtrii Mendeleev and the Shadow of the Periodic Table*. Basic Books, New York.

Gordin, M.D. 2012. The Textbook Case of a Priority Dispute: D. I. Mendeleev, Lothar Meyr, and the Periodic System. In: Biagiolo, M. and Riskin, J. (eds.) *Nature Engaged: Science in Practice from the Renaissance to the Present*. Palgrave Macmillan, New York, 59–82.

Gordin, M.D. 2013. The Table and the Word: Translation, Priority, and the Periodic System of Chemical Elements. *Ab Imperio*, 3, 53–82.

Ihde, A.J. 1964. *The Development of Modern Chemistry*. Harper & Row, New York.

Kaji, M. 1987. On Mendeleev's Path to the Discovery of the Periodic Law: Analysis of His Work Between 1854 and 1869 (in Japanese). *Kagakusi Kenkyu* (Journal for the History of Science Society of Japan), 26, 129–139.

Kaji, M. 1988. The Historical Context of the First Edition of the Mendeleev's Chemistry Textbook *The Principles of Chemistry* (in Japanese). *Kagakusi Kenkyu* (Journal for the History of Science Society of Japan), 27, 24–36.

Kaji, M. 1997. *Mendeleev's Discovery of the Periodic Law of the Chemical Elements: The Scientific and Social Context of His Discovery*. (In Japanese with English summary.) Hokkaido University Press, Sapporo.

Kaji, M. 2002. D.I. Mendeleev's Concept of Chemical Elements and the Principles of Chemistry. *Bulletin for The History of Chemistry* 27(1), 4–16.

Kaji, M. 2003. Mendeleev's Discovery of the Periodic Law: The Origin and the Reception. *Foundations of Chemistry*, 5(3), 189–214.

Kaji, M. 2007. Book review. Eric R. Scerri, *The Periodic Table: Its Story and Its Significance*, Oxford University Press, New York, *HYLE-International Journal for Philosophy of Chemistry*. 13(2), 119–121.

Kaji, M., Kragh, H., Palló, G. (eds.). 2015. *Early Responses to the Periodic System*. Oxford University Press, New York.

Kauffman, G.B. 1970. The Reception of Mendeleev's Idea in the United States and Mendeleev's Correspondence with American Scientists. *Archives International d'Histoire des Sciences* 23, 84–106.

Kedrov, B.M. 1958. *Den' odnogo velikogo otkrytia*. Izd. Sotsial'no-ekonomicheskoi Literatury, Moscow (reprinted in 2001, Editorial URSS, Moscow).

Kedrov, B.M. 1959. *Filosofskii analiz pervykh trudov D.I. Mendeleeva o periodicheskom zakone (1869–1871)*. Izd-vo Nauka, Moscow.

Kozlov, V.V. 1961. *Ocherki istorii khimicheskikh obshchestv SSSR*. Izd-vo AN SSSR, Moscow.

Krolikov, V.A. 1969. Dve oshibki v pervykh publikatsiyakh o periodicheskom zakone D.I. Mendeleeva. *VIET* (29), 129–131. Leicester, H.M., Klickstein, H.S. (eds). 1968. *Source Book in Chemistry* 1400–1900. Harvard University Press, New York.

Makarenya, A.A. 1972. *D.I. Mendeleev i fiziko-khimicheskie nauki—Opyt nauchnoi biografii D. I. Mendeleeva*. 2-e izd. Energoizdat, Moscow. (1-e izd. 1972).

Mendeleev, D.I. 1854. Rezul'taty khimicheskogo analiza selenginskoi umbry, proizvedennogo v laboratorii Glavnogo pedagogicheskogo instituta. *Vestnik imp. russkogo geograficheskogo obshchestva. St.-Petersburg.*, pt.12, section 5, Geogr. izv. i sves, 36–37.

Mendeleev, D.I. 1855. Izomorfizm v svyazi s otnosheniyami kristallicheskoi formy k sostavu. *Gornyi zhurnal* ch. 3. kn. 8, 229–400; kn. 9, 405–467.

Mendeleev, D.I. 1856a. Pyroxen aus Ruskiala in Finnland. *Verhandlungen der Russischkaiserlichen mineralogischen Gesellschaft zu St.-Petersburg.* Jg 1855–1856, 207–210.

Mendeleev, D.I. 1856b. Udel'nye ob"emy. Master's thesis.

Mendeleev, D.I. 1860a. Chastichnoe stseplenie nekotorykh zhidkikh organicheskikh soedinienii. *Khimicheskii zhumal N. Sokolova i A. Engel'garta* 3(2), 81–97; (3), 145–170.

Mendeleev, D.I. 1860b. Notiz über die Ausdebehnung homologer Flüssigkeiten. *Ann. Chem. Pharm.* 114(2), 165–169.

Mendeleev, D.I. 1860c. Sur la cohésion de quelques liquides et sur le rôle de la cohésion moleculaire dans les réactions chimiques des corps. *Compt. rend.* 51(3), 97–99.

Mendeleev, D.I. 1860d. Khimicheskii kongress v Karlsrue (Pis'mo k A.A. Voskresenskomu iz Geidel'berga ot 7 sent. 1860g *St.-Peterburgskie vedomosti*, 1860, 2 noyabrya. No. 238, 4. stlb. 2–5.

Mendeleev, D.I. 1861. *Organicheskaya khimiya*, Izd. tov-va "Obshchestv. pol'za," St. Petersburg.

Mendeleev, D.I. 1862. Essai d'une théorie sur les limites des combinaisons organiques. *Bulletin de l'Académie impériale des sciences de St.-Pétersbourg* 4, 245–250.

Mendeleev, D.I. 1865. O soedinenii spirta s vodoi. Rassuzhlenie, predstavlennoe v fiziko-matematicheskii fakul'tet I.S. Peterburgskogo universiteta dlya poluchenii doktora khimii. Tip. tov-va "Obshehestv. pol'za," St. Petersburg.

Mendeleev, D.I. 1868a. *Osnovy khimii*. ch. 1, v. 1, gl. 1–11. Tip. tov-va "Obshehestv. pol'za," St. Petersburg .

Mendeleev, D.I. 1868b. Lektsii organicheskoi khimii. chitannye D.I. Mendeleevym v Tekhnologicheskom institute. Uglerodistye vodorody. Izd. M. Matyukevich, St. Petersburg.

Mendeleev, D.I. 1869a. Opyt sistemy elementov, osnovannoi na ikh atomnom vese i khimicheskom skhodstve.

Mendeleev, D.I. 1869b. Sootnoshenie svoistv s atomnym vesom elementov. *Zh. Russ. Khim. Obshch.* 1(2/3), 60–77.

Mendeleev, D.I. 1869c. *Osnovy khimii.* ch. 1, v. 2, gl. 12–22. Tip. tov-va "Obshehestv. Pol'za," St. Petersburg.

Mendelejeff, D.I. 1869d. Über die Beziehungen der Eigenschaften zu den Atomgewichten der Elemente. *Zeitschrift für Chemie* 12, 405–406.

Mendeleev, D.I. 1870a. *Osnovy khimii.* ch. 2, v. 3, gl. 1–8. Tip. tov-va "Obshehestv. Pol'za," St. Petersburg.

Mendeleev, D.I. 1870b. Über die Stellung des Ceriums in System der Elemente. *Bulletin de l'Académie Impériale des sciences de St.-Petersburg* 16(1), 45–51.

Mendeleev, D.I. 1870c. Estestvennaya sistema elementov i primeninie ee k ukazaniyu svoistv neotkrytykh elementov. *ZhRKhO* 3(2), 25–56.

Mendeleev, D.I. 1871a. *Osnovy khimii.* ch. 2, v. 4, gl. 9–22. Tip. tov-va "Obshehestv. Pol'za," St. Petersburg.

Mendelejeff, D.I. 1871b. Die periodische Gesetzmässigkeit der chemischen Elemente. *Annalen der Chemie und Pharmacie Supplementband* 8, 133–229.

Mendeleeff, D.I. 1879. La Loi périodique des éléments chimiques. *Moniteur Scientifique*, 21, 691–735.

Mendeleeff, D.I. 1879. The Periodic Law of the Chemical Elements. *Chemical News*, vol. 40, 231–232, 243–244, 255–256, 267–268, 279–280, 291–292, 303–304.

Mendeleeff, D.I. 1880a. The Periodic Law of the Chemical Elements. *Chemical News*, vol. 41, 2–3, 27–28, 39–40–49–50, 61–62, 71–72, 83–84, 93–94, 106–108, 113–114, 125–126 (1880).

Mendelejeff, D. 1880b. Zur Geschichte des periodischen Gesetzes. *Berichte der Deutschen Chemischen Gesellschaft* 13, 1796–1804.

Mendeléeff, D. 1905. The Periodic Law of the Chemical Elements. in *The Principles of Chemistry*, 3rd English ed., vol. II. Longmans, Green & Co., London, 493.

Mendeleev, D.I. 1951. *Arkhiv D.l. Mendeleeva* tom 1. avtobiograficheskie materialy, sbornik dokumentov. Izd-vo Leningradsckogo gosudarstvennogo universisteta, Leningrad.

Mendeleev, D.I. 1953. D. I. Mendeleev. Nauchyi arkhiv. T. 1. Periodicheskii zakon. Estestvennaya sistema elemenlov. Rukopisi i tablitsy. 1869–1871. Izd-vo AN SSSR, Moscow.

Mendeleev, D.I., Kedrov, B.M. (ed.). 1958. *Periodicheskii Zakon.* Izd-vo AN SSSR, Moscow.

Mendeleev, D.I., Kedrov, B.M. (ed.). 1960. *Periodicheskii Zakon. Dopolnitel'nye materialy.* Izd-vo AN SSSR, Moscow.

Mendeleev, D.I. 1968. *Izbrannye lektsii po khimii.* Izd-vo Vyshaya shkola, Leningrad-Moscow.

Meyer, L. 1864. *Die modemen Theorien der Chemie und ihre Bedeutung für die chemische Statik.* Verlag von Maruschke & Berendt, Breslau.

Meyer, L. 1870. Die Natur der chemischen Eiemente als Function ihrer Atomgewichte. *Annalen der Chemie und Pharmarcie Supplementband*, 7, 354–364.

Meyer, L. 1873. Zur Systematik der anorganischen Chemie. *Berichte der Deutschen Chemischen Gesellschaft*, 6, 101–106.

Meyer, L. 1880a. Zur Geschichte des periodischen Atomistik. *Berichte der Delltschen Chemischen Gesellschaft*, 13, 259–265.

Meyer, L. 1880b. Zur Geschichte des periodischen Atomistik. *Berichte der Deutschen Chemischen Gesellschaft* 13, 2043–2044.

Nekoval-Chikhaoui, L. 1994. Diffusion de la classification périodique de Mendéléïev en France entre 1869 et 1934. Ph.D. Diss., Univ. Paris-Sud U.F.R. Scientifique d'Orsay.

Nye M.J. 1984. *The Question of the Atom: From the Karlsruhe Congress to the First Solvay Conference*. 1860–1911. Tomash Publishers, Los Angeles.

Paneth, E.A. 1965. Chemical Elements and Primordial Matter: Mendeleeff's View and the Present Position. In: Herbert Dingle and G. R. Martin, eds., *Chemistry and Beyond*. Interscience Publishers, New York, 53–72.

Richter, V. von. 1874. *Uchebnik neorganicheskoi khimii po noveishim vozzreniyam*. Tip. Ivanaya Yavorskogo, Warsaw.

Rocke, A.J. 1984. *Chemical Atomism in the Nineteenth Century: From Dalton to Cannizzaro*. Ohio State University Press, Columbus.

Rocke, A.J. 1993. *The Quiet Revolution: Hermann Kolbe and the Science of Organic Chemistry*. University of California Press, Berkeley.

Savchenkov, S.F. 1871. Otnosheniya mezhdu alomnuimi vesami elementov. *Goryui zhumal*. ch. 2. No.5, 234–251.

Scerri, E.R. 2000. Realism. Reduction, and The "Intermediate Position." In: Bhushan, N., Rosenfeld, S. (eds). *Of Minds and Molecules*. Oxford University Press, Oxford, 51–72.

Scerri, E.R., Worrall, J. 2001. Prediction and the Periodic Table. *Studies in History and Philosophy of Science* 32A, 407–452.

Scerri, E.R. 2007. *The Periodic Table: Its Story and Its Significance*, Oxford University Press, New York.

Seubert, K. (ed.). 1895. *Das natürliche System der chemischen Elemente. Abhandlungen von Lothar Meyer und D. Mendelejeff.* Ostwald's Klassiker der exakten Wissenschaften Nr.68. Verlag von Wilhelm Engelmann, Leipzig.

Seubert, K. 1896. Lothar Meyer. *Berichte der Deutschen Chemischen Gesellschaft* 28(4), 1109–1146.

Shchukarev, S.A., Dobrotin, R.B. 1954. Pervye nauchye raboty D. I. Mendeleeva kak etap na puti k otkrytiyu periodicheskogo zakona. *Vestnik leningradskogo universiteta* (2), 165–177.

Solov'ev, Y.I. 1984. D. I. Mendeleev and The English Chemists. *Journal of Chemical Education* 61, 1069–1071.

Spronsen, J.W. van. 1969. *The Periodic System of Chemical Elements: A History of the First Hundred Years*. Elsevier, Amsterdam.

Trifonov, D.N. 1990. Versiya-2: k istorii otkrytiya periodicheskogo zakona D. I. Mendeleevym. *VIET* (2), 25–36; (3), 20–32.

Urbain G. 1934. Comment les idées de Mendeleiev ont été accueillies en France. *Revue Scientifique*, 72, 657–661.

CHAPTER 13

Richard Abegg and the Periodic Table

WILLIAM B. JENSEN
Department of Chemistry, University of Cincinnati, USA

Who Was Richard Abegg?

The German chemist Richard Wilhelm Heinrich Abegg (Fig. 13.1), was born on 9 January 1869 in Danzig (now Gdansk, Poland) (1). He received his PhD in 1891 from the University of Berlin for work in the field of organic chemistry done under the direction of August Hofmann. He switched to the new and rising field of physical chemistry immediately upon graduation, doing post-doctoral work in the laboratories of Wilhelm Ostwald at Leipzig and Svante Arrhenius at Stockholm, as well as serving as personal assistant to Walther Nernst at Göttingen. In 1897 Abegg was appointed professor of chemistry at the University of Breslau (now Wroclaw, Poland). In 1909 he moved to the local Technischen Hochschule, where he remained until his untimely death on 3 April 1910 at age 41 in a ballooning accident near Koszalin in what is now modern-day Poland.

As might be inferred from his association with Ostwald, Arrhenius, and Nernst, Abegg's research interests quickly focused on the newly formulated theories of ionic dissociation and chemical equilibrium, where he is credited with contributing to an understanding of the theory of freezing point depression and with writing two popular introductory textbooks on the use of the ionic theory and equilibrium in reinterpreting various traditional areas of chemical synthesis and analysis (2, 3).

With the discovery of the electron in 1897 Abegg soon became interested in its use to rationalize various electrochemical phenomena and in its possible implications for both the periodic table and chemical bonding. That year he published, in collaboration with Guido Bodländer, his theory of *electroaffinity* in which he postulated that electrochemical half-cell oxidation potentials could be used as a measure of an atom's attraction for electrons and that this, in turn,

FIGURE 13.1 Richard Abegg (1869–1910).

could be qualitatively correlated with periodic trends (Fig. 13.2) in such properties as molecular polarity, solubility, and the tendency to form complex ions (4, 5).

In short, Abegg was using this parameter in much the same fashion as the modern chemist uses the concept of electronegativity, even though our modern understanding of oxidation potentials reveals that they are in fact relatively complex and contain, in addition to ionization energies, contributions from atomization and solvation energies as well (6). As a consequence we would not expect the resulting correlations to be very precise and indeed they did attract some criticism (7). However, one long-term result of this interest was the publication, under Abegg's editorship, of the first definitive reference collection of oxidation potential values, though the book in question did not appear in print until after his death (8).

Abegg's Rule of Eight and the Periodic Table

Abegg's most famous contribution—his "rule of eight"—was first published in preliminary form in 1902 (9) and in much greater detail in 1904 (10). This was an electrochemical interpretation of the 8-column, short-form of Mendeleev's periodic table (Fig. 13.3). First proposed by Mendeleev in his famous review of 1871 (11), this particular version of the table quickly became the textbook norm and remained so at least until World War II (and even longer in the case of the Soviet Union) (12).

As may be seen from Figure 13.3, Mendeleev based the group assignment of a given element on the formula of its highest oxide and included a "type" formula for this oxide immediately below the group number. For groups IV-VII he also included a second type formula for the element's known compound with hydrogen and stated the following rule connecting the two series of formulas (11):

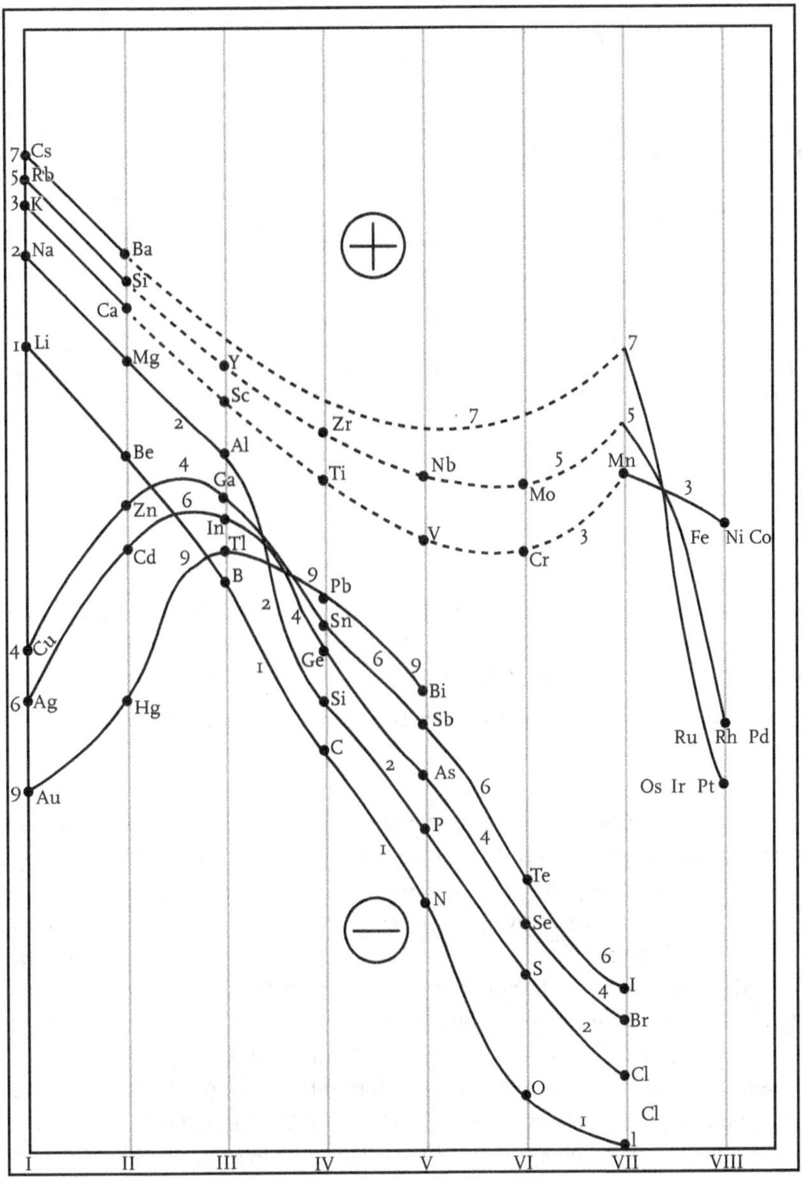

FIGURE 13.2 Abegg's plot of an element's "electroaffinity" as a function of its position in the 8-column, short form of the periodic table.

> The sum of the number of equivalents of hydrogen and oxygen which can be bound to only one atom of an element does not exceed eight.

The reason that Mendeleev used type formulas and the word "equivalent" instead of valence values, is that in 1871 he was still highly critical of the valence concept and felt that it was little more than an unnecessary relabeling of the law of multiple proportions (13).

Reihen	Gruppe I. — R^2O	Gruppe II. — RO	Gruppe III. — R^2O^3	Gruppe IV. RH^4 RO^2	Gruppe V. RH^3 R^2O^5	Gruppe VI. RH^2 RO^3	Gruppe VII. RH R^2O^7	Gruppe VIII. — RO^4
1	H = 1							
2	Li = 7	Be = 9,4	B = 11	C = 12	N = 14	O = 16	F = 19	
3	Na = 23	Mg = 24	Al = 27,3	Si = 28	P = 31	S = 32	Cl = 35,5	
4	K = 39	Ca = 40	— = 44	Ti = 48	V = 51	Cr = 52	Mn = 55	Fe = 56, Co = 59, Ni = 59, Cu = 63
5	(Cu = 63)	Zn = 65	— = 68	— = 72	As = 75	Se = 78	Br = 80	
6	Rb = 85	Sr = 87	? Yt = 88	Zr = 90	Nb = 94	Mo = 96	— = 100	Ru = 104, Rh = 104, Pd = 106, Ag = 108
7	(Ag = 108)	Cd = 112	In = 113	Sn = 118	Sb = 122	Te = 125	J = 127	
8	Cs = 133	Ba = 137	? Di = 138	?Ce = 140	—	—	—	— — — —
9	(—)	—	—	—	—	—	—	
10	—	—	? Er = 178	?La = 180	Ta = 182	W = 184	—	Os = 195, Ir = 197, Pt = 198, Au = 199
11	(Au = 199)	Hg = 200	Tl = 204	Pb = 207	Bi = 208	—	—	
12	—	—	—	Th = 231	—	U = 240	—	— — — —

FIGURE 13.3 The 1871 8-column, short form of the periodic table with Mendeleev's "type" formulas as group labels.

	Gruppe: 1	2	3	4	5	6	7
Normalvalenzen	+1	+2	+3	±4	−3	−2	−1
Kontravalenzen	(−7)	(−6)	(−5)		+5	+6	+7

FIGURE 13.4 Abegg's 1904 tabular summary of his rule of eight for groups I-VII.

Just as he had done in his various textbooks for other traditional chemistry subjects, Abegg now realized that he could also give Mendeleev's rule of eight an electrochemical interpretation based on the concept of polar valence values. The type formula for its highest oxide revealed an element's maximum possible electropositive valence, whereas the type formula for its hydrogen compound revealed its maximum electronegative valence. Instead of the sum of an element's oxygen valence and hydrogen valence being equal to eight, it was the sum of its electropositive and electronegative valence values that was equal to eight.

Mendeleev was well aware that the elements of groups I-III also formed hydrogen compounds of the types AH, AH_2, and AH_3 but did not include them on his table because their formulas violated his rule of eight. Abegg, on the other hand, realized that for these three groups the elements in question were more electropositive than hydrogen, whereas for groups IV-VII they were more electronegative. Thus their hydride formulas did not reflect their electronegative valence values and so failed to conform to the rule of eight. Instead, Abegg postulated hypothetical negative valence values of −7, −6, and −5 for these elements so that all of the groups of the periodic table would conform to his newer polar version of the rule of eight (Fig. 13.4) (14).

Whereas Mendeleev initially felt that the discovery that each element could have several valence values, depending on the kind of atom to which it was bound rather than a constant valence, invalidated the original valence concept,

Abegg embraced this variability and presented his positive and negative valence values as upper and lower bounds. Generally an element favored whichever of these limits had the smallest numerical value. Abegg called this an element's *normal valence*, whereas the higher valence of opposite polarity, which was less frequently used or not yet known in practice, was known as the element's *contra-valence* (10):

> Every element possesses a positive as well as a negative maximum valence whose sum is always eight. The maximum positive valence is always the same as the number of the periodic group to which the element belongs. Whether an element manifests its positive or its negative electrovalence depends on the differences in polarity [and thus the electroaffinities] of the elements with which it combines. The manifestation of one kind of valence appears to greatly hinder, but not completely to suspend that of the other kind. We shall call those valences which are less in number (<4), and therefore stronger than the others, the normal valences, and those which are greater in number than the others, and weaker because of contrary polarity, the contra-valences of the elements. Thus Cl possesses one negative normal valence and seven positive contra-valences, and analogously Ag, one positive normal valence and seven (hypothetical) contra-valences. The manifestation of the maximum valence is not necessary.... The contra-valences, in particular, are seldom completely employed. Increasing atomic weight [within a periodic group] facilitates their manifestation.

Abegg's confidence that he was on the right track was further reinforced by the application of his rule to the question of where the recently isolated noble gases should be placed in the periodic system. Up to this point these elements were usually placed in front of the alkali metals in a 0 group. While there was no doubt that 0 was the normal valence for these elements, the rule of eight also mandated that they should exhibit a contra-valence of either +8 or -8, thus placing them either in group 0, as before, or group VIII with the noble metals (Figs. 13.5 and 13.6). Unhappily, since a valence value of 0 has no intrinsic sign, it offered no guidance as to which of these possible choices was correct. Nevertheless, the implication that the noble gases should exhibit finite valences under proper

He	Li		F	Ne
Ne	Na		Cl	A
A	K		*Mn,*	*Fe, Co, Ni*
Fe, Co, Ni	*Cu*		Br	Kr
Kr	Rb	und	—	*Ru, Rh, Pd*
Ru, Rh, Pd	*Ag*		I	X
X	Cs		−1	−0
+0	+1		+7	+8
−8	−7			

FIGURE 13.5 Abegg's 1904 placement of the noble gases in group VIII in accord with their normal valence of 0.

8. = 0.	1.	2.	3.	4.	5.	6.	7.	8. = 0.	Gruppe
+0	+1	+2	+3	+4	+5	+6	+7	+8	+ Valenz
−8	−7	−6	−5	−4	−3	−2	−1	−0	− Valenz

FIGURE 13.6 Abegg's 1910 tabular summary of his rule of eight for groups I-VIII.

	IV. C, Si, &c.	V. N, P, &c.	VI. O, S, &c.	VII. F, Cl, &c.
+ :	4	3	2	1
− :	4	5	6	7
	—	—	—	—
Total	8	8	8	8

FIGURE 13.7 Lothar Meyer's 1884 electrochemical interpretation of Mendeleev's rule of eight. A typo in the original table has accidentally transposed the signs for the two rows.

conditions was sufficient to inspire Kossel in 1916 to correctly predict, using his own version of the ionic model and the choice of +8, that a noble-gas compound, should it ever be prepared, would most likely be a fluoride of xenon (15).

In his later writings (16), Abegg would deemphasize the concepts of normal and contra-valence and instead present his rule as the straightforward sum of an element's maximum potential electropositive and electronegative valence values (Fig. 13.6). In this form it should be noted that Abegg's electrochemical interpretation of Mendeleev's rule of eight had in fact been partially anticipated by Lothar Meyer nearly two decades earlier, when he wrote (17):

> It is not a little remarkable that in the fourth, fifth, sixth and seventh families the sum of the valencies toward electropositive and negative elements is eight, which is equal to the highest valency observed in the case of any given element (Ru, Os).

He summarized this result using the table in Figure 13.7. Unlike Abegg, however, Meyer was unwilling to extend his rule to groups I-III by boldly assuming hypothetical negative valence values, nor, in the absence of a knowledge of the noble gases, was he able to extend it to group VIII.

Had Abegg gone no further than Figure 13.6, his rule would still have been recognized as a key step in the development of modern valence theory. However, at the very end of his 1904 paper he offered—almost as a passing thought—an additional interpretation of his rule, not in terms of polar valence values, but in terms of the newly discovered electron (10):

> The sum 8 of our normal and contra-valencies has therefore simple significance as the number which represents for all atoms the positions of attachment for electrons; and the group number or positive valency indicates how many of the 8 positions of attachment must hold electrons in order to make the element electrically neutral.

In other words, for each column of the short periodic table:

$$e + v = 8 \quad [1]$$

FIGURE 13.8 Alfred Werner 1866–1919.

where e is the number of valence electrons in the neutral atom and v is the number of *valence vacancies*—a more descriptive term than either the awkward expression "position of attachment" or the later, more technical, quantum mechanical term "spin orbital." In this form Abegg's rule would go on to inspire further development of both the electronic theory of atomic structure and bond formation by such men as J. J. Thomson (18) and G. N. Lewis (19). It would also quickly enter the chemistry textbook, where it continued to be mentioned as late as the 1930s (20) and, as such, it would become an integral part of all future historical accounts dealing with the development of the electronic theory of chemical bonding (21, 22).

Most important for our purposes, however, is the fact that Abegg's rule is really the first attempt to provide, via equation 1, an electronic interpretation of the periodic table, despite the fact that the details of atomic structure were still largely unknown. Given this attempt, it is rather surprising that I have never come across an example of Abegg using a periodic table in which he replaced Mendeleev's type formulas with his own polar valence labels (especially in the form shown in Fig. 13.6).

Yet one cannot help but imagine he would have finally taken this logical step in connection with a project begun in 1908, when he assumed the editorship of a massive, multi-authored *Handbuch der anorganische Chemie* (*Handbook of Organic Chemistry*) (23). This was organized around the 8-column, short form of the periodic table with its *a* and *b* subgroups and was originally intended to illustrate in detail the application of Abegg's rule of eight. However, in true German fashion, the original projection of four volumes soon became subdivided into numerous parts and sections, many of which were 1000 pages or more in length and required separate printed volumes. In addition, the various parts and sections were published in random order as the various authors submitted their contributions. When the project finally came to a halt in 1933,

H 1.008																																He 4
Li 7.03																										Be 9.1	B 11	C 12	N 14.04	O 16.00	Fl 19	Ne 20
Na 23.05																										Mg 24.36	Al 27.1	Si 28.4	P 31.0	S 32.06	Cl 35.45	A 39.9
K 39.15	Ca 40.1																Sc 44.1	Ti 48.1	V 51.2	Cr 52.1	Mn 55.0	Fe 55.9	Co 59.0	Ni 58.7	Cu 63.6	Zn 65.4	Ga 70	Ge 72	As 75.0	Se 79.1	Br 79.96	Kr 81.12
Rb 85.4	Sr 87.6																Y 89.0	Zr 90.7	Nb 94	Mo 96.0		Ru 101.7	Rh 103.0	Pd 106	Ag 107.93	Cd 112.4	Jn 114	Sn 118.5	Sb 120	Te 127.6	J 126.95	X 128
Cs 133	Ba 137.4	La 138	Ce 140	Nd 143.6	Pr 140.5			Sa 150.3	Eu 151.79	Gd 156	Tb 160	Ho 162	Er 166		Tu 171	Yb 173			Ta 183	W 184.0		Os 191	Ir 193.0	Pt 194.8	Au 197.2	Hg 200.3	Tl 204.1	Pb 206.9	Bi 208.5			
	Ra 225	La α ?	Th 232.5						U 239.5						Ac ?													Pb α ?	Bi α ?	Te α ?		

FIGURE 13.9 Alfred Werner's 1905 long-form periodic table.

23 years after Abegg's death, it had mushroomed to 12 printed volumes and was still incomplete. Missing was volume I, which would have undoubtedly contained, had he lived, Abegg's introductory chapter on the periodic table and his rule of eight.

Defending the Short Table

As presented by Abegg, his rule of eight was, as we have seen, firmly wedded to the traditional short form of the periodic table in which all of the elements—whether main block, transition block, or inner-transition block—were wedged into just eight groups. Thus it is not surprising that Abegg felt compelled to defend this type of table when, less than a year after the publication of his paper on the rule of eight, it was challenged by no less a luminary than Alfred Werner (Fig. 13.8), the founding father of modern coordination theory (24). In its place, Werner proposed a long-form periodic table with periods of varying, rather than equal, length in which all reference to the *a* and *b* subgroups was abandoned (Fig. 13.9).

The resulting debate between Abegg and Werner would lead to a series of lengthy back-and-forths in the pages of the journal *Berichte* (25–27) in which Abegg would object to Werner's dismantling of the *a* and *b* subgroups and the resulting spatial separation of chemically related elements, his creation of many additional trivial groups of elements containing only two members each, his placement of Be and Mg above Zn rather than Ca, and especially his failure to incorporate Abegg's recent insights concerning the classification of the noble gases.

Though, in light of modern electron configurations, present-day chemists would probably favor Werner's long-form table over the short-form table championed by Abegg, Werner's table was not without its problems. He had, for example, postulated 15 rather than 14 groups for the inner-transition elements, resulting in a 33- rather than a 32-column table and, seduced by the lure of symmetry, which has misled so many purveyors of new periodic tables, he had also postulated four spurious elements near H and He in order that the final table contain two periods of each length. In the end, the debate between Abegg and Werner probably had little impact on textbook accounts of the periodic law, which continued to use the short-form table for yet another 35 years—more, one suspects, for reasons of inertia rather than sound logic.

Generalizing Abegg's Rule

The 8-column, short form of the periodic table has long since fallen out of favor and our current detailed knowledge of the electron configurations of atoms has long since invalidated the idea that all atoms have a maximum capacity of eight electrons in their valence shells. Yet Abegg's original insight that the chemical behavior of an atom, and therefore the behavior of each group within the periodic table, is best characterized by means of two rather than just one number—the first of which represents the idealized maximum number of valence electrons in the neutral atom and the second the idealized maximum number of valence vacancies—still has a certain appeal and is easily updated.

This has been done in Figure 13.10, which shows that Abegg's original contention that:

$$e + v = 8 \quad [1]$$

Valence-Manifold Chart

H-He Block
(e + v = 2)

	1+1	2+0
1	H	He

Main Block
(e + v = 8)

	1+7	2+6	2'+6	3+5	4+4	5+3	6+2	7+1	8+0
2	Li	Be		B	C	N	O	F	Ne
3	Na	Mg		Al	Si	P	S	Cl	Ar
4	K	Ca	Zn	Ga	Ge	As	Se	Br	Kr
5	Rb	Sr	Cd	In	Sn	Sb	Te	I	Xe
6	Cs	Ba	Hg	Tl	Pb	Bi	Po	At	Rn
7	Fr	Ra	112	113	114	115	116	117	118

Transition Block
(e + v = 18)

	3+15	4+14	5+13	6+12	7+11	8+10	9+9	10+8	11+7
4	Sc	Ti	V	Cr	Mn	Fe	Co	Ni	Cu
5	Y	Zr	Nb	Mo	Tc	Ru	Rh	Pd	Ag
6	Lu	Hf	Ta	W	Re	Os	Ir	Pt	Au
7	Lr	Rf	Db	Sg	Hs	Mt	110	111	112

Inner-Transition Block
(e + v = 32)

	3+29	4+28	5+27	6+26	7+25	8+24	9+23	10+22	11+21	12+20	13/19	14+18	15+17	16+16
6	La	Ce	Pr	Nd	Pm	Sm	Eu	Gd	Tb	Dy	Ho	Er	Tm	Yb
7	Ac	Th	Pa	U	Np	Pu	Am	Cm	Bk	Cf	Es	Fm	Md	No

FIGURE 13.10 The valence manifolds and (e+v) group labels for the four basic blocks of the periodic table.

is not universal but applies only to the atoms of the main-block elements, which employ ns- and np-orbitals in their various bonding interactions. To this case we must now add the following additional possibilities:

1. The atoms of the H-He block, which employ only the 1s-orbital in their bonding interactions and for which (28):

$$e + v = 2 \quad [2]$$

2. The atoms of the transition block, which employ not only ns- and np-orbitals in their bonding interactions but also (n-1)d-orbitals as well, and for which:

$$e + v = 18 \quad [3]$$

3. The atoms of the inner-transition block, which employ not only ns-, np-, and (n-1)d-orbitals in their bonding interactions but also (n-2)f-orbitals as well, and for which:

$$e + v = 32 \quad [4]$$

Rather than employing, as per Abegg's representation of 1910, a double column of idealized maximum electropositive and electronegative valence values for our group labels, we have instead used a simple binary (e+v) label, as per Abegg's suggested electronic interpretation of 1904, where the sum of e and v always corresponds to one of the above equations. Also, instead of continuing to talk of the above equations as describing a given atom's valence *shell*, we will now use the term valence *manifold* instead, since it is clear from the above orbital descriptions that the valence electron/vacancy counts for a given atom may contain contributions not only from more than one subshell (i.e., s versus p versus d versus f, etc.) but also from more than one atomic shell as well (i.e., n versus (n-1) versus (n-2), etc.).

Valence Electrons versus Oxidation States

Reflecting his commitment to the ionic bonding model, Abegg interpreted e as a measure of an element's maximum possible positive ionic charge or positive oxidation state, and v as a measure of an element's maximum possible negative charge or negative oxidation state. At present, we realize that these valence electrons and vacancies can be employed to form covalent bonds through electron sharing as well as ionic bonds through electron transfer and for this reason will avoid any further mention of polar valence. In addition, it is important to recognize, as per Abegg's original interpretation, that e and v are idealized maxima which set upper and lower bounds to an atom's observed valence values even though they themselves (especially in the case of the inner-transition block) may never be realized in actual practice.

There is, to the best of my knowledge, no generally accepted quantum mechanical criterion for unambiguously dividing the electron configuration of an atom into a core and a valence manifold. Rather such a division, which is essential to all bonding theories, is usually based on chemical experience and the range of compounds observed for the atom in question. In general, if a few electrons or vacancies of a given electronic subshell are found to significantly contribute to the bonding in at least one compound of an element in a given group, then the entire electronic subshell is counted as part of the valence manifold for the elements of the group in question.

In modern bonding theory a valence electron is not necessarily the same thing as a bonding electron or active valence. As an examination of typical Lewis dot diagrams for compounds of the H-He and main-block elements quickly reveals, some valence electrons may appear as nonbonding stereo-active lone pairs instead. Likewise, in the ligand field diagrams used to describe transition-metal complexes, both bonding and nonbonding (albeit stereo-inactive) valence electrons also make an appearance. All of this means that one must carefully distinguish between valence-electron counts and oxidation states, as these two concepts do not necessarily have the same numerical value for a given bonding situation. As we have seen, group numbers for the periodic table have traditionally been associated with maximum oxidation states and, as a result, often have many exceptions (e.g., the absence of compounds corresponding to F(VII) and O(VI)). In contrast, in Figure 13.10 we are instead employing maximum valence-electron counts and these, when one includes both bonding and nonbonding valence electrons, have, by definition, no exceptions.

The Importance of Valence Vacancies

Since we are familiar with counting valence electrons when it comes to both the periodic table and the various rules for predicting the compositions and structures of simple compounds and complex ions, it is perhaps of value to pause for a moment and review why a knowledge of the number of valence vacancies is of equal importance, using typical examples from the introductory freshman textbook.

Let us begin with the question of how to predict the ratio (b/a) of the stoichiometric coefficients for a simple binary ionic compound A_aB_b—by the adjective "simple" meaning those compounds in which the cation is in its maximum oxidation state and there is no anionic or cationic homocatenation (29). In such compounds the number of valence electrons removed from component A must equal the number of valence vacancies available on component B or:

$$ae_A = bv_B \quad [5]$$

which, upon rearrangement, gives us the desired result:

$$b/a = e_A/v_B \quad [6]$$

in which both e and v play equal roles. A glance at the group labels in Figure 13.10 will quickly verify this relationship for compounds of this type (e.g., e_A/v_B = 1/1 for NaCl, 2/1 for CaI_2, 3/2 for Al_2O_3, 4/2 = 2/1 for GeS_2, etc.)

In the case of simple covalent species of the main-block elements, in which all of the components have complete octets, the number of shared bonding electrons used for octet completion must equal the total number of valence vacancies in the component atoms or:

$$\Sigma v = e_b \qquad [7]$$

where Σv is the sum of the valence vacancies in the component atoms of the species, e_b is the total number of bonding electrons in the resulting Lewis dot diagram, and once again e and v play equally important roles (29). This equation is easily verified using the Lewis diagrams for the simple series N_2, O_2, F_2, and Ne and the v values in Figure 13.10 (Σv = 0 for Ne, 2 for F_2, 4 for O_2, and 6 for N_2, etc.). Variations of equation 13.7 have appeared many times in the chemical literature. Though Abegg was committed to an ionic bonding model in his famous 1904 paper, he did raise the question of whether there might also be a nonpolar valence in order to account for the homoatomic bonding in such species as Cl_2, S_8, and As_4 (10). Though in the end he pleaded ignorance as to the nature of such a valence, he did observe that, whatever it might be, it seemed to numerically correlate with the maximum electronegative valences of the elements in question or, in other words, with their v values.

Between 1919 and 1921 Irving Langmuir presented various elaborations of equation 7 as the basis for his octet theory of compound formation, in which he made the role of v implicit rather than explicit by substituting into equation 1 (30):

$$e_b = \Sigma(8 - e) \qquad [8]$$

It is interesting to note that Langmuir also formulated an 18e version of equation 8 for the case of the central metal atom in transition-metal carbonyls:

$$e_b = 18 - e \qquad [9]$$

In 1930 Hume-Rothery proposed his so-called 8-N rule to describe the valence values used by the nonmetallic main-block elements in forming their homocatenated standard-state structures, in which N represented the group number for the element in question in the periodic table (31). Since, in keeping with Abegg's original electronic interpretation of the periodic table,

$$N = e \qquad [10]$$

we see that this rule is just another version of equations 7 and 8.

Since 1930 many additional variations on these equations have appeared, not only for neutral molecules and complex ions, but also for the infinitely extended homocatenated anions found in so-called Zintl phases (32). The important point, however, is that all of these counting rules point to the importance, either explicitly or implicitly, of valence-vacancy counts in determining the bonding properties of an element and underscore the desirability of including this information in our group labels for the periodic table. Indeed, experience in the author's own introductory chemistry lectures has shown that this additional information greatly facilitates the students' mastery of Lewis dot diagrams (29).

Implications for the Modern Periodic Table

In principle the proposed (e+v) group labels may be used with any type of periodic table provided that it clearly separates H and He from the body of the table as a separate electronic block. However, the labels are especially effective when employed in conjunction with the updated Thomsen-Bohr step-pyramid table shown in Figure 13.11. As may be seen, this explicitly indicates the various valence manifolds corresponding to each electronic block and to equations 1-4 using different gray levels for the H-He, main, transition and the inner-transition blocks. Unfortunately, it fails to also indicate the corresponding (e+v) values for each individual group for the simple reason that the table was designed and printed before this further refinement occurred to me. Nevertheless, by referring back to Figure 13.10 when necessary, it should be possible to follow the following discussion of how these new group labels clarify our understanding, not only of this particular table, but of periodic tables in general.

Use of the (e+v) group labels leads immediately to the establishment of three kinds of possible relationships within the periodic table. Those elements having both identical e counts and identical v counts are said to be *isovalent* and are always found in both the same group and electronic block of the table. According to the terminology I first introduced in 1987 (33), such elements will exhibit a *primary relationship* to one another and are connected in Figure 13.11 either by being placed in the same vertical column or by means of black diagonal

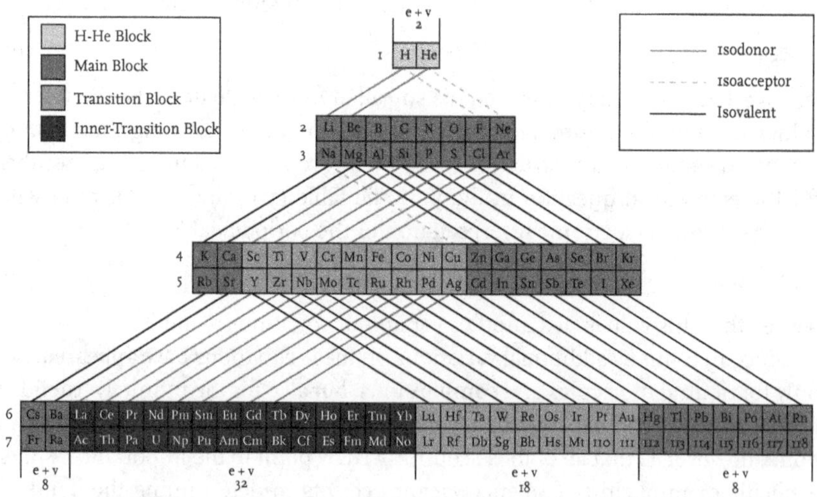

FIGURE 13.11 An updated Thomsen-Bohr step-pyramid periodic table.

tie lines. In general, isovalent elements will usually display the same trends in both their positive and negative oxidation states and can substitute for one another in compounds and reactions without altering the overall stoichiometry (e.g., Na for K or Cl for Br, etc.).

Elements which have identical e counts but differing v counts are said to be *isodonor*, since they can contribute or donate the same maximum number of valence electrons in a bonding situation. Such elements will exhibit a *secondary relationship* to one another. They are always located in both different groups and different electronic blocks and are connected in Figure 13.11 by means of gray diagonal tie lines. While elements exhibiting primary relationships to one another are found in the same group, those exhibiting only secondary relationships are said to belong to the same *family* since these secondary analogies largely correspond to the relationships traditionally attributed to the members of the *a* and *b* subgroups found in older periodic tables. In general, such elements will display stoichiometric equivalencies when in their highest oxidations states (e.g., S(VI) versus Cr(VI) or Cl(VII) versus Mn(VII), etc.) but not with respect to their intermediate and lower oxidation states (e.g., S(-II) versus Cr(0) or Cl(-I) versus Mn(0), etc.).

Elements which have identical v counts but differing e counts are said to be *isoacceptor*, since they can accept the same maximum number of valence electrons in a bonding situation. Such elements are said to exhibit a *tertiary relationship* to one another. Like those exhibiting secondary analogies, the elements in question are always located in different groups and electronic blocks and are connected in Figure 13.11 by means of broken diagonal tie lines. Since some of these tertiary analogies also correspond to the relationship between the *a* and *b* subgroups found in older periodic tables, the elements in question may also be thought of as belonging to the same family, whereas in yet other cases these relations replace what were originally assumed to be primary relationships in older tables (see below). In general, elements displaying tertiary relationships will exhibit stoichiometric equivalencies when in their lowest oxidation states (e.g., H(-I) versus Cl(-I) or He(0) versus Xe(0), etc.) but not with respect to their intermediate and highest oxidation states (e.g. H(I) versus Cl(VII) and He(0) versus Xe(VI) etc.).

Use of the proposed (e+v) group labels also resolves the various placement problems that have continued to plague the modern periodic table. The oldest of these is perhaps the question of whether H belongs in the main block as a member of the (1+7) alkali metal group or in the main block as a member of the (7+1) halogen group. Use of the (e+v) labels immediately reveals that it belongs in neither but rather in the (1+1) group of a separate H-He block. Rather than displaying a primary isovalent analogy with either of the above groups, it instead displays a secondary isodonor analogy with respect to the alkali metals and a tertiary isoacceptor analogy with respect to the halogens. The fundamental difference between elements of the (1+7) group and those of the (1+1) group is reflected in the fact that the elements of the former group, with seven valence vacancies per atom, form infinitely extended 8/8 metallic solids with partially filled conduction bands in their standard states, whereas the H atom, with only one valence vacancy per atom, forms a saturated diatomic gas.

Closely related is the question of whether He belongs in the main block as a member of the (8+0) noble-gas group, as per most conventional periodic tables, or in the main block as a member of the (2+6) alkaline-earth metal group, as per its outermost electron configuration—though I am unaware that anyone, other than Bent, has ever seriously argued for this latter placement (34). Once again the (e+v) group labels show that neither of these assignments is correct, but rather that He belongs in the (2+0) group of the separate H-He block. Rather than displaying a primary isovalent analogy with either of the above groups, it instead displays a potential secondary isodonor analogy with respect to the alkaline-earth metals and a tertiary isoacceptor analogy with respect to the noble gases.

Yet a third placement problem, which I discussed some years ago, involves the question of whether La and Ac belong in the transition block or in the inner-transition block, and vice versa for Lu and and Lr—a problem which results from the fact that there are many irregular electron configurations in this part of the periodic table (35). The (e+v) labels, which are based on total valence electron and vacancy counts, rather than on the details of irregular electron configurations, clearly show that La and Ac should be assigned to the (3+29) group of the inner-transition block, whereas Lu and Lr, with their completed $(n\text{-}2)f^{14}$ subshells, should be assigned to the (3+15) group of the transition block.

The final placement problem involves the status of the Zn group and the question of whether the members of this group should be classified as transition elements and placed in a (12+6) group or as main-block elements and placed in the (2+6) group. As I summarized in a previous paper, there is absolutely no evidence that these elements ever make use of either (n-1)d-type electrons or (n-1)d-type orbital vacancies in their chemistry and, though they are frequently lumped with the transition metals in textbooks, they are in fact main-block elements (36, 37). As a result, we encounter for the first time, as pointed out many years ago by Sanderson, a true bifurcation in the group structure of the periodic table—a bifurcation which also underlies the original debate between Abegg and Werner over whether Be and Mg should be placed above Zn or above Ca (38). This is resolved in Figure 13.10 by placing the Ca branch in the (2+6) group and the Zn branch in a (2'+6) group, and in the periodic table in Figure 13.11 by connecting Mg equally to both Ca and Zn by means of primary black tie lines.

The Question of Hypervalence

If one is to accept the valence-electron and valence-vacancy counts in Figure 13.10, there is one final issue that must be confronted and that is the so-called problem of hypervalent compounds. In order to account for such main-block species as PCl_5, SF_6, SiF_6^{2-} etc. using only traditional 2c-2e bonds, virtually all general chemistry texts invoke octet expansion or the use of outer nd-orbitals in addition to the usual ns- and np-orbitals. Such an expansion would violate the maximum valence-vacancy counts in Figure 13.10 and imply that these counts are variable within a given group of the periodic table (e.g., 2 for O but 12 for S).

The debate over this question has persisted for much of the 20th century despite the fact that quantum mechanical calculations have consistently shown that such species actually obey the octet rule through the use of 3c-4e or ω-bonds (only two electrons of which are actually bonding), rather than 2c-2e bonds—a solution first proposed by Rundle and Pimentel as far back as the 1950s (39, 40). The persistence in the general chemistry textbook of the outdated and discredited concept of octet expansion appears to be an artifact of a misguided belief that it is necessary for the application of the VSEPR model to these compounds. In fact, the VSEPR rules depend only on the number of independent repulsion units in the outermost valence shell irrespective of their bond order or type, whether these be 2c-2e single bonds, 2c-4e double bonds, 2c-6e triple bonds, 1c-2e lone pairs, or 3c-4e ω-bonds. The current consensus is that the octet rule is best viewed, as per Abegg, as an upper maximum that also allows for the existence—often for steric reasons—of stable species having less than an octet in their valence manifold (e.g., $B(CH_3)_3$).

A similar solution applies to the H-He block and the case of H-bonding, where again the bonding is best represented using a 3c-4e ω-bond rather than via duplet expansion and use of two 2c-2e bonds.

The same is true of transition-block species that supposedly contain more than 18 electrons around the central atom (41). Examination of the ligand-field diagrams for these species shows that the extra electrons reside in antibonding rather than bonding MOs. In general, as may be easily confirmed by comparing the MO diagrams for diatomics with their corresponding Lewis dot diagrams, when translating a delocalized MO diagram into the corresponding localized Lewis diagram (42):

$$\textit{1 bonding } e + \textit{1 antibonding } e \rightarrow \textit{2 nonbonding } e \qquad [11]$$

In other words, occupation of the antibonding MOs means that the extra electrons and their counterparts in the bonding MOs are transferred from the central atom to the ligands at the periphery of the complex and so do not lead to violation of the 18e rule for the central atom. This has been confirmed by the calculations of Chu and Lee for the complex $(C_5H_5)_3W^+$ and has been generalized by them under the guise of the so-called *periphery rule* (43). Since many compounds of the early transition metals have fewer than 18 electrons around the central atom, this rule, like the octet rule, is best thought of, in keeping with Figure 13.10, as an upper maximum (44, 45).

In the case of the inner-transition elements, hypervalence is not an issue. Indeed, as pointed out earlier, the chief concern is finding evidence that at least some of their many f-valence vacancies are involved in bond formation (46). However, if one examines the valence-electron counts around the central

TABLE 13.1 Electron Counts for Various Lanthanoid Complexes

OS	SPECIES	ACN	e_{CL}	e_L	e_T
La(III)	$La(NO_3)_6^{3-}$	12	0	24	24
Ce(III)	$Ce(NO_3)_6^{3-}$	12	1	24	25
Pr(III)	$Pr(napthyl)_6^{3+}$	12	2	24	26
Nd(III)	$Nd(18\text{-crown-}6)(NO_3)_3$	12	3	24	27
Sm(III)	$Sm(12\text{-crown-}4)(NO_3)_3$	10	5	20	25
Eu(III)	$Eu(15\text{-crown-}5)(NO_3)_3$	11	6	22	28
Gd(III)	$Gd(12\text{-crown-}4(NO_3)_3$	10	7	20	27
Tb(III)	$Tb(12\text{-crown-}4)(NO_3)_3$	10	8	20	28
Dy(III)	$Dy(12\text{-crown-}4)(NO_3)_3$	10	9	20	29
Ho(III)	$Ho(12\text{-crown-}4)(NO_3)_3$	10	10	20	30
Er(III)	$Er(12\text{-crown-}4)(NO_3)_3$	10	11	20	31
Tm(III)	$Tm(12\text{-crown-}4)(NO_3)_3$	10	12	20	32

lanthanoid ion for the known crown ether and other polydentate ligand complexes of these elements (Table 13.1), one quickly discovers that use of a large number of these f-vacancies is required if one is to accommodate two electrons from each of the surrounding ligand donor atoms (47). More recently Pyykkö et al have taken this a step further by theoretically studying possible complexes of the actinoids with polydentate cage π-ligands, such as Pb_{12} and C_{28} and, as a consequence, they have also postulated the existence of a 32 valence-electron maximum for the inner-transition elements, paralleling the duplet, octet, and 18-electron maxima for the elements of the other electronic blocks (48, 49).

Conclusion

In summary, replacing Abegg's original concept that each group of elements in the periodic table is characterized by an unique set of idealized maximum positive- and negative-valence values with the concept of idealized valence-electron (e) and valence-vacancy (v) counts, and applying these to the modern four-block version of the periodic table, rather than the outdated 8-column, short table used by Abegg, results in a set of consistent, albeit idealized, binary (e+v) group labels that resolve most of the residual placement problems in the modern periodic table (H, He, Zn, La, Lu, etc.). In addition, these labels provide a rationale for the existence of not only primary isovalence relationships within the table, but also weaker secondary isodonor and tertiary isoacceptor relationships. It was further shown that inclusion of the valence-vacancy counts in the group labels provides students with valuable information relevant to the application of many of the simple bonding models taught in most introductory chemistry courses.

References and Notes

1. Biographical background on Abegg may be found in S. Arrhenius, "Richard Abegg," *Z. Electrochem.*, 1910, *16*, 554–557; and W. Nernst, "Richard Abegg," *Berichte*, 1913, *46*, 619–628. This latter obituary also contains a bibliography of Abegg's publications.
2. R. Abegg, W. Herz, *Chemisches Praktikum: Experimentelle Einführung in präparative und analytische Arbeiten auf physikalisch-chemischer Grundlage*, Vandenhoeck & Ruprecht, Göttingen, 1900. Translated as *Practical Chemistry: An Experimental Introduction to Laboratory Practice and Qualitative Analysis from a Physicochemical Standpoint*, Macmillan, New York, 1901.
3. R. Abegg, *Die Theorie der elektrolytischen Dissoiciation*, Enke, Stuttgart, 1903. Translated as *The Electrolytic Dissociation Theory*, Wiley, New York, 1907.
4. R. Abegg, G. Bodländer, "Die Elektroaffinität: Ein neues Prinzip der chemischen Systematik," *Z. anorg. Chem.*, 1899, *20*, 453–499.
5. R. Abegg, G. Bodländer, "Electroaffinity as a basis for the systemization of inorganic compounds," *Am. Chem. J.*, 1902, *28*, 220–228.
6. R.T. Sanderson, "On the significance of electrode potentials," *J. Chem. Educ.*, 1966, *43*, 584–588.
7. J. Locke, "Electroaffinity as a basis for the systemization of inorganic compounds," *Am. Chem. J.*, 1902, *27*, 105–117.
8. R. Abegg, F. Auerbach, R. Luther, eds., *Messungen electromatorischer Kräfte galvanischer Ketten mit Wässerigen Elektrolyten*, Knapp, Halle, 1911.
9. R. Abegg, "Versuch einer Theorie der Valenz und der Molekülarverbindungen," *Christiania Videnskabs-Selskabets Skrifter*, 1902, No. 12; also *J. Chem. Soc. Abst.*, 1903, *84*, 536.
10. R. Abegg, "Die Valenz und das periodische System: Versuch einer Theorie der Molekülarverbindung," *Z. anorg. Chem.*, 1904, *39*, 330–380; Quote on pages 343–344; also *BAAS Report*, 1907, *77*, 480.
11. D. Mendeleev, "*Die periodische Gesetzmässigkeit der chemischen Elemente,*" *Ann. Chem. Pharm.*, 1872, *Suppl. VIII*, 133–229. Translated in W.B. Jensen, ed., *Mendeleev on the Periodic Law: Selected Writings, 1869–1905*, Dover, New York, 2002, 38–109. Though the supplement volume is dated 1872, the review was actually completed in August 1871.
12. See, for example, the inside back cover of N. Akhmetov, *General and Inorganic Chemistry*, Mir, Moscow, 1983.
13. For a discussion of Mendeleev's rule and his attitude toward the valence concept, see A.A. Makarenya, "Development of the Valency Concept in the Aspect of the Theory of Periodicity," in V.I. Kuznetsov, ed., *The Theory of Valency in Progress*, Mir, Moscow, 1980, Chapter 3.
14. Abegg would later search among the intermetallic compounds for possible examples of the elements of groups I-III employing their negative contra-valences, but without success. See R. Abegg, "Ueber die Fähigkeit der Elemente miteinander Verbindungen zu bilden," *Z. anorg. Chem.*, 1908, *50*, 309–314.
15. W. Kossel, "Über Molekülbildung als Frage des Atombaus," *Ann. Physik*, 1916, *49*, 229–362.

16. R. Abegg, "Chemische Affinität, Valenz und das natürliche System der Elements," *Scientia*, 1910, *8*, 54–64.
17. L. Meyer, *Modern Theories of Chemistry*, Longmans, Green & Co. London, 1888, 350. Translation of L. Meyer, *Die modernen Theorien der Chemie und ihre Bedeutung für die chemische Mechanik*, 5th ed., Maruschke & Berendt, Breslau, 1884.
18. J.J. Thomson, *The Corpuscular Theory of Matter*, Constable, London, 1907, 117–118.
19. G.N. Lewis, "The atom and the molecule," *J. Am. Chem. Soc.*, 1916, *38*, 762–785.
20. See, for example, H. Fenton, *Outlines of Chemistry with Practical Work*, Cambridge University Press, Cambridge, 1908, 98–100; J.W. Mellor, *Modern Inorganic Chemistry*, Longmans, Green & Co, 1927, 1053–1055; F. Ephraim, *Anorganische Chemie*, Steinkopff, Dresden, 1929, 40; J.R. Partington, *A Textbook of Inorganic Chemistry*, 5th ed., Macmillan, London, 1939, 208, 455 (The rule is still cited in the 6th edition of 1950, 415–416).
21. See, for example, C.A. Russell, *The History of Valency*, Leicester University Press, Leicester, 1971, 271–273 and A.N. Stranges, *Electrons and Valence: Development of the Theory, 1900–1925*; Texas A & M University Press, College Station, TX, 1982, 80–91.
22. W.B. Jensen, "Abegg, Lewis, Langmuir and the Octet Rule," *J. Chem. Educ.*, 1984, *61*, 191–200.
23. R. Abegg, F. Auerbach, eds., *Handbuch der anorganische Chemie*, Hirzel, Leipzig, 1908; four projected volumes. Actually at least 12 separate parts were published, some as late as 1933. The projected series was never completed.
24. A. Werner, "Beitrag zum Ausbau des periodischen System," *Berichte*, 1905, *38*, 914–921.
25. R. Abegg, "Ueber die 8. Gruppe des periodischen Systems," *Berichte*, 1905, *38*, 1386–1388.
26. A. Werner, "Zur periodischen Anordnung der Elemente," *Berichte*, 1905, *38*, 2022–2027.
27. R. Abegg, "Zur periodischen Anordnung der Elemente," *Berichte*, 1905, *38*, 2330–2334.
28. The idea of a separate H-He block was first proposed in O. Ramirez-Torres, "Stability of the 1s orbital and the chemical behavior of hydrogen," *J. Chem. Educ.*, 1955, *32*, 450–455.
29. W.B. Jensen, *Chemistry: The Science of Substance, An Introductory Lecture Course*, Oesper Collections, Cincinnati, OH, 2010, Lecture 16.
30. Langmuir gave various versions of his equations for the electron counts for covalent octet compounds and 18e transition-metal carbonyls in several papers published between 1919 and 1921. These have been conveniently collected together in C.G. Suits, ed., *The Collected Works of Irving Langmuir*, Vol. 6, Pergamon, New York, 1961.
31. W. Hume-Rothery, "The crystal structures of the elements of the b sub-groups and their connection with the periodic table and atomic structure," *Phil. Mag.*, 1930, *9 (7th series)*, 65–80.
32. For a summary, see W.B. Jensen, "Electron-orbital counting rules and bonding topology," *De Educ. Quim.*, 1978, *3*, 210–222. Originally published in Portuguese; English version available from author upon request.

33. W.B. Jensen, "Classification, symmetry, and the periodic table," *Comp. & Maths. with Appls.*, 1986, *12B*, 487–510.
34. H.A. Bent, *New Ideas in Chemistry from Fresh Energy for the Periodic Law: An Introduction to Leading Uses of the Left-Step Periodic Table*, AuthorHouse, Bloomington, IN, 2006.
35. W.B. Jensen, "The positions of lanthanum (actinium) and lutetium (lawrencium) in the periodic table," *J. Chem. Educ.*, 1982, *59*, 634–636.
36. W.B. Jensen, "The place of zinc, cadmium, and mercury in the periodic table," *J. Chem. Educ.*, 2003, *80*, 952–961.
37. There have been claims from time to time that it is possible to prepare compounds of Hg(IV) but these have always proven to be unverifiable. See W.B. Jensen, "Is mercury now a transition element?," *J. Chem. Educ.*, 2008, *85*, 1182–1183.
38. R.T. Sanderson, "A Rational Periodic Table," *J. Chem. Educ.*, 1964, *41*, 187–189.
39. An overview of the older literature on this subject may be found in W.B. Jensen, "The origin of the term hypervalent," *J. Chem. Educ.*, 2006, *83*, 1751–1752.
40. F. Weinhold, C. Landis, *Valency and Bonding: A Natural Orbital Donor-Acceptor Perspective*, Cambridge University Press: Cambridge, 2005, Chapter 3.
41. P.R. Mitchell, J. Parish, "The eighteen-electron rule," *J. Chem. Educ.*, 1969, *46*, 811–814.
42. R. Harcourt, *Qualitative Valence-Bond Descriptions Of Electron-Rich Molecules*, Lecture Notes In Chemistry, No. 30, Springer-Verlag, Berlin, 1982, 37.
43. S.Y. Chu, S.L. Lee, "Hypervalent molecules and peripheral orbitals," *J. Chem. Educ.*, 1985, *62*, 857–858.
44. W.B. Jensen, "Origins of the 18-electron rule," *J. Chem. Educ.*, 2005, *82*, 28.
45. P. Pyykkö, "Understanding the eighteen-electron rule," *J. Organomet. Chem.*, 2006, *691*, 4336–4340.
46. R. Dagani, "Uranium: beyond the 18-electron rule," *Chem. Eng. News*, 2001, *79 (Oct. 8)*, 31–32.
47. S. Cotton, *Lanthanides and Actinides*, Oxford University Press, New York, 1991, 63–66.
48. J. Dognon, C. Clavaguéra, P. Pyykkö, "Towards a 32-electron principle: $Pu@Pb_{12}$ and related systems," *Angew. Chem. Int. Ed.*, 2007, 46, 1427–1430.
49. J.P. Dognon, C. Clavaguéra, P. Pyykkö, "A predicted organometallic series following a 32-electron principle: $An@C_{28}$ (An = Th, Pa^+. U^{2+}, Pu^{4+})," *J. Am. Chem. Soc.*, 2009, *131*, 238–243.

CHAPTER 14

The Chemist as Philosopher

D. I. Mendeleev's "The Unit" and "Worldview"

MICHAEL D. GORDIN
Department of History, Princeton University, USA

THE PERIODIC SYSTEM OF chemical elements is almost certainly the most widely recognized scientific object in the world today, even though extensive debates persist about what it exactly *is*. Is it a theory, a collection of empirical data, a tabular arrangement of that data, a particular (best) tabular arrangement, a "paper tool," or something else besides? Precisely because the periodic system has over close to 150 years remained so significant to the training and practice of scientists, the broader field of science studies has devoted considerable attention to it, most prominently in the philosophy of science.[1]

Among the many different approaches to articulating a philosophical foundation for the periodic system, one central strand is historicist, which places great emphasis on the individual (or individuals) to whom one attributes its discovery (Gordin 2012). Almost universally, credit for the formulation of the periodic system is assigned to St. Petersburg chemist Dmitrii Ivanovich Mendeleev (1834–1907) for his 1869 table of elements, which he later used to predict the properties of three yet-undiscovered elements.[2] Although the philosophical justification of the periodic system by no means requires engagement with Mendeleev's own views about the periodic system—or, as argued in Gordin (2004, 182–189), how those views changed over the course of his lifetime as periodicity became more central to chemical practice—nonetheless it remains of interest to understand precisely what Mendeleev thought he was about in constructing his system, as well as his post hoc justifications of it.

---

[1] For an introduction to the field, see Scerri 2007, as well as almost any issue of the journal *Foundations of Chemistry*.

[2] For a biographical account of Mendeleev's many activities beyond the periodic system, see Gordin 2004.

There is, however, an obstacle to the full development of this line of inquiry: the Russian language itself. There is a substantial body of Imperial Russian, Soviet, and post-Soviet scholarship that would be of interest to the international community of philosophers and historians of chemistry, but it remains locked in a language not widely read by Western scholars. Even more problematic, only a very slender selection of primary sources are accessible in Western European languages (most widely cited are those available in English, although the corpus is larger if one includes French and German). The most significant of these are a Dover edition of translations into English of Mendeleev's chemical publications about the periodic law (Mendeleev 2002); translations of the later editions of his textbook, *The Principles of Chemistry* (*Osnovy khimii*); and Mendeleev's pamphlet (1904) on the chemical conception of the luminiferous ether (although this last is only rarely referenced). Despite some limitations,[3] these sources have doubtless been helpful in deepening scholars' awareness of Mendeleev's own thought process, although the chemical emphasis of these pieces has obscured the fact that Mendeleev did produce philosophical writings. My goal here is to make editions of two such texts available to the English-reading public.

I selected these two pieces—"The Unit," from 1877, and "Worldview," from 1905—because in them Mendeleev most explicitly attempted a philosophical point of view, along the lines of the style of philosophy common in Russia at the time. (They will surely fall short in terms of rigor for an analytic philosopher today.) Neither text is, however, straightforward: the first was printed under a pseudonym, and the second he declined to publish. They are, therefore, reflections Mendeleev felt uncomfortable sharing openly. They also come from different contexts, both later than his earliest writings on the periodic system, but contemporaneous with some of his mature publications on it. (Much of the scholarly literature tends to flatten the timeline, with pieces from different periods cited side-by-side without regard to Mendeleev's temporal development.) Neither piece translated here directly addresses the periodic system, although in each Mendeleev presents a nuanced and sophisticated philosophy of nature, with significant implications for his thoughts on epistemology and ontology.

"The Unit" (Popov 1877) was published in under the pseudonym "D. Popov" in *Svet* (which means either "light" or "world" in Russian), a hybrid science-art-philosophy journal edited by his zoologist colleague at St. Petersburg University, Nikolai Petrovich Vagner (1829–1907).[4] It appeared a year after Mendeleev's public debunking of spiritualist séances (Gordin 2004, chapter 4),

[3] Principal among these is the reliance on the later editions of the *Principles*. The periodic system was produced in the process of writing this textbook in 1868–1869, as has been widely recognized (Gordin 2004, chapter 2, and references therein), but the versions that are translated are only the substantially revised later editions (all editions after the fifth edition of 1889), which include substantial revisions about the nature and significance of the periodic system, responding to scientific developments. These are therefore unreliable sources in elucidating the context of discovery.

[4] On Vagner, see Gordin 2011.

in direct opposition to Vagner, who was a major participant in those debates on the spiritualist side. In Mendeleev's annotated bibliography of his writings, produced in his final decade, he commented on "The Unit" and its byline thus: "It was written by me and signed with a pseudonym (which I did only once in my life)—D. Popov—from the name of my wife, A. I. Popova" (Mendeleev in Shchukarev and Valk 1951, 83). (Actually, Anna Popova was not yet his wife; Mendeleev's divorce was not finalized—or even proposed—at that date, and he married Popova only in 1882.) Mendeleev added:

> This was a transitional time for me: much was changing in me; at that time I read a great deal about religions, about sects, about philosophy, economic articles. Something is expressed [in this piece]. I took a pseudonym—for the reason that then the certainty in the veracity of the path chosen by me was still weak in me. And now I would write exactly the same directly with my last name; everything that was said, reading it anew, I would sign.
>
> (Mendeleev in Shchukarev and Valk 1951, 82–83)

There are several features of this article that prove interesting, such as his engagement with units of measurement (on Mendeleev's career-long involvement in metrology, see Gordin 2004, 164–171) and hints of his abstract conception of chemical elements (Scerri 2007; Gordin 2009). "The Unit" is also Mendeleev's most extended discussion of the "naturalistic fallacy," although he did not use that term.

"Worldview" (Mendeleev 1948) is a much more difficult text in almost every respect: its convoluted prose style, its provenance, and its interpretation. The translation published here is based on the autograph manuscript from the Mendeleev Archives at St. Petersburg State University, edited by E. Kh. Fritsman. I have modified this version slightly in accordance with a more recent edition in Mendeleev (1995, 408–413). "Worldview" was intended to be an afterword for his collection of political and social essays, *Cherished Thoughts* (*Zavetnye mysli*), published while Russia was in the midst of the Revolution of 1905, which resulted in Tsar Nicholas II issuing the "October Manifesto" that provided for a constitutional monarchy with an elected parliament (Duma). On the original manuscript, Mendeleev wrote "Decided not to publish," with the date 28 September 1905 (Mendeleev 1948, 157).[5] He eventually produced an afterword for *Cherished Thoughts*, composed on 4 October 1905, in which he mentioned the existence of a drafted statement of his worldview that he had refrained from publishing "because the exposition seemed to me insufficiently full" (Mendeleev 1995, 406). The afterword replicated in abbreviated form some of the thoughts about his tripartite metaphysics—also present in later editions of the *Principles* and in his ether pamphlet, as discussed in Gordin (2004, chapter 8)—as well as his ruminations about Don Quixote. It is clear from the picture presented in this document that earlier scholarly claims that

[5] All dates here are presented in the old style Julian calendar, which lagged twelve days behind the new style Gregorian calendar in the nineteenth century, and thirteen days in the twentieth.

Mendeleev was some variant of anti-metaphysical positivist (as in Vucinich 1967, Stackenwalt 1976, and Bensaude-Vincent 1986) are impossible to sustain.

One final point: As readers of Mendeleev have long been aware, the chemist adored footnotes, often presenting some of his most intriguing revisions and second thoughts in this manner (as seen in *Principles of Chemistry*). Most of the footnotes to these texts, however, are my own, always marked by my initials. There are two original Mendelevian footnotes in "Worldview," one in "The Unit."

"The Unit"

In the 6th issue of *Svet* Mr. Alenitsin proves that in nature there is no zero, that it is a fiction.[6] One can consider such a point helpful for certain gentlemen; but if one can ascribe existence to ideas, if a word corresponds to what exists, and if every word is already an abstraction, then the word and idea (or abstraction of this sort) which is called *zero* exists in consciousness. That is why to speak about zero means to speak not about nature, but about an idea, an abstraction, a generalization.

The unit is another matter. It appears to be not only an idea, but also a reality. And people definitely consider it to be such: they compose from it, it is the alpha and omega of the philosopher who seeks "I" in capital letters,[7] it is the foundation of creation, subdivisions end with it, all the sense of individualism is in it, in a word, it without a doubt exists. Isn't it true that its existence is entirely true, unquestionable? And meanwhile, think and you will come to the conviction that the unit is simply even unthinkable in nature. The units of length, weight, time, all kinds of strengths of forces—in a word, all units used in the sciences—are known to be conventional. They do not exist, they are thought up by us, i.e. they are fictional. Let us take, for example, our Earth—even that is not a unit, because there are other earths: planets. There are just as many suns, because the stars are all exactly the same kind of thing as our sun, and if the sun does not have plural declensions in general use, then this depends only from the fact that language was formed at a time when one believed in unitarity more than one should. Give it a try, and you can say: suns, *of* suns, *with* suns, etc.[8] It is true that these words don't sound pretty, but that is only because we did not grow up with the idea of the plurality of worlds and suns, we learned it, we were not born with it, it was imbibed only through the long path of study—and, chiefly, because we grew up with the concept of unit, we learn to count from units, we even think in units. Because of this, most likely, people affirm to us, at one time (occasionally, not more) a single *unit* for

[6] "Positive and Negative," p. 129. (Alenitsin 1877–1878—MG)

[7] The Russian first-person pronoun is not ordinarily capitalized.—MG

[8] In the original: "солнцы, солнцев, солнцами." Russian is a case language, and certain forms conventionally exist only in the singular, as is the case for "sun." The words Mendeleev uses here strike native speakers as extremely awkward.—MG

success, although they are prepared to give high marks for diligence and behavior.[9] The whole misfortune is that we are already very confident in the autonomy of the unit and we forget or do not know that in nature a unit is impossible and, moreover, a unit in nature is even unthinkable. We think up the zero, but not the unit. Look and consider. Let's imagine a single, say, ram. He doesn't exist at all. He will die and will not remain a ram and will become a zero and will forever remain a zero. And remember that one person or one ram is very close to zero. Two—a man and a woman—are thinkable in nature as the beginning of a genus, as the embryo of a collective, development, a consciousness, self-consciousness, isolation; but one male or one female or a unit—that won't even lead to a conception of anything. Behind each of us, around every one, beside every one, together with every one, in every word, sound, concept—in everything, in everything one feels the aggregate, the complexity, the mass of units, the collective. Individualism, or the entire essence of our education, is a ripe and even rotting fruit of the concept of the unit which exists autonomously in nature. But from this fruit, when it rots, a seed will certainly remain, and this seed will give a new, splendid development.

"*I* am the king of nature, it is *mine*, *I* recognize myself, *I* will live, *I* will begin to create, *I* will be in a state of bliss, *I* found..." These are all concepts, words, and thoughts, supported by the firm confidence in the unit—not thought through entirely—and it will reconstruct itself, it will change with the ages, it will fade away into thoughts. Thus a great deal has already changed since the time when Homer, or even Virgil, wrote.

"And you are not the king of nature," they will say to us, "and if you rule that is only because you received and use the legacy of your predecessors, because you were formed in a family, in a society, in a state. Alone, you are simply nature's slave. Your *individual* is only zoological, animal, and all your *humanness* and all that you revere, all of this is from others, with others, and not for you alone, not personal but general. Understand this and you will cease boasting about yourself. Yes! This is not yours, but was given to you by someone. Thus the right hand is not the property of the right arm, but of the entire person. Wool from a sheep, thread from a spinner, cloth from a weaver, a seam from a tailor—this is the matter and the property not of one, but of many people and rather a lot of people at that. You are conscious of yourself," they will also tell us, "only because your thought developed from your father and mother, sisters and brothers, teachers, and colleagues, in a word, from the fact that you are not a unit in nature but a part of a whole, a cell in a large organism. And your laudatory *I* is just as senseless as would be your arm's boast that it draws or writes, that the hand belongs to it."

People will speak thus when they come to realize that there is no unit in nature. And then a new phase will begin, then individualism will fall, then the Slavic communal idea will replace the idea of the unit, and the matter will go

[9] This is a pun. Russian grading runs on a scale of 5 to 1, with the lowest number ("unity") being a failing grade.—MG

much further than at present.[10] New epochs, we propose, will begin both in thoughts and deeds, and in beliefs and popular fates, but—never fear—there will be less nonsense, and neither the ancient idea of the state, nor the newer, free unit, will disappear: things will be better for them because there will be more truth and sense. For centuries the affairs of life will grow more complex because people will know more, and better—they will preserve, they will adapt [to life] consciously and unconsciously. Buddhist self-abnegation and contemporary self-aggrandizement are both bad extremes. The good is more complicated than them, possibly more desirable.

There is no zero in nature, and the unit is unthinkable in nature, although it appears to exist. Both the zero and the unit are words, ideas. Taking the word—for in the beginning everything was and will be a word—apart from the rest, we at first understand it as corresponding to something real, singular. But meanwhile it never has this character. The word is already a generalization. The "horse" is not this one and not that other horse, but the general, abstract concept. The word "unit" is also an abstraction, but furthermore an abstraction of a higher order than that which is connected with the word "horse." We cannot say anything, not a single word, without falling into abstraction and this abstraction about the unit as something which exists in nature is precisely the same kind of abstraction, the same kind of fiction, as the concept of zero. There is no difference. Both abstractions are inevitable, lawful, helpful; the thought becomes tangible, living, fruitful in the word. We usually even think in words. But say that in the beginning there will be just a word, as something real, in the end there ought to be a true idea which corresponds to it as something general; in the contrary case the word is nothing more than empty noise. The idea of the unit will only be true when people understand that in nature it is the same kind of meaningless thing as the zero, that the unit is nothing in and of itself, that it is only a creation of our minds similar to the ones resorted to in geometry when one imagines that a curve is composed of a large number of straight lines. People have understood the unit poorly up to the present, they have become enamored of it, and because of the parts they did not see the whole; but it has become high time to recognize that we all separately consider ourselves significant units, but in essence each of us is in and of himself a zero. We go a degree higher and then the unit will be of a higher order—the family, society, state, humanity. And on these degrees our life will become better and we will begin to seek bliss not personally, seek delight not as units, we will not mistake the good with delight but will begin to understand ourselves as nothing more than microscopic cells in the entire organism. Thus we will go ever higher, beginning from the conventional zero and unit into the nonconventional infinite. This path of thought is also perfectly clear and simple, inevitable and helpful, as that path of thought according to which everything

10 The "Slavic communal idea" is a reference to the commune (*obshchina*) which dominated Slavophile thought. On the full political and philosophical implications of Slavophilism, the heyday of which was in the 1840s and 1850s, see Walicki (1975).—MG

visible is formed from indivisible atoms, everything living from cells, as the first and simplest units to which our contemporary gaze has reached. Only the lowest organism is a single cell. And just as how for the cell all its functions are mixed up, so will ours all remain mixed up until we come to terms with the concept of the social organism, until we cease being individualists, until we recognize that we are units of the lowest order. And so we dreamed that everything is conditional and relative, only our philosophical *I* is unconditional, original, and existing. Speaking truthfully: in its idea our proper, personal unit is rarely different from zero and in no form is it identical, in its actual nature, to thought free from conditional forms and worlds—this *I* of the individualists is no more, no less, than a zero.

The idea of zero, in my opinion, was harmless, but the concept of the unit has caused a lot of misfortune. People's sense in nature will be clear only when they come into consciousness together with the idea of infinity. If one recognizes in nature a unit, it is impossible to deny either the zero or infinity. In life only the finite is possible and desirable. It is constrained by the concepts of the greatest good and the smallest evil.

D. Popov

"Worldview"

I cannot—I even simply don't have enough of such audacity—to close the exposition of my *Cherished Thoughts* without having tried to communicate my own fundamental principles, which were worked out through the entire totality of what I have experienced and learned in life, since my entire exposition was essentially determined by these principles, albeit not directly. I consider this all the more necessary in our time, which is openly occupied by "reevaluation" and the concentrated striving to find anew the supposedly lost "origin of all origins," setting forth at times from a subjective autonomous point of view, at times from some kind of abstract unity, whether it be energy in general, or electricity in particular, or something other—only not the ancient origin point, which we name God. They are now trying to establish the reciprocal distance from physics to metaphysics as vanishingly small, so that in physics (especially after the discovery of radioactivity) they directly transition into metaphysics, and in the latter they are trying to achieve the clarity and objectivity of physics. The old gods are overturned and they seek new ones, but they have not arrived at anything at all permissible or integral; and so skepticism is legitimated, remaining content with aphorisms and denying the possibility of an integral general system. This is very sadly expressed in philosophy, setting out from Schopenhauer and Nietzsche; in the natural sciences we are trying "grasp the ungraspable"[11] following the example of [Wilhelm] Ostwald or even [Johann

[11] In Russian, "нельзя объять необъятное"—"it is impossible to grasp the ungraspable"—is an aphorism attributed to Koz'ma Petrovich Prutkov, a fictional author invented by Aleksei Nikolaevich Tolstoy and three of his cousins.—MG

Heinrich] Ziegler (in Switzerland, viz., in his "Die wahre Einheit von Religion und Wissenschaft" by J. H. Ziegler, Dr. philos., Zürich, 1904, and still better in his: "Die wahre Ursache der heller Lichtstrahlung des Radiums", 1905)[12]; in the whole intelligentsia, accustomed to cling to "the last word of science" but which cannot understand anything about what is being done now in the sciences; and saddest of all the reigning skepticism is expressed in our straying youth, since they themselves, as they know, often have to sort out open contradictions between what they read and hear in the various auditoriums of one and the same faculty, which sets up the youth to consider themselves as judges and their teachers (either one or all of them) as backward, seeking support for themselves, and the youth only value "freedom," understood in the form of a flowing robe. It is well known that this skepticism ruined the foundations of the ancient world, which appeared so strong, and not a few thinkers think the same about the foundations of modernity. Not thinking this way, I will try, as much as I know how, to express my point of view because, first of all, I hope "not to ruffle the geese's feathers"[13] and all the same to some degree express those foundations on which skepticism of a scientific or philosophical character builds itself, and, second, I will begin directly with the results, of which I advise[14]—for the sake of brevity—to be adhered to also in the resolutions now being prepared in our State Duma.[15]

Contemporary scientific-philosophical skepticism draws its origin from the everlastingly existing striving—and which ought to exist eternally—of people to recognize the unity of all the internal and external world, which is conveyed in the recognition of the One God, and in the striving to achieve as much as possible this original concept of "unity" or to come to know it more closely. In my opinion it is absolutely necessary to recognize the former as correct and the second in all respects incorrect, unrealizable, and leading to skepticism itself. Some saw this unity in the sun, others in an autocratic, imaginary, and eternal old man, a third group in the individual reason of persons, a fourth in some abstracted higher reason, a fifth see it in some kind of unified matter, a sixth in energy or force, a seventh in the will, an eighth in individualism, a ninth in humanity—you never know what else they will come up with. The striving to achieve in one way or another "the unit" or "unity" is a natural consequence of inquisitiveness, and recently it has exerted especial pressure when the successes of the real sciences began not only to clearly increase, but also to be visible even in everyday life. Formalism, attributed usually to all religious faiths—not excluding the shakers, the Bábists,[16] or the Protestants—is also the well-known realization of the fact that the realistic demands of reason account

12 Ziegler 1904, 1905.—MG

13 An idiom attributed to Russian fabulist Ivan Krylov.—MG

14 Here in the manuscript Mendeleev intended to insert a page reference, presumably to *Cherished Thoughts*, but never completed it.—MG

15 The Duma, Russia's first parliament, was formed as a consequence of the Russian Revolution of 1905, taking place precisely as Mendeleev composed this document.—MG

16 The Bábí faith was a religious movement that flourished from 1844 to 1852 in Iran.—MG

for very little, because the eternal, the general, and the unified in any event is logically higher than the real, which is understood only in the temporary, the partial, and the heterogeneous, only graspable by reason in abstraction, which indeed comprises the domain of the sciences, including philosophy (if it does not become—on stilts—the science of sciences). The sciences essentially abstract from direct realism, and if they are either real in subject or real in utility (because they give useful predictions), then this only underscores the necessity of abstractions, their significance and utility. It must [be] a great confusion of thought when one wants with scientific examples to find a realization of higher unity, to express using only the real the multiplicity of realities or abstractions. This ends up running around in circles.[17] And when one sees this they throw it out immediately, they immediately fall into skepticism relating to any and all generalizations—of course except for words, which are in themselves nothing more than the original generalizations. The realization, in whatever form it might be, of a generalization so abstract as a general "unit" or "unity" just simply contradicts the entire spirit of the sciences and cannot lead to anything other than the doubts of skepticism. The flaw here is not at all in the very idea of unity but only in the striving to realize it in images, forms, and particular conceptions. This will never achieve anything by the very logic of the matter, and one must not attempt to imagine the general "unit" either in such materialities as matter or energy, or in such realities as reason, will, the individual, or all humanity, because both the one and the other must be contained by this general "unit" and both the one and the other comprise only a topic of the generalizing sciences.

And thus I explain skepticism by the unreasonableness of forcing science, which generalizes realism and in the form of predictions conquers it for human use and by this means returns to reality, forcing science to relate these very same topics to its extreme generalizations. And one must not do this because scientific generalizations are not an already-changing limitlessness or reality, but are limited by the fact that one can only study (and only "pebbles on the banks of the ocean of the unknown" are studied, as Newton said) up to the point that it became possible to predict something and these scientific generalizations ought to remain unchanging until the very study of reality compels changing, expanding, and perfecting them. From this all of metaphysics, upon which rests all of skepticism, has not given anything sensible and useful.[18]

But enough about that. In any event it is absolutely necessary to admit the enormity of the mass of the completely unknown. Whether there is or there isn't, in this or that given area of knowledge, a certain limit which is impossible to transgress, I am not going to explore because it is entirely inappropriate to resolve this for the communication of what comprises the subject of my original thoughts. The issue at the present time is only about what a reasonable

---

[17] Literally, "a squirrel in a wheel."—MG

[18] But I beg you to notice that I do not say "will not give"— because it is still impossible to know that because it is impossible to foresee the border of scientific knowledge and prediction.

generalization may now achieve, about what one ought to or is able to agree (although one must be personally satisfied only temporarily), entirely besides the "origin of all origins," for which the ground will be created not through study but through what is called faith and is determined by instinct, will, feeling, and heart. After all, where indeed will the generalizations of reason cease? Won't all of the work of centuries gathered in science reduce to a single formulation of particulars? Where is the limit of contemporary reasonable generalizations if not in a general "unit"? Precisely there is the question of worldview, the task of that array of thoughts by which from long ago such directly applied sciences—such as the medical, engineering-technical, and juridical sciences—differ from the philosophical sciences, to which relate not only philosophy, philology, and history but also all the mathematical and natural sciences. The first are bound with the second so tightly that many minds have been confused by this tightness, but simple common sense easily recognizes that the applied sciences are moved by the philosophical sciences and, at the same time, that the philosophical sciences are worked out only because they (albeit only with a dim light) illuminate all the same the path of life—i.e. they serve utility both directly and essentially—by means of the applied sciences. Already one primary and clearly never-ending search for new parts of truth, which distinguishes science, directly points to science's striving to perfection and to the recognition of the bottomlessness of the unknown; more briefly: serving science teaches modesty united with persistence, and weans one from premature arrogance and slavery to prejudices. And since science, setting out from actuality or reality, gradually all the same reaches certain propositions or affirmations, doubtlessly justified by observations and experiments, one has a right to consider these partial truth or "laws." No one, it seems, will take this away from science. But since in the republic of science "freedom" of opinions is guaranteed to that degree that there are no attempts (either secretly or openly) to ask the majority, so everyone is free to speak in the name of science—not only everyone who has studied a little, and any author, scribbler, or humorist, but even a simple passerby, and therefore it is extraordinarily or extremely easy to get lost concerning "the last words of science." And one will not seek here, perhaps, any sort of signs—except rather negative ones—for differentiating all forms of usurpation from the actual voice of science, since the sense capable of leading here is not innate but is gained only through long and bitter experience. It shows, however, that the truly scientific is usually accompanied by a calm modesty, and there is no true science where people try sharply and with judgmental reception to shut the mouth of any contradiction, although there sometimes appears an artistic virtuosity and many footnotes to "the last word of science." Read a bit of how Copernicus and Newton introduced the truths they found and you will be convinced. It's as if true science speaks or advises "please don't believe my words, and try only to verify them," from which on the one hand I cannot but express this advice: consider as authentic science only that which is affirmed after doubts and every type of trial (observations and experiments, numbers and logic), and you will not trust very much "the last word of science," which has not tried, and does not expect, ever newer proofs. The new search for truth, only that is science, but from that it does not at all follow that it will lead to "last words." Working in science more than 50 years, you become convinced of the need for this carefulness. I will not prove this here, although I do not foreswear returning to this subject in another place under another circumstance. There are a lot of circumstances, now a good many, more than at any other time. Yes, some want to do a "reevaluation" even in science, such is already the time, everywhere—not just among us—the brew is fermenting and one is called on to write *Cherished Thoughts* clearly, as though for escaping wholesale misunderstandings, albeit only partially. Thus for this goal I consider it necessary again[19] to say that, in my judgment, the edge of the sciences has scarcely been achieved and by all appearances ought still for a long time to serve as the edge of scientific knowledge, the edge beyond which the nonscientific realm already begins. This edge always should come into contact with reality, from which it sets forth and to which it returns, and it will lead (I repeat again—to avoid misunderstandings—in my opinion) to the acceptance of the fundamental trinity of those which do not meld into each other, do not combine, are eternal (as far as we are able to recognize in reality), and which determine everything: substance (or matter), force (or energy), and spirit (or psyche). The recognition of their confluence, origin, and division lies already outside the scientific realm, constrained by actuality or reality. It is affirmed only the fact that in everything real one has to admit either substance, or force, or spirit, or, as is always the case, their combination, because substance without force, force (or motion) without substance, or spirit without flesh and blood, without forces and matter, are equally unthinkable in real phenomena. I do not at all intend to develop this theme here, I even prefer to leave it unclear but to express it in the *Cherished Thoughts* I consider necessary, necessary because not only Count D. A. Tolstoi but an entire band of people with him propose—from ignorance, of course—that in studying substance and the forces that are proper to it, scientists do not recognize spirit, they attribute everything to substance and forces.[20] There were and are such people, I don't deny it—but only [note] that they primarily grew up with classicism, which proves it is a frightening, crushing bore. This was explained a long time ago, although it is often forgotten.

---

[19] The thought set forth here (I do not insist about the degree of its originality and even directly propose that it is very widely distributed in scholarly circles) was expressed more clearly in 1902 in the article "An Attempt at the Chemical Conception of the World Ether," placed in the *Vestnik samoobrazovaniia* (Messenger of Self-Education), and recently published by me as a separate brochure (Mendeleev 1904). I might be mistaken but all the same I propose that in the epoch of "reevaluation of values" it is helpful to set forward what you consider as a general value, let them reevaluate it, and perhaps they will think that even scientists don't have anything valuable, and that the entire matter goes under the hammers of "auctioneers."

[20] Count Dmitrii Andreevich Tolstoi (1823–1889) was Minister of Popular Enlightenment, and therefore in charge of education and the universities, under the strongly conservative Tsar Alexander III. See Sinel (1973) for a more sympathetic account of his policies than Mendeleev's.—MG

With these thoughts I conclude the book, knowing or, better said, understanding, that now is not the kind of time for gradualist thoughts like mine, capable only somewhat of influencing the disturbed minds of today's youth, for whom this book was principally written. It is possible to be effective in this matter only through images, as Cervantes did with his Don Quixote. One is vicariously sorry for him, and it is impossible not to recognize the purity of his convictions, and yet people have stopped imitating him because they already very clearly saw how people like him produce only nonsense and ridiculousness. One must know how to write about how, in searching for freedom, one acts against freedom. Alas, I do not have these talents; I did not summon and nurture them. However, regardless of my attitude to this current book, I won't repent of writing it, but delight in the fact that I finished it, because, no matter what, all the same I hope it will be clear from my book what kinds of thoughts saturated the professors of the times of the late Count D. A. Tolstoi—who, I will admit, I consider the first cause of many contemporary Russian misfortunes and an exemplary and skillful mischief-maker and confounder.

27 September 1905 D. Mendele[ev]

References

Alenitsin, V.: Polozhitel'noe i otritsatel'noe. *Svet* 1, no. 6, 129–134, no. 10, 227–233 (1877); 2, no. 5, 172–178 (1878).

Bensaude-Vincent, B. 1986. Mendeleev's periodic system of chemical elements. *British Journal for the History of Science* 19, 3–17.

Gordin, M.D. 2004. *A Well-Ordered Thing: Dmitrii Mendeleev and the Shadow of the Periodic Table*. Basic Books, New York.

Gordin, M.D. 2009. Die Periodentafel oder Atombilder wider Willen. In: Bigg, C. and Hennig, J. (eds.), *Atombilder: Ikonographien des Atoms in Wissenschaft und Öffentlichkeit des 20. Jahrhunderts*, 43–50. Wallstein Verlag, Göttingen.

Gordin, M.D. 2011. Seeing is believing: Professor Wagner's wonderful world. In: Daston, L.J. and Lunbeck, E. (eds.), *Histories of Scientific Observation*, 135–155. University of Chicago Press, Chicago.

Gordin, M.D. 2012. The textbook case of a priority dispute: D. I. Mendeleev, Lothar Meyer, and the periodic system. In: Riskin, J. and Biagioli, B. (eds.) *Nature Engaged: Science in Practice from the Renaissance to the Present*, 59–82. Palgrave Macmillan, New York.

Mendeleev, D. I. 1904. *An Attempt towards a Chemical Conception of the Ether*, tr. George Kamensky. Longmans, Green, and Co., New York.

Mendeleev, D.I. 1948. *Mirovozzrenie*, ed. E. Kh. Fritsman. *Nauchnoe Nasledstvo* 1, 157–162.

Mendeleev, D.I. 1995 (1905). *Zavetnye mysli: Polnoe izdanie*. Mysl', Moscow.

Mendeleev, D.I. 2002. *Mendeleev on the Periodic Law: Selected Writings, 1869–1905*, ed. and tr. William B. Jensen. Dover Publications, Mineola, New York.

Popov, D. [D.I. Mendeleev]. 1877. Edinitsa. *Svet* 1, no. 11, 247–249.

Scerri, E.R. 2007. *The Periodic Table: Its Story and Its Significance*. Oxford University Press, New York.

Shchukarev, S.A. and Valk, S.N. 1951. *Arkhiv D. I. Mendeleeva: Avtobiograficheskie materialy, sbornik dokumentov*. Izd. Leningradskogo Gosudarstvennogo Universiteta imeni A. A. Zhdanova, Leningrad.

Sinel, A. 1973. *The Classroom and the Chancellery: State Educational Reform in Russia under Count Dmitry Tolstoi*. Harvard University Press, Cambridge, MA.

Stackenwalt, F.M. 1976. The Economic Thought and Work of Dmitrii Ivanovich Mendeleev. Ph.D. dissertation, University of Illinois at Urbana-Champaign.

Vucinich, A. 1967. Mendeleev's views on science and society. *Isis* 58, 342–351.

Walicki, A. 1975. *The Slavophile Controversy: History of a Conservative Utopia in Nineteenth-Century Russian Thought*, tr. Hilda Andrews-Rusiecka. Clarendon Press, Oxford.

Ziegler, J.H. 1904. *Die wahre Einheit von Religion und Wissenschaft*. Kommissions-Verlag Art. Institut Orell Füssli, Zürich.

Ziegler, J.H. 1905. *Die wahre Ursache der hellen Lichtstrahlung des Radiums*, 2d. ed. Kommissions-Verlag Art. Institut Orell Füssli, Zürich.

CHAPTER 15

The Philosophical Importance of the Periodic Table

MARK WEINSTEIN
Educational Foundations, Montclair State University, USA

THE CENTRALITY OF THE periodic table to chemistry is beyond dispute. What seems just as obvious to me is that the table should be seen to play an equally central role in the philosophical understanding of scientific inquiry. This may be a minority opinion; if we look at philosophical discussions of scientific issues broadly, such a view seems unsupported by philosophical practice. Philosophers have been exercised by the problematic aspects of science: revolutions rather than normal scientific practice; aspects of science that are conceptually problematic, for example, quantum mechanics; areas of science that include explanatory accounts that deviate from standard models, for example, evolutionary theory; or aspects of science that raise moral or social issues, such as the biomedical sciences. Chemistry, with a long track record of unsurprising growth, with myriad of applications taken for granted, and with a strongly supported and unifying theory may seem to be just too boring to exercise philosophers interested in resolving puzzles, developing surprising theories, and engendering novel insights. But as I will attempt to show, the most normal of normal sciences, physical chemistry with the periodic table at its core, offers a view of science relevant to central philosophical concerns. In what follows I will offer an overview of three philosophical areas for which the periodic table is salient, while indicating a logical image of a scientific structure of the sort that the table exemplifies. I look first at methodology, and in particular the role of counterevidence in evaluating generalizations. Second I look at how the table permits a reinterpretation of foundational epistemological notions of truth. Finally, I will look at ontology, how the table supports our commitment to the fundamental nature of reality.

The basis of my analysis is a model of emerging truth (MET). This metamathematical model is available in a number of publications and I will include only its most basic elements in a technical appendix. In place of the formal construction I will offer the philosophical intuitions it encodes, intuitions that draw upon the structure of chemistry with the periodic table at its core. The metamathematical construction reflects the tradition of formal accounts of scientific practice commonly referred to as rational reconstructions. Its purpose is to offer an idealization of a scientific structure so as to make the logical underpinnings transparent. Given the audience and the purpose of this volume the appendix gives just enough of the technical apparatus to offer a heuristic image, intuitive set-theoretic constructions that hopefully offer some sense of what the formalism provides. The details of the construction of the MET are found in Weinstein (2013) and in a somewhat abbreviated version with clear references to episodes in the history of the periodic table in Weinstein (2016). What I offer here is the intuition behind the MET and the philosophical insights it affords. These insights both reflect the periodic table and its history and offer a logically transparent image of the system of explanatory relationships that the table provides. The explanatory structure with the periodic table as the keystone offers a framework for human understanding at the highest level of epistemological adequacy. It is one of the most durable, useful, and detailed constructs in the history of science. It accounts for an enormous proportion of everything we know and can do and it provides, both to common sense and to deep theory, the basis for ontological commitment: an informed sense of what constitutes reality.

It is the unparalleled success of the periodic table as a scientific project that supports the relevance of the periodical table for philosophy. A description of the underling logical structure the periodic table supports shows how truth emerges, how speculative theory engages with a developing set of experimental practices and instrumentation, moving toward an increasingly adequate sense of the evidence: what I call *consilience*. The logical structure captures what is perhaps the most salient aspect of the periodic table, its ability to account for an increasingly divergent set of phenomena: what I call *breadth*. And perhaps more important, it illustrates the power of successive layers of theoretic interpretation to increase the scope and adequacy of explanations: *depth*. Finally, it shows how a commitment to a view of reality can be sustained and progress despite theoretic vagueness and experimental indecision.

These three intuitive properties, consilience, breadth, and depth are given formal treatment in the MET as follows: consilience is constructed as a chain of increasingly adequate models in light of theoretic expectations (MET, 1.1-1.2); breadth, by increasingly many model chains (MET, 1.3); and depth, the number of higher order reducing theories that offer theoretic reinterpretations of lower-order generalizations (MET, 2.1-2.4). The construct offers an account of ontological commitment (MET, 3-3.4.2) and a metric for resolving claims and counterclaims based on the three properties (MET, Part II).

With that much in place we move on to the philosophical importance of the periodic table, looking first at methodology, then epistemology, and finally ontology. Each section will include quite general comments about the table and about the chemical theories with which it engages. I hope to state the obvious in my descriptions of chemistry and support some possibly controversial claims relying on the work of Scerri (2007). When discussing philosophy, I will refer to fundamental positions and support more technical allusions with references to online encyclopedias. The MET will be described in, hopefully, intuitive ways with technical detail referenced in the appendix.

Part I: Methodology

There are two competing foundational images of scientific methodology, each reflecting a side of the logical divide in philosophical accounts of science. The first is deductivist, best exemplified by the classical deductive nomological (D-N) model (Hempel & Oppenheim 1948) and its heir, falsificationalism as championed by Karl Popper (1963). The second is inductivist, ranging from the early attempts at inductive logic in John Stuart Mill to the modern concern with Bayes' theorem, seen as the basis for what Charles Sanders Peirce first called abduction, the process by which we generate theories from available evidence (Magnani 2001). We turn first to deductivist accounts and then offer a few brief indications about the insights the MET offers to Bayesian accounts of hypothesis confirmation.

The idea behind the D-N model is simple and goes back to Aristotle. A hypothesis seen as a generalization, the *explanans*, is instantiated in some empirical array, the *explanandum*, given a set of initial conditions. In model theoretic terms there is model of the *explanandum* that is a submodel of the *explanans*. Strictly speaking, and this is crucial to the construction in the MET the model of the *explanandum* is isomorphic to some intended model of the *explanans*. In more intuitive logical terms if the *explanandum* is a logical consequence of the *explanans*, given some initial conditions, the hypothesis explains the phenomena instantiated and to that extent the hypothesis is confirmed.

This view has notorious difficulties. The most basic difficulty is built into the idea of logical consequence itself, since an implication is trivially true if the antecedent is false, the validity of the relation between a hypothesis and its implications is indifferent to the truth of the hypothesis: this is the notorious problem of induction from David Hume onward. One solution for deductivists is the falsificationism of Popper. This influential view highlights the dialectical nature of scientific inquiry. Hypotheses are put forward and tested; rejections settle the case, confirmations are at best tentative, at worst logically unsupported. This result is sufficient to show the inadequacy of the approach. It seems utterly fanciful to think the periodic table is possibly false, its apparent success the result of some massive coincidence over time. Even more telling, it is historically implausible to see the persistence of the periodic principle in the face of inconsistent evidence as a methodological weakness. This requires an examination of the dialectics of explanations put forward.

The core relation that governs dialectical argument is contradiction. Even more than inference, the basic idea of inconsistency has maintained its centrality in logic of all sorts. This persistence of the classical account of contradiction is clear from the history of syllogism, where—whatever the status of the immediate inferences indicated by the sides of the square, contraries, subcontraries, and subalternation—the role of the diagonals and the relation of contradiction, persisted as the unaltered core of logical relations (Parsons 2015). The reason was supportable on logical terms; accepting contradictions seemed fatal since contradiction between the premises and conclusion was the ultimate test for what followed inferentially and what did not. Essential to inquiry, contradiction seems vital in any dialectical account; the force of refutation by contradiction seems fundamental. In particular, consistency of evidence with a generalization put forward as a hypothesis was essential. Contradictory evidence is the crucial challenge that a hypothesis must confront (Weinstein 2012).

An examination of the dialectical use of contradictions in the history of science offers a more complex picture. Successful processes of inquiry include long periods when contradictions are disregarded in the interest of continuing inquiry. An obvious example is the law of definite proportion by volume, expressed in 1809 by Gay-Lussac as, "the volume of gases entering into a chemical reaction and the gaseous products are in a ratio of small integers" (Scerri 2007, 37). Held as almost a regulative principle, the law was confronted with countless counterexamples, recalcitrant, yet often roughly accurate, measurements reflected the lack of knowledge of the time (Scerri 2007, 29–33). More common than is apparent from discussions in philosophy of science, the tolerance of inconsistent evidence reflects the competing pulls of empirical adequacy and theoretic commitment: not one to the exclusion of the other, but both in an uneasy balance.

This is readily seen in the history of the periodic table of elements. Throughout the history of the introduction, elaboration, and extension of the basic insight of periodicity—the organization of known elements into an organized array based on essential properties evaluated by acceptable experimental procedures and reflecting generalizations informed by available theoretic images—attempts at constructing periodic systems were confronted by bodies of inconsistent evidence (Scerri 2007, 42–48). This reflected the ongoing situation in early chemistry. From problems with Dalton's formula for water to the placement of argon in the periodic table, theoretic expectations were often confounded by empirical results. Contrary to the definitive role Popper sees falsification to play, contrary evidence did not result in the abandonment of the many hypothetically postulated generalizations; rather, such generalizations were often used as the basis for additional experiments and theoretic elaborations. And as was often the case, this led to continuing empirical and theoretic advances. If such inquiry procedures are to be seen as reasonable a rather different account of dialectics within inquiry must be attempted.

The key to such an account is to see truth as an emergent property across inquiry, rather than as a simple correspondence relation, exemplified as truth based on the consistency of hypothesis and evidence. The locus of argument in inquiry needs to be moved from specific arguments conjoined in a consistent array as the basis for more elaborate arguments (the scientific analogue of recursive argumentation as in formal logic) to a holistic notion of arguments seen embedded in fields: networks of generalizations. Rather than seeing arguments as no stronger than their weakest link, arguments in inquiry need to be seen as overlapping threads of argument, mutually reinforcing and deriving their explanatory power from analogies across similar lines of experimental inquiry, buttressed by increasingly adequate overarching theories. Truth, as we shall see in the following section, is not assigned to elements as in the metaphors that drive formal logic, but may be more aptly seen as a property of the field from which these elements derive their explanatory power. Truth is a field property defined on emergent relations across a network, as in Quine's metaphor of the "web of belief" (Quine 1953).

Such an account of truth requires a dialectical image that is available from the theory of argument as first developed by Stephen Toulmin (1969). In my reconstruction of the Toulmin model, generalizations (*explanans*) are seen as warrants, connecting grounds (initial conditions) and a claim (*explanandum*) (Weinstein 2006). The use of warrants to account for the dialectical role of contradiction in inquiry requires a notion of warrant strength in support of acceptable of inferences within disciplined inquiry. What is required in light of the historical facts of theory development is a principled account of the power of warrants to resist counter-evidence. This is readily available in terms of the basic properties of the MET, which affords an intuitive level of warrant strength as a function of consilience, breadth, and depth (MET, Part II). The idea is simple: the strength of both a generalization and evidence needs to by contrasted, given the depth of commitment to both. In logical terms, such inferences are non-classical; they are paraconsistent, adaptive, and defeasible (Priest et al. 2015). Logical accounts of inquiry must be sensitive to the range of support for both generalizations and evidence. The stronger the support for the generalization the more resistant it is to counter-evidence. The more deeply the counter-evidence is supported by generalizations including accepted methodological principles (analogous to Toulmin's backing) the more the counter-evidence needs to be taken seriously (Weinstein 2012).

My use of Toulmin points to the inherently naturalistic and historical perspective of my work. My concern is with pragmatics, how actual arguments are rationally grounded in a coherent logical perspective. My starting place is not with argument in the abstract sense of entailment familiar from logical theory, but with a limited notion of warranted entailments that begins with concepts and procedures in use. This move permits me to disregard the classic arguments against earlier formalist accounts in philosophy of science based on construing scientific inferences (explanations and reductions) as open to the entire range of logical implication (Omer 1970). Rather a hypothesis and the evidentiary basis (both pro and con) need to be limited to actual claims put forward.

The pragmatic stance, seeing counterexamples to generalizations as functioning dialectically, rather than purely logically as in the standard view, requires a rethinking of the logical status of generalization, generally viewed as universal propositions. This raises a fundamental methodological question that can be viewed from a logical perspective: What is the meaning of universal quantification in scientific contexts? If counterexamples to a generalization that serves as a warrant must be seen within the context that determines how deeply a warrant is embedded in a system of other warrants, if the epistemic force of the counterexample needs to be determined in terms of the robustness of the warrant seen within a larger theoretic structure, how can we construe the universal quantifier? In intuitive terms, what does the word "all" mean in a sentence that purports to describe the physical realities uncovered by science?

The history of the developing periodic table offers both insights and challenges for a purely logical account of universal generalizations. Take as a salient example Prout's hypothesis, a cornerstone of the periodic table, and in 1817, a bold and ultimately fruitful conjecture. In syllogistics terms, "all elements are composed of hydrogen atoms." The evidence that prompted the conjecture was an outgrowth of a deep explanatory principle in early atomic theory, that molecules could be described in terms of whole number ratios among their constituent parts. This was the underlying intuition of atomic theory in its original form (that atoms, being indivisible, would only enter into combination as discrete individuals). But more significantly, there was a growing body of evidence showing whole number ratios resulting from both the experimentally ascertained weights and volumes after chemical decomposition, and more importantly, as the correlative theoretic notion of atomic weight was applied to the empirical results obtained by measuring weights on increasingly sensitive balances (Scerri 2007, 33–37). Unfortunately, in 1825 Jacob Berzelius, "compiled a set of improved atomic weights that disproved Prout's hypothesis" (Scerri 2007, 40).

Prout's hypothesis remained inconsistent with the evidence for at least a century. Nevertheless, Prout was correct in seeing hydrogen as the basis of the elements, since hydrogen with one proton serves as the basic underlying structure as we move across the periodic table, each element adding protons in whole number ratios. The number of protons yield the final organizational principle of the table, once atomic number is distinguished from atomic weight by including the presence of neutrons, which were unknown until the mid-20th century. All of this was based on warrants that supported inferences and that function, inferentially, as universal generalizations.

This raises deep issues about the applicability of concepts drawn from mathematical logic to chemical inquiry. For a universal quantifier to have logical meaning a domain must be stipulated. Concepts are then defined extensionally in terms of subsets of elements in the domain. so, unless a domain is stipulated, chemical generalizations cannot be interpreted as a universally quantified sentence. And it seems clear from the history of chemistry that no domain for its generalizations was specifiable except in the most programmatic sense. The early chemists had no clear sense of the domain across which their tentative generalizations were to be applied. Rather theoretic models were elaborated from small hard-won experimental regularities, beginning with the earliest experiments of Lavoisier and Dalton. Although in the most optimistic theorists, the range of such regularities was broadly construed to cover all chemical phenomena, just which phenomena were chemical remained unsettled until the 20th century. The argument about the nature of heat, and the interpretation of Rumford's classic experiment on heat generated by mechanical means, is a clear example of how difficult it was to draw the line between chemical and other sorts of physical properties, absent an adequate theory of the sort that the periodic table exemplifies (Cardwell 1971). And since there is no clear candidate for the domain, and hence no extensional definition on the domain, the paradigm drawn from the role of logic in the foundations of arithmetic as the guiding intuition has no purchase. Modern logic could rely on the natural numbers as an available domain, for it was provable that every logical expression could be so interpreted (Kleene 1950). But without the prior determination of a domain the universal quantifier remains a metaphor at best. This is obvious for extensional definitions, that is, definition in terms of a domain of objects. It is less obvious perhaps, that there were no clear intentional definitions, no clear meaning, for the elements that were the focus of chemistry until the meaning of the chemical terms that named them were settled by inquiry.

The extent of the chemical elements was unknown in the early decades of the inquiry that was to result in the periodic table. The prevailing notion of decomposition into elements through chemical means was in its childhood (if not its infancy), and there was no reason to believe that any current conception was adequate. Without the understanding of isotopes, the placement of chemical elements was confounded. Until the discovery of neutrons, the experimental evidence based on atomic weights could not be reconciled with the underlying theory. Analogous situations persisted to varying degrees well into the 20th century as chemical understanding evolved. As essential theoretic developments were reflected in the increase in experimental competence, progress was expected and changes in the underlying conceptions of the nature of the chemical elements were welcomed. But that is to say that the domain needed to define the scope of generalizations was not available in principle. Philosophical hand waving about all possible elements, or about where we would end up in the long run offers small comfort. It is more reasonable to say that "all" is a vague statement of hope and intention and leave it at that. But that leaves us nowhere, for without some idea as to how generalizations comport with their evidence no logic of science is possible.

Science when sufficiently mature and theoretical, as in physical chemistry, requires clear and often mathematical theories. In such formal theoretic contexts, as in mature scientific theories, predicates often refer to an unbounded, yet definable domain as specified by a particular theory. The paradigmatic example in arithmetic is the Peano axioms, which define the natural numbers.

Within a scientific theory predicates are given clear and explicit definitions in respect of an equally explicit domain of theoretic entities, and so universal quantification makes clear sense as applying to the entire range of possible instances. That is, science, like mathematics, takes "all" seriously. But when the theory reaches out to the world, to the models of data that stand as its confirmatory basis, such clarity is often obscured by the empirical facts. For although experimental data is interpretable in terms of the theory and its predicates, the world has something to say about the specifics. And the specifics, more often than not, point to ambiguities in the theoretic model in response to the empirical regularities the theory strives to explain. What is unclear is the range of the natural uniformities, construed as models of data, to which the generalizations, couched in theoretic terms, may be seen to apply (even in approximation). And so, although the domain for universal quantification is clear in theoretic models, the range of its application to empirical models remains indeterminate, and so in that crucial sense, "all" is not logically definable. This requires a more nuanced view of falsification—one that focuses on the emerging sense of the domain as it becomes available through the advance of inquiry, and that views falsification as a decision based on the epistemic surround.

What is both relevant and clear in the history of chemistry is that generalizations are reasonably insensitive to refutation by particular counterexamples. And so, understanding the methodological dialectic requires an account of scientific structure that underlies generalizations and their resistance to counterexample in perfectly appropriate ways. The MET offers a logical structure in terms of the progress of a theory in terms of the three intuitive characteristics, consilience, breadth, and depth that weights the embeddedness of a generalization as against the force of the counterexample. Examples abound in the history of the table. Mendeleev's attempts at applying the basic structure of periodicity, the placement of beryllium, the incorporation of newly discovered elements, the placement of gallium, and his notable predictions of scandium and germanium indicated successes. Equally noteworthy from our perspective are that Mendeleev's less-successful predictions, that newtonium as the analogue to ether and the postulations of two elements lighter than hydrogen by analogy with the noble gases, were not fatal to the project (Scerri 2007, 140).

As the principle of periodicity developed not only were new substances found to conform to the requirements of the periodic table, but the ability to articulate the chemical properties of these substances based on underlying theoretic analysis resulted in the understanding of the physical world around us at a level of precision that was barely imaginable by the early chemists. The result of the willful suspension of empirical disconfirmation and even the suspension of the need for experimental confirmation in light of theoretic commitment often results in an eventual increase of goodness of fit with theory as the empirical properties of new substances are ascertained through newly developed instrumentation and technical skills. An example of the latter is the theoretic understanding of semiconductors, first envisioned in the 1920s, but only experimentally applied in the 1950s, leading to both theoretical and practical advances including the enormous range of applications of transistor technology and the unification of chemistry and electronics at a micro level (Riordan & Hoddeson 1997).

The give-and-take between the theoretic model and the models of its supporting evidence is required for the theory to be empirical, that is, falsifiable. The history of theoretic generalizations in relation to their empirical database, however, is not the simple one of refutation by experimental counterexample. Theories often resist anomalous data in light of the robustness of the theoretic contexts within which the interpretation of empirical data occurs. Theories are subject to modification or even disconfirmation in light of recalcitrant facts interpreted within the domain and predicates of the theory—facts that fail to support its theoretic generalizations. But inconsistent empirical data may also be resisted in the name of the power of the theory, measured by its overall empirical basis, and as important, its place in a network of other theories, each one of which is supported by its own empirical basis and its own place in the network of related theories. It is this give-and-take between theoretic embeddedness on the one hand, and risk of modification or falsification in light of recalcitrant empirical data on the other that the discussion that follows attempts to illuminate. Such an account relies on the structure of the MET, which looks at a theory in terms of its underlying model structure. The array of lower-order empirical models and their adequacy in the face of experimental and instrumental sophistication (MET, 1.1-1.2); the scope of the theory, that range of new phenomena that fall under that theory (MET, 1.3), and most important its connections with deep explanatory theories that permit a uniform reinterpretation (reduction) of a range of theories using higher order explanatory theory, which reduce generalizations in terms of deep theoretic principles with powerful explanatory force (MET, 2.1-2.3). Such a structure explains how counterexamples are to be considered on their merit, rather than serving as automatic refutations as in classical logic and Popper, based on the embeddedness of supporting warrants in scientific structures (MET, Part II).

The ability to envision a structure that supports the strength of warrants has consequences for the positive role of evidence in supporting hypotheses. Without going into the technical details of statistical theories of confirmation, the MET yields a surprising and intuitive solution to the problem of prior probabilities as required for the application of Bayes' theorem as a model of abductive reasoning from evidence to hypotheses. Bayes' theorem, a logical consequence of standard probability theory, asserts that the probability of a hypothesis H conditional on a given body of evidence E is the ratio of the unconditional probability of the conjunction of the hypothesis with the evidence to the unconditional probability of the evidence alone:

$$\mathbf{P}_E(H) = \mathbf{P}(H \,\&\, E) / \mathbf{P}(E), \mathbf{P}(E) > 0.$$

The classic concern is characterized in terms of what may be called "confirmational relativity"—that is, probabilities reflect individual estimations of degrees of belief (Joyce 2008), most essentially, the probability of the hypothesis prior to the evidence and the probability of the evidence itself. This subjectivist interpretation of probability, when contrasted to frequency interpretations based on available empirical data, has resulted in the rejection of Bayesian statistics as inadequate for the needs of objective inquiry. Given the need for an affirmative model of hypothesis verification this is problem for understanding scientific methodology. The brute facts seem to be that scientists do increase their estimation of hypothesis based on evidence, just as they frequently reject counter-evidence in light of the centrality of a hypothesis in a particular inquiry. We need some sense of how this is supportable and the MET offers a possible answer.

A classic problem in philosophy of science is that of entrenchment. Nelson Goodman had a persuasive argument that showed that evidence could confirm the most unlikely hypotheses as well as plausible ones. The classic puzzle was that all evidence that emeralds were green was also evidence that emeralds were "grue" (green until some arbitrary date in the far future and blue thereafter, Goodman 1965). This mirrored problems with the D-N model, which also permitted the construction of logically equivalent but strange predicates as alternative available explanations. The MET meets such fanciful constructions head on by requiring that for an explanation to be included, it must be seriously presented and accepted by members of the community that governs inquiry (Weinstein 2013, 85–86). Although this is a pragmatic interpretation of the relevance of a hypothesis in terms of a decision of a community, the MET, by looking at inquiry over time, moves beyond the narrowly pragmatic vision of a community decision to the larger pragmatic vision that sees explanatory progress as an indicator of the objective warrant for the decision. The model structure of MET permits of a metric of entrenchment in terms of the scope of the scientific structure within which it sits, its evidentiary base and its theoretic connections: the history of a theory in terms of its realizations over time (MET, 1).

Looking at the scientific structure that is at the core of the MET yields a similar resolution of the subjectivity of the assignment of probabilities needed for Bayes' theorem. The model chains in the MET are all finite, each consists of the confirming models put forward (MET, 1.1) and if these chains prove to be progressive (MET, 1.2) they are acceptable, meeting the test of consilience. If in addition the set of progressive model chains is increasing (MET, 1.3) the evidence supports the theory as meeting the test of breadth. Given that the MET is limited to model chains put forward and that the set of these at any time is finite we can assign prior probability to a hypothesis as a function of the number of confirmatory empirical models in a chain and an even greater increment based on the number of confirmatory model chains it supports. The probability of the evidence is a function of how closely it approaches the standard of the acceptable model chains. This creates a metrical basis for the dialectical evaluation of evidence and counter-evidence (MET, Part II; See also Weinstein 2013, chapter 4). The probabilities are still confirmational as in the standard interpretation of prior probabilities, but they are no longer relative to an individual belief, but rather reflect the acceptance of evidence by the community engaged in the inquiry within which the hypothesis sits. The power of this conception will become clearer as we move to the consequences of the MET for epistemology and finally to ontology.

Part II: Epistemology

The core of epistemology is truth and the philosophical analysis of truth can be seen as including three foundational perspectives. The first version of foundationalism, based on the apparent availability of abstract knowledge, can be generalized as coherence. That is, claims to truth must withstand a logical test at a very high level of rigor. The paradigmatic example for the Greeks was geometry, and in the modern era mathematics. The second version, based on the primacy of sense perception and the availability of the world as it appears can be generalized as correspondence. Truth is seen as a relation between our beliefs and the world as it becomes available to us. Correspondence in a formal sense was also given a central role in modern logic since the availability of antecedently known truths of arithmetic enabled standard semantic theory, which sees truth as a function mapping sentences unto a domain whose structure was understood (Hodges 2014). The third view is generalized from the practical effectiveness of trial and error reasoning, and our ability to use both sense and reason to accommodate our purposes. This is characterized as a pragmatic notion of truth. These are all deeply compelling images and are often dialectically used against each other, but the very durability of the philosophical discussion points to the insight that each includes. Looking at the logical structure of chemistry through the lens of the MET offers epistemological insights as to how these foundational metaphors may be elaborated and combined. Given our concern is with empirical science, we begin with correspondence.

The obvious problem with seeing the truth of the periodic table as correspondence in any simple sense should be obvious. The world the table describes, although independent of the table, is a creature of the table. What we see through the table is the structure that the table provides. The comparison with logical theory is telling. A logical semantics is a mapping from sentences unto a structured domain within which truths and entailments can be ascertained. In this standard view the domain under consideration is most often the natural numbers, a productive source of logical theory for decades (Mostowski 1966). But unlike logical theory, the domain to which the principles, generalization and descriptions that constitute the periodic table is not available, for it was not known, and cannot be known, except as the table evolves. That is not to diminish chemistry's basis in experimentation and the human senses. As with all human knowledge, including the logical, chemistry begins in intuition. Sense perceptions were essential, including such relatively recondite sense percepts as taste; acids were initially defined in terms of their feel on the tongue. And experience with manipulating ordinary substances, such as refining metals or brewing beer, supports the centrality of experimentation in human experience (Toulmin & Goodfield 1963).

Understanding and controlling the world of objects requires the identification of natural kinds, classifications that enable useful distinctions to be made. But the natural division into physical substances based on practical manipulation proved an unreliable basis for the developing chemical theory. The identification of natural kinds in chemistry was fraught with error and subjected to extensive revision as the field progressed. The obvious distinctions that constituted the pre-theoretic knowledge of chemical elements and compounds were incomparably poorer than the distinctions among kinds of substances that the developing theory would expose. Deep principles of organization, such as the conservation of mass determined the key procedure of careful weighing, and of course, various speculative ideas about atomic theory, were essential to the development of chemistry. These theoretic principles, as applied to the sorting of physical substances, offered increasingly productive chemical distinctions, ranging from the early study of gases to the sophisticated analyses that led to organic chemistry. Chemistry developed as the result of the application of new concepts, procedures and technology, and over a century led to practical applications understood both in terms of their experimental outcomes and the availability of increasingly coherent theories.

The early experimental and theoretic efforts formed a roughly stable body of generalizations that permitted analogical application to similar cases as chemistry advanced. Thus, not only are there levels of description, experimental and theoretical, but accepted descriptions at a particular level reach out to neighboring phenomenon on the same level creating networks of similar chemical knowledge—for example, the differentiation of acids and bases, identifying metals as a chemical class, the analysis of the family of substances formed by carbon rings, the structure of crystals, and the like. Major theoretic advance, however, occurs when these networks of analogous chemical generalizations are subsumed under higher levels of laws and models, as in the development of molecular theory based on the understanding of chemical reactions. This started in the 19th century with advances in chemical analysis, but the power of unification is best seen in the periodic table of elements. The table is a substantive model, a comprehensive image that has driven chemistry as it opened the door to the deeper connections between chemical understanding and the physics of the molecule, the atom, and ultimately theories of elementary particles. The yield: modern physical chemistry with all of its riches.

The image should be clear. A scientific knowledge structure, its end points in experience increases in depth as higher order theories offer new understanding of aspects of the lower order generalizations. It expands it breadth, both through analogy, as similar discoveries form broader networks of similar classifications, and through common causal and other functional connections. This breadth is substantive when groups of similar generalities are subsumed under a higher order theory. Such reductions of the generalizations in lower-order theories, in terms of the concepts of higher-order reducing theories, create major advances in both theory and practice. These are the grand unifications of physical sciences: classes of similar and even apparently dissimilar phenomena are seen to be the result of the same underlying forces. Such unifications, for example, electromagnetism, physical chemistry, organic chemistry, biochemistry, and astrophysics, not only unify understanding but also yield measurements and predictions of increasing precision. But the yield is not immediate, nor is it easy to come by. Chemical understanding, for example, required that predicted relationship be better borne out by the details of the experimental process. Driven by deep principles such as the conservation laws, and supported by layers of physical models (molecules, atoms, elementary particles), the theoretic demands set ideal values which experiments were to increasingly approximate.

As indicated above, chemistry as a knowledge structure can be seen to exemplify three essential criteria of epistemic adequacy: breadth of applicability (at all theoretic levels), depth of theoretic understanding, and progressive improvements of measurement and other relevant descriptions as understanding progresses, yielding greater consilience. These criteria, as reconstructed in the MET give an abstract and formal image of this complexity. And in the style of philosophy within which my conception was formed, the availability of a metamathematics serves as an indication of the complexity with which the deep structure that constitutes the logic of science needs to be reconstructed.

This is evident from the reconceptualizations of the three foundational metaphors for truth in terms that the MET provides. In terms of coherence, the logical health of the system, a strict deductivism using logical implications result in falsification, since few experiments come out exactly as planned. But the dialectics of the acceptance of generalizations, warrants for decisions of inclusion or exclusion from the body of working hypothesis within the field on the basis of available evidence, whether experimental or theoretic, is not promiscuous. It is well managed in terms of the estimations of warrant strength (MET, Part II). On an epistemological level, the coherence between the evidentiary basis and the intended model is relaxed to include approximations, goodness of fit between evidence and theory and between linked theories (MET, 1.1, 2.1) But the approximations are well managed, again in terms of the pragmatic decisions by working scientists. This is not something that should be defined a priori as in deductivist models, but is the outcome of deliberations by participants engaged in the inquiry and operationalized by journal editors and reviewers. The MET offers the logical structure. The MET moves from truth in a model to the history of approximations of models in model chains seen over time. We look backward at forward-looking scientists, and see how the accumulation of better empirical data reinforced their project for which the periodic table became the keystone (MET, 1.1). And, how better theoretic models drove new experiments that yielded better approximations and new chains of models (MET 1.2). We see increasing coherence between evidence and theory as an index of the emerging truth of physical chemistry. But even such brief and idealized remarks about the history of chemistry point further. The key notion of breadth marries coherence to correspondence. More than the coherence of theory and evidence, chemistry is characterized by the ever-widening increase in the scope of its application (MET, 1.3). The gradual adjustment of the theoretic parts, evolving experimental prowess and increased understanding, permits each theoretic part of the developing structure to be increasingly seen in terms of some others (MET, 2.1). The growth of theory connects with an expanding vision that encompasses vast reaches of the world as we come to understand it. And so, increased coherence results in elaborated correspondence as more of the known universe is included within our theoretic grasp.

The MET has a deep affinity with pragmatic notions of truth. It, however, sees pragmatic concerns on two distinct levels. In terms of the construction of key aspects of the MET—the degree and kind of approximation relations that indicate the progressive nature of a sequence of models (MET, 1.1)—are determined in light of the practice of the field; levels of acceptable deviation from theoretic expectations are determined by those engaged in the inquiry. This also holds for the choices of models to be included in the expression of the theory: the identification of model chains is subject to the actual decisions made by those engaged in the inquiry. The MET looks broadly at the scientific structure that results from such decision (MET, 1). It is an invitation to historians of science to see the interchanges that they chronicle as part of a larger and logically definable image. The second level of pragmatic concern reflects the undeniable accomplishment of chemistry as a practical tool with countless applications, and with many more to come. But it is not only the practical success that drives its truth, rather it is the increasing adequacy of its description and most important the depth and saliency of its explanations, the coherence and fruitfulness of its theory, that drives it pragmatic value. The pragmatic theory of truth is not limited to the transient concerns of practical men and women but must speak to the deep insight of Charles Sanders Peirce who saw truth as the end toward which successful inquiry tends.

Part III. Ontology

Ontology is the oldest and deepest of philosophical concerns. It rests on the distinction between appearance and reality. And as such, it seeks to grasp being in itself. A simple version of the question that drives ontology is: what is the world really made of? This was a basic question for Greek philosophy in its beginning. My concern is with how we commit to an ontology, to a vision of the reality that constitutes the world. It is worth considering that for a sane, moderately informed person chemistry offers the answer to the simple question. My claim is that chemistry offers the answer to my concern as well.

Chemistry shows us the world as it appears through our deepening understanding and search for truth. Physical chemistry, with the periodic table at the center, shows us how we commit to an ontology: what the world is made of on the basis of how we come to know that world.

The ontological commitments of chemistry rest on the stability and success of the inquiry through which the world as understood by chemistry has been constructed. In what follows I will offer reminders of aspects of chemistry that are indicators to its epistemological stability: its increasing consilience, its enormous breath, and its profound depth. The task here is to show how the MET offers a formal image of these three properties leading to ontological commitment.

The periodic table stands at the center of an amazingly complex joining of theories at levels of analysis from the most ordinary chemical formula in application to industrial needs to the most recondite particle physics. The range of these ordinary things—electrical appliances to bridges—has been interpreted in sequences of models, developed over time, each of these responding to a particular need or area of scientific research. Examples are no more than a listing of scientific understanding of various sorts: the understanding of dyes that prompted organic chemistry in Germany in the late 19th century; the elaboration of the biochemical basis of life; the smelting of metals and the improvement of alloys (e.g. stainless steel); the synthesis of new substances (notably bakelite and nylon); the work of Faraday and the consequences for the development of electrical apparatus; the development of the transistor and the exploration of semiconductors; and the discovery and utilization of radioactive substances. This multitude of specific projects is linked empirically by clear operational concepts and explained in the terms the periodic table provides. Chemistry had combined the independent areas of study of gases, liquids, solids, and organic substances in a unified theoretical structure. Each of the areas began as an independent field of inquiry, whose power was both evident and evidently increased bringing together disparate areas of empirical concern under a unified approach, and increasingly understood in terms of the deeper unifying assumptions of particle physics and electromagnetic wave theory.

The deep work in science is to unify theories. The more mundane work is to clarify and extend each of the various applications and clarify and modify existing empirical laws. This is accomplished in two ways: first, by offering better interpretations of empirical and practical understanding as the underlying theories of their structure become clearer; and second by strengthening connections between underlying theories so as to move toward a more coherent and comprehensive image of physical reality, as underlying theories are modified and changed.

How to describe that structure? In modern logic, the notion of truth is tied to the notion of satisfaction, that is, true in a model (Tarski 1983). A true sentence or set of sentences is true, just when it refers to aspects of an array correctly; the traditional formulation is: "Snow is white" is true only when snow is white. Mathematically, there is a function mapping the terms in the sentence onto elements in some relevant model. This is supported by the intuition that truth is a correspondence between a sentence and the world. And the notion of correspondence is reinforced in modern logic since its concern with the foundations of mathematics yielded an available model in terms on which logical theory could develop. Truth in logical theory was essentially tied to what was true of the natural numbers, since any expression, formula, predicate, or sentence in standard logic could be interpreted (satisfied) in that domain (Kleene 1950). Arithmetic was the model of choice; it was available and understood and became the paradigm in terms of which modern logic was constructed (Mostowski 1966).

Chemistry, of course, can support no such model. There is nothing in physical science that, like true in the natural numbers, serves as the obvious anchor for a formal semantics. The history of chemistry does not offer a model. Rather it offers a succession of models: models of experimental data, explanatory models, ideal types, graphic images, mathematical formulations, and illuminating metaphors. It is the growth and relationship among these models that gives the structure that supports the claim that chemistry arrives as close to truth as any reasonable person would require. To capture this requires that truth be defined in terms of relationships across models. The MET shows how truth emerges from the progress of chains of models over time, and how the relationships among them yield ontological commitment.

Truth in a model is an essential concept. Without it we have no logic. But the identification of truth in a model with truth just reflects the metaphysical and epistemological biases of standard logic in light of the univocal character of mathematics as traditionally understood. If I am right, it is not truth in a model that is the central issue for truth in science, but rather the choice of models that indicate truth. And this cannot be identified with truth in a model for it requires that models be compared.

To look at it another way, if we replace mathematics with science as the central paradigm from which a logical theory of truth is to be drawn, the identification of truth with true in a model is severed. For there is no particular model in which scientific theories are proved true. Rather, science shows interlocking models connected in complex ways. The principles that support reduction between theories are enormously difficult to find and invariably include all sorts of assumptions not tied to the reduced theory itself. The classic example is the reduction of the gas laws to statistical mechanics. The assumption of equiprobability in regions is just silly as an assumption about real gases, but the assumption permits inferences to be drawn that explain the behavior of gases in a deeply mathematical way, and in a way that gets connected to the developing atomic theory at the time, much to the advantage of theoretical understanding and practical application (Nagel 1963). That is to say, reductions, like explanations, include gaps and mismatched models (Cartwright 1983). This requires a rational reconstruction that has a lenient view of model relations, approximation rather than isomorphism as required in standard logical theory.

The MET looks to functional relations among models of a theory, the history of their relations to each other over time, and in relation to their logical surround. Truth, in the final analysis, becomes a property of the field of models, rather than of a model per se, as in the standard notion of truth as satisfaction. Two basic preliminaries: first, the crucial empirical dimension. There is a set of privileged models, empirical models of the data put forward relevant to the confirmation of the theory. Such data sets are, as usual, based on observations, whether direct or mediated through instrumentation. But like all empirical data, they must be objectively ascertainable and open to all competent observers. What makes a scientific theory empirical is the constraint that all models of the theory have connections with empirical models (MET, 1). What makes it a potentially true theory is that the empirical evidence is increasingly adequate to the theory (MET, 1.2) and that the theory is increasingly fruitful, yielding new explanations (MET, 1.3). Second, and essential for ontological commitment, for any sequence of models whether theoretic or empirical—but with the exception of the highest order theoretical models—there may be found higher level models. For empirical models, higher order models are explanatory. For explanatory models, there are higher order theoretic models that are reducing models, models that explain theoretic models by reinterpreting the theoretical elements in terms of a more abstract domain (MET, 2.1). Powerful theories expand the reinterpretation with layers of higher order theories (MET, 2.2). More powerful theories expand the reinterpretation across a wide range of theories (MET, 2.3). And the most powerful offer a reinterpretation that increases the yield of all of the theories within its scope (MET, 2.4).

The MET attempts to characterize the process of truth emerging as evidence of the epistemic adequacy of the warrants becomes available, supporting theoretic explanations and governing theory driven inferences. Truth will ultimately be defined as an ideal outcome of a process of model consolidation. The ultimate image is that of a field of models that exhibit maximally desirable epistemic properties over time. This is supported by an ongoing evaluation of truthlikeness and is reflected in the dialectic of claim and counterexample (MET, Part II). At the level of empirical generalization, truthlikeness reflects the increase in the number and degree of articulation of empirical models over time: that is, consilience. It is often seen in the increasing precision with which measurements and calculations can be made, congruent with the expectations as specified in an intended model (MET, 1.1). Thought of in visual terms, consilience speaks to the grain of an explanation, the level at which detail can be elaborated and explained as both measurement and available explanatory concepts are applied at a time and as they evolve over time. Examples of the increase of precision characterized by better and more detailed empirical measurements is evidenced throughout the history of chemistry, and most recently in its application to both inorganic and organic substances now described with precision down to the atomic level.

Another measure of empirical adequacy is breadth (MET 1.3), visualized as a horizontal measure in which an explanatory model is represented in the field by an increasing array of lower level—more empirical—models at a time, and as a function of time. Breadth can be applied to theoretic models as well as empirical models, and is a measure of the range of application of a theory. Examples of broad explanations abound in mature science. Crude atomic models and the formula with which they were represented were first applied to explain properties of gases, then to liquids (fluid dynamics) and solids (crystallography), until by the 21st century the scope of chemical understanding is virtually comprehensive of the entire range of available empirical knowledge of the material universe.

The empirical basis is a necessary requirement for scientific theories seen as adequate. But there is a more deeply theoretic dimension, the depth with which any theory is supported by other theories (MET, 2.1–2.4). This can be visually construed as the vertical measure as higher order theories offer deep reinterpretations of lower level models in terms of more abstract and general theoretic concepts and constructions. This last *desiderata*, characterized as reduction in the philosophy of science literature, is one of the most essential characteristics of mature physical science and is arguably the basis for the enormous rapid advance of physical chemistry in the 20th century. Deep reductions not only explain phenomena more adequately, but such theoretic reconstruals of known generalization open the door for vast areas of new applications. An obvious example is the use of scientific theories to first deepen and then expand empirical knowledge almost beyond the reaches of human imagination. In the early 19th century, Auguste Comte argued that we could never know what stars are made of since we could never visit them. By the 1860s William and Margaret Huggins applied spectrum analysis, then used to analyze chemical substances on Earth, to analyze starlight as an indication of their chemical composition. A more telling example is the enormous advance of organic chemistry as applied to the understanding of human physiology, with applications as broad as modern medicine, the modification of living organisms at the level of genes and, in the foreseeable future, prosthetics that are integrated into the neural network and manipulated by thought alone. A similar case could be made for sciences as diverse as material science, geology, paleontology, and anthropology. Each of these has advanced far beyond the initial range of empirical generalization to include theory-driven understanding that reaches to the stars, to the beginnings of the universe in time, as well as into the distant past with the analysis of the chemical composition of ice cores reaching back into Earth's history countless millions of years, the examination of DNA for dating fossils, and the prevalence of particular isotopes in determining the history of the surface of Earth, including the understanding of cosmic events such as the destruction of dinosaurs.

As we look at the history of the periodic table of elements, the key notions of consilience, breadth, and depth are apparent as the theory evolves over time. Historically, this is no simple linear progression. The history of the periodic table is fraught with unresolved issues and reevaluations of empirical and theoretic understanding. A theory of truth in inquiry needs to account for the changing epistemic weights assigned to models as they interact. The goal is a metric that correlates evidence of varying degrees of robustness, flowing from different sources. The MET gives an abstract account of this and points to the possibility of a linear metric based on the history of the various models over time. This is a simplification, reconstructing the complex reality into a more tractable logical image. A statistic is more reasonable than the linear metric that the formalism indicates, but does not require. Such a statistical account, however, would complicate the model in a manner that obscures the noetic clarity of the idealized linear account. But, however the sequence of models is characterized, truthlikeness becomes a function of the scientific structure itself. Truth in an ontologically significant sense is defined in terms of an optimal outcome (MET, 3-3.4.32).

The periodic table, at the center of the structure envisioned by the MET, connects with the vast domain of chemistry, physical and organic, which in association with roughly parallel theoretic clusters, mechanics (statics and dynamics), and electromagnetic wave theory, explains just about everything we do and can do in the physical world. This cluster of theories has increased in its explanatory power over the last 150 years as individual theories are expanded and refined, and inter-theoretic connections made. Scientists develop translation procedures to and from observable phenomena and to and from related theories. The strongest connections are often the result of higher order theories. Above the periodic table stands particle physics, quantum theory, quantum electro-dynamics, and general relativity offering the possibilities of the deepest connection and the most profound understanding. But this is not logically straightforward.

To take an example, quantum mechanics has a complex relationship to chemistry. Developments of the early model of Bohr have led to better theoretic understanding and more adequate models. Scerri asserts "more accurate calculations can now be carried out in quantum chemistry than were even dreamed of in the old quantum theory" (Scerri 2007, 229). But advances do not meet the requirement of deductive certainty. Many applications result in approximation; Scerri offers the example of the Hartree-Fock method (Scerri 2007, 231). Approximation relationships among models are a basic element in the construction of the MET, which does not require that empirical models be isomorphic to theoretic models (MET, 1). But this does not lead to methodological license since we require first, that approximations met the standard set by those working in the field and second, that sequences of models be progressive, that approximation relations between empirical and theoretic models improve over time.

Weakening the conditions on empirical models is not enough when we look at the reduction of aspects of chemistry to quantum mechanics. Scerri cites the basic principle of electronic configurations including the Hund and Pauli principles as "essentially empirical and none of them have been strictly derived from the principles of quantum mechanics" (Scerri 2007, 233). These are complex issues and differences of opinion abound, but the MET offers a principled description of the logic of such reductive disputes. A successful reducer does not have to account for all of a theory, it merely has to show an increase in its reductive scope over time (MET, 2.2). The empirical basis is there; the deep understanding in terms of reduction to higher order theories must be sought after.

The practice of science both reflects and supports the logical procedure of drawing appropriately modified inferences based on pragmatic decisions by members of the field. A metamathematical model must support the range of flexibility of such inferences, while retaining enough integrity as a model structure so that a formal semantics is indicated and its inferential structure exposed. From an ontological point of view the issue is how the management of inquiry over time leads to a commitment to a model that encapsulates the realties that the inquiry exposes. The MET describes this in several steps. The key notion is that of the intended model. The core intuition is captured in Quine's slogan: to be is to be the value of a variable (Quine 1953). That is, within formal reconstructions ontological commitment is made through the choice of domain through which the extensions of terms are fixed.

As we have seen, unlike traditional logic which has the natural numbers as an available domain for its models, there is no universal model for chemistry, and so there is no available domain. The MET looks at the history of models over time. This requires that we look at the history of realizations of a *scientific structure* (MET, 1): a theory, its models and the reducing theories ordered in time. Within such a structure we look at a model that persists across the array (MET, 3). We call the elements in the domain of such a model an *ontic set* (MET, 3.1). It is a putative ontology for the structure. If in addition it is the intended model of a progressive reduction chain (MET, 2.2) that is progressively branching (MET, 2.3), then it is a *favored ontic set*; that is to say it is the presumptive ontology for the theory that generates the structure. If, in addition, the theory meets the constraints on empirical adequacy (MET, 3.3) and the theory at the head of the structure meets the constraints of reductive adequacy (MET, 3.3.1) the intended model of the theory is the most favored candidate for the ontology of the structure (MET, 3.3.3); its ontic set becomes the favored ontic set for the array (MET 3.4.1) and truth within the scientific structure is standardly defined in terms of true in the domain of that theory (MET, 3.4.2).

Of course, as in all rational reconstructions, this is an idealization of the scientific realities it purports to describe. The ontology of physical chemistry is open to dispute. How high in the chain of abstract theories, the microphysics behind the relatively settled aspects of the periodic table, should we go in search of a comprehensive ontology? Quantum mechanical models are clearly available and as science advances choices for understanding the deep ambiguities of quantum mechanics will be advanced and debated. But what the MET tells us is that the physical understanding of the basis of chemistry is not subject to logical and philosophical constraints alone. Just how to characterize the hard won understanding of quantum physics, in logical terms its intended model, is not the only philosophical concern. If we are to understand the power of such deep scientific theory we must address how it meets the needs of the durable chemical structure that it seeks to explain. The ontological task is to decide which model of the microphysical theories we propose can serve as the reinterpretation of the array of chemical theories whose explanatory power is beyond dispute. The intuition in relation to ontology should be increasingly apparent. A theory, whatever the dispute as to its intended models takes its ontological commitment in light of how the theory fairs in relationship to other theories whose models it must accommodate under reduction. That is, we fix the reference of our theoretic terms in light of the facts of the matter, the relevant facts being how a theory may be redefined in light of its place in inquiry as inquiry progresses. But this is not arbitrary; for the philosopher, the power of a rational reconstruction must rest ultimately on the noetic clarity of the philosophical criteria that it exposes as it encapsulates the flow of inquiry. The scientist sees this from a different perspective.

The awareness on the part of inquirers of the history of success of scientific structures enables participants in the inquiry to rationally set standards for model choice in terms of plausible standards, based on successful practice. These are internal criteria, scientific standards of empirical and theoretic adequacy that govern the dialectic of claim and counterclaim and are reflected in the tenacity with which claims are maintained and the epistemic force of counterexamples brought forward. Scientists must make educated guesses as to relevance and truthlikeness; this is reflected in the decisions of editorial boards and funding agencies. But this is not methodological anarchy as in the notorious views of Feyerabend (1975). And the MET shows why.

The MET requires us to look at the history of approximations within the models of a theory as they meet or approach intended models, and most essentially, goodness-of-fit relations between models donated from above, from reducing theories, which reinterpret intended models of the reduced theory. The philosophical yield is both in the clarity of the criteria for epistemic force (the various notions of progressive as applied to chains of models and to reducing theories) (MET, 1.2-2.4); an ontologically significant definition of truth internal to the structure seen as a whole (MET, 3-3.4.2). And finally, an account of inferential force (the hierarchy of warrants; MET, Part II). Truthlikeness is seen as a quantifiable metric as a theory is increasingly progressive, and is markedly increased, as reducing theories donate intended models to reduced theories unifying the structure in terms of a common ontology through the reinterpretation of intended models (MET, 2). Truth is defined as an ideal outcome, identified with the strongest possible degree of truthlikeness (MET, 3.4.1).

The guiding intuition is fairly standard: true theories ramify. The force of the MET is in its logical clarity, and what that permits. The construction displays what one can mean by "ramify" and indicates how a metric might be defined. The structure could be modeled with any finitary assignment, and consequences drawn. It offers an adequate metaphor for truthlikeness as the outcome of inquiry of the sort found in physical chemistry, the most plausible candidate for a naturalist ontology.

Technical Appendix

The Model of Emerging Truth (MET)

Part I

1. A scientific structure, TT = ⟨T, FF, RR⟩, where T is a set of sentences that constitute the linguistic statement of TT closed under some appropriate consequence relation, Con(T), and where FF is a set of functions F, such that for each F in FF, there is a map f in F, such that f(T) = m, for some model or near model of T (Weinstein 2013, 85–86). And where RR is a field of sets of representing functions, R, such that for all R in RR and every r in R, there is some theory T* and r represents T in T*, in respect of some subset of T (Weinstein 2013, 88).

A scientific structure is first of all, a set of nomic generalizations, the theoretic commitments of the members of the field in respect of a given body of inquiry. We then include distinguishable sets of possible empirical models (or appropriately approximate models) and a set of reducing theories (or near reducers). What we will be interested in is a realization of TT, that is to say a triple ⟨T, F, R⟩ where F and R represent choices from FF and RR, respectively. What we look at is the history of realizations, that is, an ordered n-tuple: $\langle\langle T,F_1,R_1\rangle,\ldots,\langle T,F_n,R_n\rangle\rangle$ ordered in time. The claim is that the adequacy of TT as a scientific structure is a complex function of the set of realizations.

1.1. Let T´ be a subtheory of T in the sense that T´ is the restriction of the relational symbols of T to some subset of these. Let f´ be subset of some f in F, in some realization of TT. Let $\langle T'_1,\ldots,T'_n\rangle$ be an ordered n-tuple such that for each i, j (i < j,) T'_i reflects a subset of T modeled under some f´ at some time earlier than T'_j. We say the T is *model progressive under* f´ iff:

(a) T´k is identical to T for all indices k, or
(b) the ordered n-tuple $\langle T'_1,\ldots,T'_n\rangle$ is well ordered in time by the subset relation. That is to say, for each T'_i, T'_j in $\langle T'_1,\ldots T'_n\rangle$ ($i<j\leq 2$), if T'_i is earlier in time than T'_j, T´i is a proper subset of T'_j.

1.2. We define a *model chain* C, for theory T, as an ordered n-tuple $\langle m_1,\ldots,m_n\rangle$, such that for each mi in the chain $m_i = \langle d_i, f_i\rangle$ for some domain d_i, and assignment function f_i, and where for each d_i and d_j in any m_i, $d_i = d_j$; and where for each i and j (i < j), m_i is an earlier realization (in time) of T then m_j.

Let M be an intended model of T, making sure that f(T) = M for some f in F (for some realization ⟨T, F, R⟩) and T is model progressive under f. We then say that C is a *progressive model chain* iff:

(a) for every m_i in C, m_i is isomorphic to M, or
(b) there is an ordering of models in C such that for most pairs m_i, m_j (j > i) in C, m_j is a nearer isomorph to M than m_i (Weinstein 2013, 94).

The requirement in condition (b) is an idealization, as are all similar conditions that follow. We cannot assume that all theoretic advances are progressive.

Frequently, theories move backward without being, thereby, rejected. We are looking for a preponderance of evidence or where possible, a statistic. Nor can we define this a priori. What counts as an advance is a judgment in respect of a particular enterprise over time best made pragmatically by members of the field. We are engaged in rational reconstruction where, logically, clarity trumps descriptive adequacy, in presentation, but where descriptive adequacy is still at the heart of the intuition.

1.3. Let $\langle C_1,\ldots,C_n\rangle$ be a well ordering of the progressive model chains of TT, such that for all i, j (i > j), C_i is a later model chain than C_j. TT is *model chain progressive* iff $\langle C_1,\ldots,C_n\rangle$ is well ordered in time by the subset relation. That is to say, each later model includes and extends the models antecedent to it in time.

2. We now turn our attention to the members of some R in RR. The members of RR represent T in T* in respect of some subset of T, k(T). Let $\langle k_1(T),\ldots,k_n(T)\rangle$ be an n-tuple of representations of T over time, that is, if i > j, then $k_i(T)$ is a representation of T in T* at a time later that $k_j(T)$. We say that TT is *reduction progressive* iff

(a) k(T) is identical to Con(T) for all indices, or
(b) the n-tuple is well ordered by the subset relation.

2.1. We call an n-tuple of theories RC = $\langle T_1,\ldots,T_n\rangle$ a *reduction chain,* and $\langle T_1,\ldots,T_n\rangle$ a *deeper reduction chain* than j-tuple $\langle T'_1,\ldots,T'_j\rangle$, iff n > j and for all i, j there is a r_i in R_i such that r_i represents T_i in T_i+1 and similarly for T'_1, and further for all T_k ($k \leq j$) T_k is identical in both chains Note, the index i must be different from the index j, since if i = j, there is no T_{i+1}.

2.2. We call a theory *reduction chain progressive* iff T for an n-tuple of reduction chains $\langle RC_1,\ldots, RC_n\rangle$ and for each RC_i (i < n), RC_{i+1} is a deeper reduction chain than Rc_i.

2.3. T is a *branching reducer* iff there is a pair (at least) T´ and T* such that there is some r´ and r* in R' and R*, respectively, such that r´ represents T´ in T and r* represents T* in T and neither T is represented in T* nor conversely.

2.3.1. B = $\langle TT_1,TT_2,\ldots,TT_n\rangle = \langle\langle T_1, F_1, R_1\rangle,\langle T_2, F_2,R_2\rangle,\ldots, \langle T_n, F_n, R_n\rangle\rangle$ is a *reduction branch* of TT_1 iff T_1 is a branching reducer in respect of T_i, and T_j ($i \geqq 2$; $j \geqq 3$ for i, $j \leq n$)

2.4. We say that a branching reducer, T is a *progressively branching reducer* iff the n-tuple of reduction branches $\langle B_1,\ldots,B_n\rangle$ is well ordered in time by the subset relation, that is, for each pair i, j (i > j) B_i is a later branch than B_j, that is, the number of branching reducers has been increasing in breadth as inquiry persists.

3. Let TTT be an n-tuple $\langle TT_1,\ldots,TT_n\rangle$ of scientific structures seriously proposed at a time. Let $\langle\langle T_1,F_1,R_1\rangle,\ldots, \langle T_n,F_n,R_n\rangle\rangle$ be their respective realizations at a time. We say that a set of models M, M = $\{m_1, m_2,\ldots, m_n\}$ is a *persistent model set* iff

(a) $M = \langle\langle m_1 = \langle d_1, f_1 \rangle, m_2 = \langle d_2, f_2 \rangle, \ldots, m_n = \langle d_n, f_n \rangle\rangle$ and for all i, j $d_i = d_j$, or
(b) M is a model in a set of ordered subsets of TTT, such that the sequence is well ordered in time by the subset relation.

3.1 M is an *ontic set* for TTT, that is, M is a set of models that are putative ontologies for M in that their common domain persists under reduction relations over time.

3.2. We say that an ontic set **O** is a *favored ontic set* iff

(a) **O** is the set of intended models of a theory T is the first member of a progressive reduction chain. (O is thus the ontic set of all of the theories in the chain.)
(b) the members of the reduction chain are themselves reduction progressive
(c) T is a progressively branching reducer.

3.2.1. Notice that the set consisting of an ontic set and the sets that it generates (the set of sets under the reduction relation), form a persistent model set.

3.3 TT is *progressive* if

(a) TT is model chain progressive
(b) TT is model progressive
(c) TT is reduction progressive.

3.3.1. We call T a *progressive reducer* if

(a) T is reduction chain progressive
(b) T is a progressively branching reducer.

3.3.2. We say T is a *favored reducer,* if

(a) TT is progressive
(b) T is a progressive reducer.

3.3.3. T is a *most favored reducer* if T is a maximally progressive reducer, that is, T is the nth member of a reduction chain such that for all T_i, $(i < n)$ T_i is a favored reducer. (Notice, T is not reduction progressive, since it stands at the head of the longest reduction chain).

3.4. The set **O**, of ontic models of T, is thus, a favored ontic set in respect of every T_i $(i < n)$ in the reduction chain.

3.4.1. If T is a most favored reducer, and **O** is its favored ontic set than **O***= $\{m_1 \cup, \ldots, \cup m_n\}$ of models m_i in **O** is the ontology of scientific structure TT.

3.4.2 A truth predicate for TT can then be constructed in fairly standard Tarskian as "s is true" for s in T and T in TT, iff **O***||-s where **O*** is the ontology of TT.

Part II

The core construction is where a theory T is confronted with a counterexample, a specific model of a data set inconsistent with T. The interesting case is where T has prima facie credibility, that is, where T is at least model progressive, that is, is increasingly confirmed over time (Part I, 1).

A. The basic notion is that a model, cm, is a *confirming model* of theory T in TT, a model of data, of some experimental setup or a set of systematic observations interpreted in light of the prevailing theory that warrants the data being used. And where

(1) cm. is either a model of T or
(2) cm is an approximation to a model of T and is the nth member of a sequence of models ordered in time and T is model progressive (1.1).

B. A model interpretable in T, but not a confirming model of T is an *anomalous model.*

The definitions of warrant strength from the previous section reflect a natural hierarchy of theoretic embeddedness: model progressive, (1.1), model chain progressive (1.3) reduction progressive (2), reduction chain progressive (2.2), branching reducers (2.3) and progressively branching reducers (2.4). A/O opposition varies with the strength of the theory. This defines levels of warrant strength, W1 through W6, respectively. Anomalies face a dialectical burden as a function of the strength of warrants. We require that a warrant be model progressive at least, that is there is evidence that is either stable or getting better.

So, if T is merely model progressive, W1, an anomalous model is type-1 anomalous, if in addition, model chain progressive, W2, type-2 anomalous etc. up to type-6 anomalous for theories that are progressively branching reducers, W6.

P1. The strength of the anomaly is inversely proportional to dialectical resistance, that is, counter-evidence afforded by an anomaly will be considered as a refutation of T as a function of strength of T in relation to TT. In terms of dialectical obligation, a claimant is dialectically responsible to account for type 1 anomalies or reject T and less so as the type of the anomalies increases.

P2: Strength of an anomaly is directly proportional to dialectical advantage, that is, the anomalous evidence will be considered as refuting as a function of the power of the explanatory structure within which it sits.

P*: The dialectical use of refutation is rational to the extent that it is an additive function of P1 and P2

References

Cardwell, D.S.L. (1971). *From Watt to Clausius: The Rise of Thermodynamics in the Early Industrial Age.* London: Heinemann.
Cartwright, N. (1983). *How the Laws of Physics Lie.* Oxford: Oxford University Press.
Feyerabend, P. (1975). *Against Method.* London: New Left Press.

Goodman, N. (1965). *Fact, Fiction and Forecast.* Indianapolis, IN: Bobbs-Merrill
Hempel, C.G. and Oppenheim, P. (1948). Studies in the logic of explanation. *Philosophy of Science*, 15:2, 135–175.
Hodges, W. (2014). "Tarski's Truth Definitions," *The Stanford Encyclopedia of Philosophy* (Fall Edition). Edward N. Zalta (ed.). <http://plato.stanford.edu/archives/fall2014/entries/tarski-truth/>.
Joyce, J. (2008). "Bayes' Theorem," *The Stanford Encyclopedia of Philosophy* (Fall Edition), Edward N. Zalta (ed.). <http://plato.stanford.edu/archives/fall2008/entries/bayes-theorem/>.
Kleene, S. (1950). *Introduction to Metamathematics.* Princeton, NJ: Van Nostrand.
Magnani, L. (2001). *Abduction, Reason and Science.* New York: Kluwer.
Mostowski, A. (1966). *Thirty Years of Foundational Studies.* New York: Barnes and Noble.
Nagel, E. (1963). *The Structure of Science.* New York: Harcourt Brace and World.
Omer, I.A. (1970). On the D-N model of scientific explanation. *Philosophy of Science*, 37, 417–433.
Parsons, T. (2015). "The Traditional Square of Opposition," *The Stanford Encyclopedia of Philosophy* (Summer Edition), Edward N. Zalta (ed.). <http://plato.stanford.edu/archives/sum2015/entries/square/>.
Popper, K. (1963). *Conjectures and Refutations.* London: Routledge and Kegan Paul.
Priest, G., Tanaka, K., and Weber, Z. (2015). "Paraconsistent Logic," *The Stanford Encyclopedia of Philosophy* (Spring Edition), Edward N. Zalta (ed.). <http://plato.stanford.edu/archives/spr2015/entries/logic-paraconsistent/>.
Quine, W.V.O. (1953). *From a Logical Point of View.* New York: Harper and Row.
Riordan, M. and Hoddeson, L. (1997). *Crystal Fire.* New York: Norton.
Scerri, E.R. (2007). *The Periodic Table: Its Story and Its Significance.* New York: Oxford University Press.
Tarski, A. (1983). *Logic, Semantics and Metamathematics.* Indianapolis, IN: Hackett.
Toulmin, S. (1969). *The Uses of Argument.* Cambridge: Cambridge University Press.
Toulmin, S. and Goodfield, J. (1963). *The Architecture of Matter.* New York: Harper and Row.
Weinstein, M. (2006). A metamathematical extension of the Toulmin agenda. In: D. Hitchcock and Verheif, J. (eds.), *Arguing on the Toulmin Model: New Essays on Argument Analysis and Evaluation.* Dordrecht: Springer.
Weinstein, M. (2012). A mathematical model for A/O opposition in scientific inquiry. In: Beziau, J-Y. and Jacquette, D., *Around and Beyond the Square of Opposition.* Basel: Birkhäuser/Springer Basel.
Weinstein, M. (2013). *Logic Truth and Inquiry.* London: College Publication.
Weinstein, M. (2016). "The periodic table and the model of emerging truth." *Foundations of Chemistry*, 18:3, 195–212.

INDEX

Figures, notes, and tables are indicated by f, n, and t following the page number.